DYNAMICS OF INSECT BEHAVIOUR

DYNAMICS OF INSECT BEHAVIOUR

T. N. ANANTHAKRISHNAN
K. G. SIVARAMAKRISHNAN

Published by:
Scientific Publishers
5A New Pali Road, P.O. Box 91
Jodhpur 342 001 (India)

E-mail: info@scientificpub.com
Website: www.scientificpub.com

ISBN: 978-81-7233-740-7

Typeset by Rajesh Ojha

Printed in India

PREFACE

Insect behaviour is a multifaceted discipline encompassing such diverse areas as ecology, conservation, insect-plant interactions, pest management, vector control, ecosystem engineering, chemical ecology and genetics. Insects, as in the case of other animals are known to endure behavioural diversities, overcoming difficulties in escaping from predators, gathering food, fighting rivals, selecting mates and nurturing the young. The diverse ways in which insects accomplish these functions successfully are of considerable interest and this volume presents a nearly vivid picture of the behavioural diversities of several groups of insects – aerial, terrestrial and aquatic. The phenomenal migration of the monarch butterflies, ants hunting in swarms of thousands, shrill sounds of the cicadas, songs of the grasshoppers, flashing of glow worms and fireflies communicating in darkness go to show that behaviour is an exciting part of the natural history of insects. Adaptive regulation has promoted the continued existence of insects in a wide variety of environments.

Many insects such as locusts show migratory movements of a prolonged nature that carry these insects outside familiar habitats. Persistence, linearity, undistractability and stored energy distinguish migration from other movements. The ceaseless underground foraging of ants aerating the soil, efficacy of pollination of bees, wasps and flies playing an essential role as scavengers are well known. The diversity of behaviour of some aquatic beetles and bugs, the skating behaviour of some water bugs, gall inducing behaviour of some insects and vector behaviour of some mosquitoes are proverbial. In trying to explore the behavioural diversities of some of these insects the authors have kept in mind the essential roles that insect play in nature.

The editors wish to thank the contributors who have taken pains to assess and document the behavioural diversities of several groups of insects, knowledge of which will go a long way towards contributing to our knowledge of the behavioural diversities of insects. The editor thanks the co-editor, Prof. K.G. Sivaramakrishnan for his painstaking efforts in contacting the authors to submit chapters on time. The editors are equally thankful to the publishers for their patience and interest in this publication.

T.N. ANANTHAKRISHNAN

CONTENTS

Dynamics of Insect Behaviour
T.N. Ananthakrishnan & K.G. Sivaramakrishnan
Scientific Publishers (India), Jodhpur, 2012, pp. 1-10

Chapter - 1

DYNAMICS OF INSECT BEHAVIOUR: AN INTRODUCTION

T.N. Ananthakrishnan

INTRODUCTION

Insects abound in all kinds of terrestrial and aquatic ecosystems, presenting an endless variety of form and function. In view of their small size, short generation time, sensory abilities and interactions with plants and other organisms, they have dominated every type of environment. Their immense diversity, environmental adaptabilities and behavioural intricacies need increased assessment and understanding. As early as 1954, Dethier suggested that the first barrier to overcome in insect-plant relationships is a behavioural one. For a proper assessment of their behavioural dynamics, the need for a better appreciation of the wide variety of ecological patterns such as distribution, abundance and population dynamics cannot be overemphasized. Among the noteworthy behavioural trends are aspects such as death feigning, foraging, migration, nesting and oviposition. Besides, behavioural plasticity in response to host range, abundance and competition with host-associated groups involve multiple adaptations. Such levels of herbivory as leaf mining, fruit boring, stalk and wood boring are common. In moving to a new host, species tend to modify their own niche and occupy formerly unoccupied space. In so doing, they tend to modify the environment facilitating production of niches for other species. The distance travelled to locate a host may vary depending on adult behaviour and life stage of the concerned species. Some species tend to modify the environment facilitating production of new niches for other species, enabling habitat provision and new niche opportunities – an aspect termed environmental engineering (Erwin, 2005). Needless to indicate that plant species composition is affected by competitive interactions of insects. Insect pollination is the basis for the evolutionary history of flowering plants. Insects have provided the best examples for commensalism, mimicry, mutualism, structure and dynamics of food webs by the movement of insects from one habitat to another. Dispersal influences food web structure and insect dispersal tends to add to our knowledge of community ecology (Thompson, 2006).

DISPERSAL BEHAVIOUR

In a newly developing habitat, insects such as macropterous plant hoppers colonize at low densities, mate soon after arrival, leaving offspring that develop mostly into fecund adults, followed by population growth. Dispersal is important for the success of insect species exploiting temporary habitats (Zera and Denno, 1997). Macropterous adults of planthoppers are known to migrate long distances before they colonise a habitat. When a species arrives at a new site, its ability to survive there would be determined by its ability to fit into the local ecosystem – an aspect referred to as ecological fitting (Wilkinson, 2004). Female choice and male signaling behaviour tend to lead to what is termed as assortive matings among populations of different hosts, resulting in the diversification of sexual communication. Increased understanding of evolution of oviposition behaviour is important and adult feeding preferences can influence oviposition preferences. Consequences of host shift are well known and are very evident in membracids such as *Enchenopa notata*. A shift from tree feeding to diverse galling sites such as leaf, petiole, stem and buds is common and such new adaptive zones tend to be invaded through oviposition mistakes. Choice of oviposition sites is partly dependent on the trichomatous nature of the plant. Many lepidopterans show a high preference for the abaxial surfaces of plants over adaxial surfaces, more because of the protection afforded to eggs (Ramasamy, 1994). In many species structure of the ovipositor correlate with increasing trichome density. Plasticity in male searching behaviour of females such as those of membracids tend to contribute to assortive meetings among populations of different hosts. If the normal host is locally rare, adaptations to a new plant, even if inferior, may be favoured by its availability. In short, behavioural adaptation tends to be more pronounced than physiological adaptation.

VIBRATIONAL COMMUNICATION

Insects are also vibration sensitive species which not only monitor vibrations to detect predators and prey, but also introduce vibrations to communicate with other individuals (Cocroft and Rodrigues, 2005). A wide variety of plants are used as transmission channels for insect vibrations. They play a major role in pair formation and other stages of reproduction. For many plant feeding insects, vibrational communication is important for meeting the challenges of life on the planet, like access to feeding sites and also help herbivores to avoid predation. In response to group signals from offspring, mothers move into the aggregation and attempt to drive off predators. Vibrational signals in concert with chemical signals attract mutualistic ants that provide protection from predators (Travassos and Fierce, 2000). Signals of some butterfly larvae such as those of lycaenids also function in avoiding predators (De Vries, 1991). In general, a wide variety of plants are used as transmission channels for insect vibrations. Given the large number of vibration orienting signals, the need for increased information on interactions between vibrational species and natural enemies, cannot be overemphasized.

ANALYSIS OF BEHAVIOURAL DIVERSITY

Success of insects' behavioural plasticity also includes their ability to modify behaviour through learning. Though most insect behaviour trends tend to be innate, many are environmentally and physiologically modified. Cryptic behaviour in mantids mostly looking like barks of trees, caterpillars as twigs feigning death, lying motionless (thanatosis) are well known. Butterflies with eye marks like sphingids with larvae assuming a threatening posture and with eye-spots giving a snake-like appearance, are equally well known. Covering with lichens detritus acts as barriers to predators in some lace-wings and reduviids. Defence by hiding is common in geometrid moths, and looper caterpillars resembling twigs, membracids imitating thorns, stick insects like sticks, Lepidoptera with venation matching of leaves, dead or alive, mottled with fungus and looking partially eaten, add to the complexity of behavioural diversity. Sit and wait predators within striking range are flower mimicking mantids, Odonate nymphs remaining concealed in vegetation are good instances. A very good example of sit and wait is offered by the antlion, *Myrmeleon* larvae which dig triangular pits with steep sides, the larvae lying buried at the base of the pit, waiting for the passing prey to fall in. Raptorial legs of mantids and labial masks of dragonflies are equally good examples.

Also well known is the functional diversity relating to such aspects of behaviour as saltatorial (jumping), gressional (walking), natatorial (swimming), fossorial (digging) and raptorial (prey-seizing) ones. Functional feeding groups in aquatic ecosystems are referable to shredders (living in decomposing plant tissues), collectors feeding by filtering particles, scrapers feeding on attached algae by grazing surfaces and piercers feeding by piercing and sucking contents. Whirligig beetles which skim the surface of ponds and lakes have flipper-like middle and hind legs enabling propulsion at high speed. Behavioural patterns of damselflies differ from dragonflies in that the former do not feed in the air and prefer to search for insects in foliage and are able to fold the wings and are less aggressive, not mating in flight unlike dragonflies.

PARENTAL CARE

Care of eggs and adults is common in social insects but tending solely by males is very unusual and is known in giant water bugs such as *Belostoma* and *Lethoceros*. Back brooding by males, carrying eggs on their back is common in these insects, besides *Sphaerodema*. Active brood tending behaviour is called 'brood jumping' involving underwater jumping.

Parental care is a post hatching behaviour including provision and protection of food resources. Tending of eggs and young is common in orthopteroids, Dermaptera, blattids, thrips, embiids, Hemiptera, Coleoptera and Hymenoptera. Quasisocial behaviour is seen in communal nesting consisting of members of the same generation, all of which assist in brood rearing. Subsocial behaviour is seen in thrips galls, besides bark dwellers. In semisocial behaviour, there is a reproductive division of labour with some females laying eggs, others acting as workers. Eusociality is common in insects, with division of labour including a caste system involving –queens, soldiers and workers. In eusocial wasps and bees, the queen has an enlarged abdomen producing only workers in the first brood. As the colony matures, subsequent broods result including increasing populations of males and workers. Honey bee workers exhibit polytheism with young workers tending to be hive bees engaged in within-house activities, while older workers are foraging hive bees. Quality and quantity of larval diet determine caste differentiation, the food for the future queen being the 'royal jelly', with high sugar content. The dance language of the bees is well known, with nearby available food being communicated by the 'round dance' and the more distant food resources identified by the 'waggle dance' with abdomen in the figure of '8' circuit. In vibration dance, workers vibrate indicating communication of a suitable site for a new host. Trophallaxis is typical of termites wherein by feeding on oral secretions, transfer of pheromones occur.

Most insect behaviours are innate, programmed genetically, with the ability to modify behaviour through 'learning'. A good example is the digger wasps which can learn nest location through markers such as making a short flight to mark the terrain. Successful colonization and establishment by invasive species provide novel opportunities to study life cycles in novel habitats. Many butterflies and moths evince a behaviour called 'puddling', with clusters of hundreds being attracted to copious quantities of liquids including substances like sodium. Cannibalistic mating is evident in mantids with males being eaten up by females. Adults of a few species of noctuid moths are known to frequent the eyes of humans and cattle, feeding on lachrymal secretions. Leaf cutter ants dominate the ecosystems where they occur, with adults highly polyphagous, and larvae monophagous feeding on fungi. Polymorphic workers exhibit an elaborate range of behaviour including cutting living tissues, earning the name 'leaf cutter ants'. They excise secretions of live vegetation with their mandibles and transport the pieces to the nest, where other individuals macerate the plant tissues with their mandibles. Other specialized workers tend the fungus gardens by inoculating new substances with fungal hyphae. Ants and fungi cooperate to breakdown plant defenses, with ants removing protective leaf waxes that deter fungi which convert indigestible cellulose carbohydrates. Ant colonies have the uncanny ability to calculate the shortest path to different food sources using pheromone trail communication very typical of recruitment behaviour in ants. Among diverse defensive roles of insects, caddisfly larvae construct cases of sand grains and organic matter. Bagworms construct a portable case of silk-bound vegetable materials, the cases affording protection during pupation. Chrysomelid beetles decorate themselves with their faeces while the larvae of lacewings and reduviids cover themselves with detritus and lichens enabling disguise. Death feigning or thanatosis is a behaviour common to several beetles, notably weevils. Click beetles or elaterids have the ability to propel in the air.

The host plant is a clear locus of ecological adaptations. Host-associated adaptive divergence is largely driven by the presence of particular chemicals. Adaptation to specific host plants involves shifts in the time of mating, ovipositional site and biochemical preference for a site. Several phytophagous insects show a wide host range and small scale changes in host range tend to reflect behavioural plasticity or local adaptation in response to differences in host quality and abundance. Host range expansion results from what has been termed 'Oscillation hypothesis' of host plant range and speciation. Opportunistic colonization of host plants is typical of membracids with host shifts resulting in reproductive isolation and speciation. Male location and courtship behaviour involve sex attractants and many aggregating insects tend to produce alarm pheromones as in termites, bees and ants. Cannibalistic mating is typical of mantids, males being consumed by larger females during or after mating (Gullan and Cranston, 2010). Parental guarding is also known in reduviids and belostomatids. Behavioural intricacies such as that of the dung beetles are of interest. From the time the dung is dropped, invasion of diverse species of scarabid beetles is of common occurrence. Most dung beetles excavate networks of tunnels immediately beneath or beside the pits. Some excise a chunk and move it some distance to a dug-out chamber. Head butting or rolling spherical balls is also of common occurrence. Scarab grubs alter their microenvironment by creating a temporary chamber in the soil. Several scarab grubs show seasonal patterns of vertical movement associated with soil temperature (Villani and Wright, 1990).

Trophobiosis is very typical of ant - homopteran relationship. Aphids tended by ants are able to feed more rapidly, resulting in a higher rate of reproduction. Aphid aggregation tends to increase when tended by ants and also result in the increase of life span of aphids. Aphid honey dew also contains specific ant attractants and may transport aphids to new feeding sites. By defending aphids, these ants tend to alter the structure of arthropod communities associated with host plants. Tandem running involves a behaviour with chemical trail communication very typical of recruitment behaviour in ants.

Collective defense is common among gregarious and social insects such as some wasps crowding outside their nests in orderly clusters. Some chrysomelids form defended circles (cycloalexy); termite soldiers may block nest entrance with their modified heads, a process called phragmatism. Females of fireflies indicate their location through light flashing by their luminiscent organs, containing luciferin which is oxidized by the enzyme luciferase enabling flashing. Many aggregating insects produce chemical releases of alarm behaviour resulting in rapid dispersal or aggressive defense. Meloids or blister beetles exude skin-blistering irritants, a technique called reflex bleeding.

Transport by larger individuals of another species results in phoresy as in hippoboscid flies. Egg parasitizing Hymenoptera like trichogrammatids attach themselves to adult females of host species gaining immediate access to eggs at oviposition. Cryptic insects display what is known as startle display, opening the forewings, revealing the brightly coloured "eyes". Behavioural strategies evident in running wastes are the use of currents to drift from an unstable location to locate aggregation of prey. Burrowing deep into the substrate is also a lotic behaviour and those that live in strong currents are called rheophilic. Lotic insects being smaller with flexible bodies live among cracks and crevices of rocks, avoiding currents. Gerrids or water striders scavenge or predate at the water surface as exemplified by *Notonecta*, while nepids and belostomatids which are predatory, also feed on detritus.

Seasonal migratory tendencies are very typical of danaid butterflies and of these the most outstanding is the monarch butterfly, with often roosts of over ten million individuals. They are "behaviourally wired" moving upwards and towards light. The caterpillars show a defensive behaviour, curling up and dropping off their host plant even when mildly stimulated. Their penchant for basking in the sun is great – a behaviour tending to reduce the duration of the caterpillar stage. Adult monarchs can soar and glide for considerable distances, typical of migratory species. Males use pyrolizzidine alkaloids from certain plants to produce sex hormones which entice females to cease flying and be receptive for mating. Interestingly enough 20-30 million butterflies are found in some large Mexican roosts (Schappert, 2004). Some migratory insects show ectothermic behaviour or rely on internally generated heat mostly from metabolites associated with flight (endothermy). The basic cause of a specialized host association is the insects' behaviour in the selection of certain species over others as food or oviposition substrates (Bernays and Chapman, 1994).

Display of elaborate armature such as horns or teeth varying in size, the larger ones restricted to males entirely is evident in species showing sex-limited polymorphism evident in the extreme sized males of several mycophagous tubuliferous Thysanoptera such as *Ecacanthothrips*, *Tiarothrips*, *Kleothrips* besides a host of other species displaying behavioural intricacies. Extreme sized males of *Tiarothrips* show excessively enlarged parts like head and very long third antennal segment or excessively enlarged forefemora in *Kleothrips* or well armed and enlarged forefemora as in *Ecacanthothrips*, besides several others (Ananthakrishnan, 1969, 1973) playing essential roles in the protection of eggs and early instars.

BEHAVIOURAL RESPONSES OF PHYTOPHAGOUS INSECTS TO PLANT SECONDARY CHEMICALS

Behavioural studies have substantially contributed towards an understanding of the possible function of diverse receptors involved in host finding and resultant discrimination of feeding and oviposition. The choice of oviposition sites is to some extent dependent on the degree of trichomes present. A better protection is afforded to eggs, moths prefer abaxial surfaces for oviposition, especially because of higher densities of trichomes, leading to increased survival of immatures. Behavioural responses of phytophagous insects to plant secondary compounds notably to secondary compounds, many of which tend to act as stimulants and deterrents to oviposition, in particular to Lepidoptera. Acceptance or rejection is related to various internal excitatory factors. In short, how natural enemies search for herbivores depend to a great extent on behavioural, ecological, genetic and developmental constraints of both natural enemies and the concerned hosts. Since each plant produces a different blend of plant chemicals when damaged, natural enemies make use of generalized plant volatiles. Insects adapt to plant chemical quality evincing behavioural, ecological and physiological adjustments. Insects can sequester plant defensive substances store them in reservoir and use whenever needed.

CONCLUSION

The adaptive spectrum of insects to their host plants involves behavioural as well as metabolic changes which reflect the overwhelming diversity of phytophagous and mycophagous insects. Insect behaviour and adaptation to a host plant are oriented by physical and chemical components. Behaviour modifying compounds tend to deter or repel insects and interfere with normal development of a given insect species. Several physiological, ecological and behavioural processes mediated by environmental factors are very typical of insect-plant interactions, especially subtle interactions of orientational, feeding and ovipositional responses to behaviourally active components of plants. Polymorphism and structural allometry in thrips are associated with competition for resources resulting in diverse ecological and behavioural intricacies. Fighting behaviour in males is typically characteristic of polymorphic males with egg guarding and larval defense behaviour. The need for intensive studies on the physiology and behaviour of as many species as possible, cannot be overemphasised.

Nowhere can the behavioural diversity of a mosquito be better described than the account by Michael Auft (2004). "She comes out night, flying out of the woods and swamps in search of a suitable landing spot. When she finds one, she plants all six legs on uncovered flesh and bends her body forward. She hunkers down and manouveres her spike of snout into position. She bores that spike, a complex package of micromachines – serrated tools for cutting, pumps for removing liquids through the epidermis then punctures it through a thin layer of fat. In short, she drills the tip down into tiny capillaries, then taps them, sucking blood in the name of future generation".

REFERENCES

Ananthakrishnan, T.N. 1969. *Indian Thysanoptera,* CSIR Zool. Monograph 1: 1-171.

Ananthakrishnan, T.N. 1973. *Thirps, Biology and Control*. Macmillan, India, pages 120.

Bernays, E.A. and Chapman, R.F. 1994. Host *plant selection by phytophagous insects* Chapman & Hall

Cocroft, R.B. and Rodrigues, R.L. 2005. Behavioural Ecology of Insect's vibrational communication Bioscience 55: 323-334.

De Vries, P.J. 1991. Ecological and evolutionary patterns in riodinid butterflies. In: *Ant-Plant Interactions*. C. Huxley and D.F. Cutler (eds.) Oxford Univ Press, pp. 143-156.

Dethier, V.G. 1954. Evolution of feeding preferences in phytophagous insects Evolution 8: 33-54.

Erwin, D.H. 2005. Seeds of diversity Science 308: 1752-1753.

Gullan, P.J. and Cranston, P.S. 2010. *The Insects: An outline of Entomology* Wiley, Blackwell, IV Ed.

Michael, Auft 2004. The forever enemy *In*: John Hopkins Magazine. pp. 40-47.

Ramaswamy, S.B. 1994. Physiological basis of feeding and oviposition behaviours in moths. In *Functional dynamics of insects* edited by TN Ananthakrishnan, Oxford and IBH, New Delhi, pages 54-78.

Schappert, O. 2004. *The Last Monarch Butterfly*. Firefly Books (US) Inc. Buffalo, New York.

Thompson, R.M. 2006. Embedding food webs in ecological landscapes. Global Ecology and Biogeography 15: 109.

Travassos, M.A. and Pearce, N.E. 2000. Acoustics, context and function of vibrational signalling in a lycaenid butterfly-ant mutualism Animal Behaviour 60: 13-26.

Villani, M.G. and Wright, R.J. 1990. Environmental influences on soil micro-arthropod behaviour in agricultural Systems. Annual Review of Entomology 35: 249-269.

Wilkinson, C. 2004. Status of coral reefs of the world: 2004. Australian Institute of Marine Science, Townsville, Queensland. 547p.

Zera, A.J. and Denno, R.F. 1997. Physiology and Ecology of dispersal polymorphism in Insects Annual Review of Entomology 42: 207-230.

Dynamics of Insect Behaviour
T.N. Ananthakrishnan & K.G. Sivaramakrishnan
Scientific Publishers (India), Jodhpur, 2012, pp. 11-18

Chapter - 2

BEHAVIOURAL STRATEGIES AND ADAPTIVE TRENDS IN INSECTS

T.N. Ananthakrishnan

INTRODUCTION

The problem of structural patterns in insects has fascinated biologists over the years, in particular the possible mechanisms by which new patterns could arise during development. Many species have inherited the capacity to react to fluctuating environments by varying body structure as well as population size. Extreme intraspecific diversity involving an array of characters tend to provide 'release mechanisms' enhancing the variability of species. Variations are magnified to a very large extent by the incidence of heterogeny or allomorphosis or absolute allometry of insects (Ananthakrishanan, 1969). Correlation of microclimatic data such as humidity, temperature, rainfall, light and windspeed among vegetation influence the abundance of insect species at different times of the day. Patterns in the activities of the concerned insect provide clues regarding their specific periodicity patterns. Needless to indicate that every species has its own limits of climatic tolerance, with some endowed with greater adaptability to a wide range of host plants. Many species on emergence from the soil and other hiding places like the bark of trees, migrate over short distances in search of host plants. As such, the synchronization of a complex of environmental, visual, tactile and chemical factors mediating the behaviour and physiology of the concerned insects tend to be vital factors in the interaction of an insect with its host plant. The multiplicity of interacting factors and the adaptive dynamics of the concerned insects hold the key for successful interaction between insects and plants. Thus the host plants form an important 'island' for the concerned insects to determine survival, growth, and rate of increase. In most polyphagous insects, non-preference due to oviposition and feeding generates a series of complex responses depending on the host responses and environmental characteristics.

The chemical composition of the plant is of fundamental significance in the acceptance or rejection of food or oviposition by insects (Ananthakrishnan, 1992). Morphological features interfere physically and physiologically, the latter involving host selection, feeding, ingestion, digestion, mating and oviposition (Norris and Kogan, 1980). Besides, the nature and dimension of trichomes on the laminar surface of the leaves influence the feeding of many insects. The choice of the oviposition site depends on the presence or absence of trichomes. Many species of moths prefer the abaxial surfaces of accepted plants for better protection afforded to eggs and because of high density of trichomes (Navasero and Ramaswamy, 1994). Preference to hirsute surfaces for oviposition tends to increase survival of immatures. All the same, many like *Chilo partellus* oviposit preferentially on the lower surfaces which are smooth and hairy.

Oviposition is the culmination of a series of behavioural events, the first component being the orientation of the gravid females to the prospective host plants. Following orientation as a whole to the plant, insects orient to different plant parts in the selection of suitable oviposition site. Oviposition to a prospective host plant involves normal as well as chemosensory stimuli. The acceptance or rejection of a plant usually depends on contact with the plant surface or through probing after landing. Feeding behaviour therefore is an important factor in determining the resistance of the host plant to insect attacks. In some plants with virulent poisons like the milkweed with bitter taste, insects have found a way of consuming leaves. On landing, an insect immediately cuts through the midrib, munching beyond the cut, not only severing the vein, but also carving out a circular trench on the underside of the leaf enabling feeding only in this area. Similarly develops immunity to poisoning, storing the poisons in the tissues, so that birds find them distasteful. Recognition and preference of host plants therefore involve integration of diverse behavioural factors. Many caterpillars show patterns of integrated behaviour, several being gregarious as larvae. They are obligatory gregarious species with a foraging behaviour. Aggregation of social caterpillars vary in size from a few individuals to assemblages of hundreds as in the caterpillar larvae of *Eupterotes mollifera*. Caterpillars that forage together may aggregate as a community as in the case of the tent caterpillar, *Malacosoma* (Ananthakrishnan, 2009).

Behavioural events involved in the acceptance of a host plant for oviposition are: orientation → encounter → landing → surface evaluation → acceptance and oviposition. The behavioural responses of phytophagous insects involve stimulants and deterrents to ovipositing insects such as Lepidoptera. Dethier (1982) has suggested that the ultimate factors affecting behaviour are more internal, such as motivation and degree of satiety. In achieving this, various species vary in their sensitivity to the positive and negative signals provided by the plant. On alighting from a potential host plant, acceptance or rejection depends on the insects' assessment of the balance between stimulants and deterrents (Renwick and Huang, 1994). After feeding, many caterpillars move away from the damaged leaves or partially eaten leaves enabling the damaged laves to fall to the ground so that they tend to distance themselves from the natural enemies (Heinrich, 1979). Reducing rates of predation by natural enemies, swallow tail caterpillars and other insects quickly move away from the host plant, tending to reduce action by predators and other natural enemies (Whitman and Nordlund, 1994).

As for the mechanism of pollen collection bumble bees have developed complex methods of collecting pollen. Hairs on the furry body are endowed with microscopic hairs which pick up the sticky pollen and the bristly comb that fringes the lower half of each hind leg, help to sweep the pollen. Another example relates to army ants soldiers which enter the nests of wasps driving at the occupants.

Temperature and photoperiod tend to affect host seeking behaviour by altering the nutritional state of the insect. The orientation of the red cotton bug *Dysdercus koenegii* changes as they become water-stressed (Saxena, 1969). Some ovipositing butterflies avoid ovipositing on plants that bear previously laid eggs. Crowding on host plants tends to trigger the development of winged forms facilitating dispersal. Generally thrips and aphids are wind-born over long distances so that eggs are randomly deposited on hosts (Dixon, 1973; Price, 1976). In order to enhance adaptive fitness intense searching behaviour is adopted by several insects (Vinson, 1981). Insect populations confront a wide array of environmental resources and as each population attempts to achieve maximum fitness, adaptive variation in host finding behaviour becomes common (Ananthakrishnan, 1994). Both genetic and non genetic sources of variation could influence behavioural responses, but only genetically based variation can serve as a raw material for natural selection, which can bring about host shifts (Agarwala, 1996).

After feeding, many caterpillars move away from the damaged leaves or partially eaten leaves at the petioles allowing damaged leaves to fall to the ground enabling the caterpillars to distance themselves from natural enemies (Heinrich, 1979). Reducing rates of predation of natural enemies swallow tail caterpillars and often insects, quickly move away from the host plant, tending to reduce action by predation and other natural enemies (Whitman and Nordlund, 1994).

Ant tending is well known in aphids associated with increased rates of honey dew production. Aphids tended by ants are able to fed more rapidly, enabling increased reproduction rates. Ants allow tending to encourage aphids to move to new feeding sites by 'herding' them. By releasing hardening wax from cornicles, aphids tend to defend themselves against natural enemies. The most obvious behavioural adaptation in aphids is the secretion of honey dew in response to ant palpation. Tended aphids are also more gregarious than untended aphids. Ants respond to aphid alarm pheromone behaviours with antipredation behaviour. They may also transport the aphids to new feeding sites and new host plants. By regurgitating droplets, ants may also feed the aphids. Ants tend to forage on honey dew secreted by homopterans, an aspect termed as 'Trophobiosis'. Ant colonies have also the uncanny ability to calculate the shortest path to different food sources using pheromone trails. Behaviour of alate aphids enables dispersion and finding a suitable host, to locate special secondary compounds of nutritional value, apteral on the other hand move only when necessary (Sovenson, 2007). Ants have diverse communication system and can deposit chemical trails to recruit nestmates to discover food. May produce many volatile chemicals to signal alarm and can produce recruitment pheromones (Franka, 2003).

ADAPTIVE BEHAVIOUR IN THRIPS

Thrips are endowed with diverse feeding strategies and also show subsocial behaviour as evident in some mycophagous species. While the majority of thrips species are phytophagous and mycophagous, predatory species feeding on thrips, mites, coccids and psocids are not uncommon. Some species occur within galleries of termite mounds. Very rarely thrips are subaquatic live *Organothrips bianchii* infesting cultivated and wild colocasia plants growing in stagnant water. Some like in the axils of needs. Incidence of thrips on bryophytes, pterido-phytes and gymnosperms exist, besides moss, lichen and fern infecting species. Species infesting dead wood are associated with early stages of fungal attack. Very rarely some terebratian species feed on the stores of wheat rusts, coffee leaf rust and red rust of bananas. Massive invasion of thrips to the extent of overflowing populations are known in *Mallothrips Indicus*, besides overcrowding of *Isoneurothrips* in Eucalyptus with massive migratory tendencies.

The co-adaptation of flowers and thrips involved in the mechanisms of pollen transfer is based on the behavioural patterns involved in pollination. In short, the adaptive signific-ance of thrips-flower association and the pollination potential of thrips mainly depend on the dynamics of pollen transfer mechanism (Ananthakrishnan, 1994). By and large, the leaf surface dominates the environment of an insect on it, ant the nature of the surface is of primary importance in its behaviour ecological strategies.

Galling behaviour is very common thrips, mostly phlocothripids from simple leaf folds or leaf rolls to highly contorted masses of leaf tissues including several hundreds of adults and larvae as in *Calycopteris floribundus* by *Austrothrips cochinchinensis*. An interesting substance of eusociality is seen in Acacia galls on phyllodes in Australia, the phyllodes glued together, thrips breeding within (Crespi and Mound, 1997).

ECOLOGICAL STRATEGIES OF SPECIES

Host plants play a central role in determining numbers, and life history strategies are dictated by the habitat or areas accessible to the feeding stages. The plant species is not a homogenous entity, but a heterogeneous one with microchemical environment changing in time and space, with changing condition of climate, growth and soil (Dethier, 1954). The individual leaf is always considered as heterogeneous habitat forming the smallest resource units (Withers, 1980). The conditions of the host plant are a major factor in determining population density. Southwood (1977) has shown how ecological strategies of a species are evolved in response to the habitat as expressed by the "reproductive success matrix". In assessing insect-plant interactions, the stage of development of the host plant, time of attack, source of infestation, build-up of insect population and host plant strategies and responses are vital. In an attempt to understand the source of infestation it would be necessary to study population fluctuations of insect species, both within the crop and weeds around (Ananthakrishnan, 1984).

Floral morphology and behaviour are typical, with many flowers attracting a wide variety of insects such as the bee flowers with release fragrance attracting thrips, beetles and flies. Carpenter bees like *Xylocopa* which emits fragrant chemicals, besides honey bee by marking flowers with scents, inform others of the same species. In bumblebees, the hocked hairs on the furry body pick up the sticky pollen and the bristly comb fringing the hindlegs, help to sweep the pollen,

Natural enemies not only respond to chemicals released from plants, but also respond to chemicals produced or released by herbivores themselves. In this way plans influence the volatile emissions thereby influencing the interactions between insects and their natural enemies. Some plants tend to actively release natural enemy attractive chemical signals in response to attack by insects. Thus plants tend to communicate with members of the third trophic level (Whitman and Nordlund, 1994). Plants are therefore a vast storehouse of bioactive compounds, plant chemical diversity producing different blends of volatile chemicals when damaged. As such, insects tend to adapt to plant chemical quality evincing behavioural, ecological and physiological adjustments. Adaptation to hosts is manifest the successful colonization and regular emergence of newer generations. Needless to indicate that lower concentration of varying essential free amino acids also tend to influence the degree of acceptance of plants by insects like thrips and heavy aggregation result if all the nutrient are present in the right proportions.

There is an increasing need for a deeper understanding of the chemical perception of plants by insects with different host plant selection strategies. Generalist or polyphagous species use different chemical cues to recognize their preferred hosts, while the specialists perception of hosts is mediated through key stimuli in their preferred hosts (Renwick, 1983; Visscher, 1986). By comparing the oviposition behaviour and sensory physiology of adult insects with different host specificities, it is possible to infer that the same compounds which are stimulatory to specialists, tend to be repellent or deterrent to generalists. Needless to emphasise that larval host plants by adults is a prerequisite for oviposition, especially specialists. In short, whether a potential host is selected or not is decided by a balance between physical stimuli, attractants and repellents, feeding deterrents and oviposition stimulants and deterrents.

CONCLUSION

Discussing the intra-and interplant behavioural dynamics of the cotton bollworm complexes, Uthamasamy (1994) concludes that it is very difficult for an individual insect to locate efficiently the best site for ovipositor or feeding. Further, differences 'among individuals and populations contribute to a change in behavioural processes which decide the successful survival of a species. A combination of physiological, ecological and behavioural characters, enable pests to survive in a notable habitat.

ACKNOWLEDGEMENTS

I wish to take this opportunity in thanking Prof. Ralph Holzenthal, Professor of Entomology, University of Minnesota, St. Paul, for the ready loan of reference books which have greatly helped in the preparation of the first two chapters.

REFERENCES

Agarwala, B.K. 1996. Host related behavioral responses in insect populations and their genetic basis. In *Biotechnological perspectives in the Chemical Ecology of Insects* edited by T.N. Ananthakrishnan, Oxford & IBH, New Delhi, pages 102-115.

Ananthakrishnan, T.N. 1969. *Indian Thysanoptera*. CSIR Zoological Monograph 1, 171 pp.

Ananthakrishnan, T.N. 1982. Thrips and pollination biology. Current Science 51(4): 168-172.

Ananthakrishnan, T.N. 1984. *Bioecology of Thrips*. Michigan Indira Publishing House, pages 232.

Ananthakrishnan, T.N. 1992. *Dimensions of Insect-Plant Interactions.* Oxford and IBH, New Delhi, pages184.

Ananthakrishnan, T.N. 1994. Viewpoints in the chemical ecology of insects. Phytophaga 6: 69-72.

Ananthakrishnan, T.N. 2009. Ecodynamics of Insect Communications. Scientific Publishers (India), Jodhpur. pages 132.

Crespi, B.J. and Mound, L.A. 1997. Ecology and evolution of social behavior among Australian gall thrips and their allies. In: Choe, J.C. and Crespi, B.J., eds. The Evolution of Social Behaviour of Insects and Arachnids. Cambridge: Cambridge University Press, 166–180.

Dethier, V.G. 1954. Evolution of feeding preferences in phytophagous insects. Evolution 8: 33-54

Dixon, A.F.G. 1973. *Biology of Aphids.* Edwin Arnold, London, pages 258.

Heinrich, B. 1979. Foraging strategies of caterpillars: leaf damage and possible predator avoidance strategies. *Qecologia* (Berlin), 42, 325-337.

Navasero, R.C. and Ramaswamy, S.B. 1994. Morphology of leaf surface trichomes and their influence on oviposition by *Heliothis virescens.* Crop Science.

Norris, D.M. and Kogan, M. 1980. Biochemical and morphological basis of resistance. In *Breeding plants resistance to insects* edited by FG Maxwell and PR Jennings, John Wiley and Son, New York, Pages 24-60.

Price, P. 1976. Colonization of crops by arthropods: Non equilibrium communities.

Renwick, J.A.A. & Huang, X.P. 1994. Interacting chemical stimuli mediating oviposition by Lepidoptera. In: T.N. Ananthakrishnan (ed.), Functional Dynamics of Phytophagous Insects. Oxford & IBH Publishing, New Delhi, pp. 79–94.

Saxena, K.N. 1969. Patterns of insect-plant relationships determining susceptibility or resistance of different plant parts to insects. Entomologia Experimentalis et Applicata 12: 751-766.

Southwood, T.R.E. 1977. Habitat, the template for ecological strategies. Journal of Animal Ecology 46: 337-365.

Uthamasamy, S. 1994. Intra- and interplant behavioural dynamics of the cotton bollworm complex. In *Functional dynamics of phytophagous insects* edited by T.N. Ananthakrishnan, Oxford and IBH, New Delhi, pages 57-78.

Vinson, S.B. 1981. Habitat location. In *Semiochemicals: Their role in pest control* edited by D.A. Nordlund, R.L. Jones and W.J. Lewis, Wiley, New York, pages 51-57.

Whitman, D.W. and Nordlund, D.A. 1994. Plant chemicals and the location of herbivorous arthropods by their natural enemies. In *Functional dynamics of insects* edited by T.N. Ananthakrishnan, Oxford & IBH, New Delhi, pages 133-159.

Dynamics of Insect Behaviour
T.N. Ananthakrishnan & K.G. Sivaramakrishnan
Scientific Publishers (India), Jodhpur, 2012, pp. 19-27

Chapter - 3

ECOLOGICAL DIVERSITIES AND BEHAVIOURAL INTRICACIES IN INSECTS

T.N. Ananthakrishnan

Insect are major constituents of terrestrial ecosystems in terms of their biomass and interactions with other organisms. Ecologically and behaviourally they tend to differ and in spite of their morphological diversity they are functionally diverse and species specific in their habitats. In terms of their biomass and interactions with other organisms, they play an essential role in our understanding of ecology, developmental biology and evolution. Food provision in communities and ecosystems and removal of waste products and dead organisms are very typical of insects. In many ecosystems they play a major role in regulating litter dynamics and soil organic matter. Insects like Lepidoptera and Hymenoptera play a significant role in plant pollination and biological control of weeds indicating the need to further inventories and monitor them. Insect defoliators have considerable effect on the growth and survival of forest vegetation, with the capacity to alter the function of forest ecosystems. Soil insects are essential for letter breakdown and provide nutrient inflow. Insects serve as facilitators for inter-specific interactions through proresy, transmission of pathogenic organism, seed dispersal and alteration of microhabitat structure (Miller, 1993). In short, entomological research has progressed by leaps and bounds and such aspects as population dynamics and regulation of populations by parasites and predators have contributed towards a litter appreciation of the factors enabling these aspects. In general, insects like ants and termites are considered to be good ecosystem engineers. They provide the basic tools for the study of genetic and molecular markers which have commercial value.

A predatory ant

Ants milking aphids

Plate 1.

A pollinating bee

A butterfly in the act of pollination

Plate 2.

A pyrrhocorid bug with its offspring

The monarch butterflies in flight

Plate 3.

Different species tend to adopt different foraging strategies adapted to the exploration of resources. Ecologically and behaviourally insects tend to differ and inspite of their diverse morphological traits, they are very species specific in their habitat. In ecologically sensitive areas insects like the Lepidoptera play a significant role in plant pollination and biological control of weeds. Diptera pollinate several crops of importance, not to mention of some flies which also involve in pollen gathering (Mitra and Banerjee, 2007). Different species of ants are functionally grouped into harvesters, scavengers, engineering ants, hunters and mushroom growers. Trail laying ants by dint of their pheromones tend to have both recruitment and orientation effects (Foottit and Adler, 2009).

ORIENTATION DYNAMICS

Insect orientation maneuvers involve changes in body position and related functional aspects. The long distance migration of thousands of kilometers by monarch butterflies to their overwintering site, not to mention of the migratory movement of locusts involving pheromones enhancing the tendency to group, are typical examples. Accessory glands of gregarious females promote gregarious behaviour. Facilitating movement towards or away from bright light involves *klinokinesis*, while *lelotaxis* is a direct form of orientation in which the intensity of the stimulus is processed. Monarchs and honey bees adopt a 'compass orientation' called *menotaxis* which is also responsible for the attraction to lights in nocturnal insects. The end points of orientation are attraction and aggregation. Other types of behavioural orientation are positional, object orientation, home range orientation and geographic orientation. Orientation that enables insects to fly upwind as in male moths flying over distances to a pheromones releasing female, and mosquitoes flying hundreds of meters to a prospective vertebrate host are common instances. Walking insects use mechanosensory inputs to determine wind direction. Gryllotalpids or mole crickets use well developed forelegs for digging deep galleries for foraging plant roots. In short, insects create the biological foundation for all terrestrial ecosystems through their cycling nutrients, pollinating plants dispersing seeds and maintaining soil structure and fertility (Majer, 1987).

Some honeybee species like xylocopids, scent mark flowers, informing others of the same species of a recently visited flower.

Some establish a scent trail to inform nestmates as to where a good floral source can be located. Lepidoptera undergo a sequence of behaviour, including searching, orientation, encounter, landing, evaluation of surface and acceptance. During searching behaviour, visual clues such as plant colour and shape are significant. Besides volatile chemicals produced by the plant, initiate the orientation and encounter behaviour that attract females from a distance. To enhance dispersal, phoresy occurs through the attachment of an organism to the host, enabling colonization of a new habitat. In addition to transport, the phoretic host provides suitable shelter.

IMMIGRANT INSECTS

Organisms that colonize a new area tend to affect native communities and ecosystem. Immigrant insects continue to the principal threats to native biodiversity. Climate change also tends to affect their abundance and distribution, affecting ecosystem structure and function. Establishment of immigrant mosquitoes also tend to serve as disease vectors. Pollination of native plants also tends to be disrupted when introduced honey bees compete for floral resources with native plants. White flies such as *Bemisia tabaci* species complex are transported with commerce and transmit several viruses, not mention of thrips like *Frankliniella*, the main vector of topsovirus causing epidermis of tomato spotted wilt. In so far as mosquitoes are concerned, introduction of vector species tend to cause epidemic malaria. Other disease viruses like dengue virus and viral agents causing encephalitis and also introduced by immigrant mosquitoes (Juliano and Lounibos, 2005). Immigrant ants tend to affect seed dispersal and pollination. The behaviour of alien insects in novel environments and their spectacular success in such environments in exemplified by the introduction of *Rodolia cardinalis* to suppress the extensive invasion of prickly gear caches. While regulating the abundance of angiosperms, insects create the biological foundation for all terrestrial organisms. Exclusion, detection, eradication and control are the principal means of dealing with invasive species (Wheeler and Hoeebeke, 2009) and the need to know the extent of distribution of invasive species cannot be overemphasized.

Aposematism wherein insects like stick insects, leaf mimicking moths and caterpillar patterned and coloured like the vegetation they consume or twigs or bark they inhabit. Some resemble dead objects as in cryptic insects. Aposematism is used by insects for warming colouration sporting bright colour patterns. Warming coloration boldly signals that the insect is toxic. Aposematic signals are produced by a wide range of insects as in many groups of beetles and bugs. Other cryptic insects resemble dead objects as in bird droppings and these insects tend to match in their life style behaviour. Cryptic insects tend to select resting and matching backgrounds. Warming colouration boldly signals that the insect is toxic or injurious. Aggregation as defense is useful if individuals are toxic as in coccinellid and milkweed bugs.

Soil texture and structure tend to have a direct impact on insect behaviour and adaptation. Soil insects and other arthropods more vertically in response to soil temperature, moving downwards in winter and resume feeding with temperature increase (Villani *et al.,* 1999).

APOSEMATIC BEHAVIOUR

Aposemation is used by insects for warning colouration sporting bright colour patterns. Warning colouration boldly signals that the insect is toxic. stick insects, leaf mimic insects such as moths and many caterpillars are patterned and coloured like the vegetation they consume or twigs or bank they inhabit. Some resemble dead objects as in many cryptic insects Aposematic signals are produced by a wide range of insects as in many groups of beetles, reduviids, to mention a few other cryptic insects resemble dead objects as stones, bird droppings and tend to select resting and matching backgrounds. Warning colouration signals that the insect is toxic or injurious.

GALLS AND GALLING BEHAVIOUR

Abnormal outgrowths develop on vascular plants are termed galls which are defined as complex biological expression involving the inducing organism and the host plant. Galling behaviour is well known in different groups of insects such as Thysanoptera, Hymenoptera, Diptera, and has been defined as a series of complex and coordinated interactions between the concerned plant and the insect. As defined by Raman *et al.,* (2005) gall is an expression of suite of adaptation achieved by the host plant for the inducer, the insect displaying diverse feeding behaviour, the biology and behaviour of the inducing insect being a critical factor, along with other; life history performance, feeding, oviposition behaviour, host plant selection being key factor facilitating gall induction. Raman's objective view that the earlier publication of Ananthakrishnan on Biology of Gall Insects (1984) concentrating on the biological and behavioural aspects of galling and from the entomologists view point behaviour of insects often relevant. To quote Raman *et al.,* (2005) "The inducing arthropod also demonstrates several adaptation in its life history. Gall inducing arthropods are highly host specific and have the ability to select the right host plant and the light site on that host plant, in a natural environment where different plant species coexist... and often display a sophisticated biology and physiology, rendering more efficient users of host plants". Gall-inducing arthropods display an interaction demand for a particular host plant to complete their life cycles as evident in highly specialized feeding and reproduction behaviours and a high level of specificity to particular host plants. From leaf curls, leaf folds, leaf rolls, leaf surface projection and convoluted pouches Tubuliferan thrips revel in gall formation. This diversity in thrips species and their functional involvement have made the study of cecidology unique.

The structure and dynamics of food webs can have profound effects as a result of movement of insects from one habitat to another. Studies on insect dispersal from this angle could contribute towards our knowledge of community ecology (Thompson, 2006). Movement of insects from one habitat to another tends to have a profound effect on the structure and dynamics of food webs (Polis *et al.,* 2004). Insect colonies are highly structured and respond to changing needs and this plasticity constitutes to the reproduction success of colonies. An insect colony in likened to a super organism as a result of the behaviour of individual workers. In advanced ensocial species there is also plasticity in the division of labour and this is achieved by diverse mechanism of behavioural integration.

ACKNOWLEDGEMENT

The author wishes to thank his son A. Ramdas Iyer, New Jersey for the photographs of insects in their natural set-up (Plate 1-3). My thanks are due to Prof. Ralph Holzenthal of the Department of Entomology, University of Minnesota, St. Paul, for loan of reference books and personal discussions.

REFERENCES

Foottit, R.G. and Adler, P.H. 2009. Adaptive strategies of adaptive arthropods. Ann. Rev. Entomol. 44: 233-256.

Juliano, S.A. & Lounibos, L.P. 2005. Ecology of invasive mosquitoes: effects on resident species and on human health. *Ecology Letters*, 8: 558-574.

Majer, J.D. 1987. The conservation and study of invertebrates in remnants of native vegetation. In D.A. Sounders, G.W. Arnold, A.A. Burbridge and A.J.M. Hopkins (eds.): Nature Conservation: Role of remnants of native vegetation, Survey Beatty and Sons, Sidney.

Miller, J.C. 1993. Insect natural history, multispecies interaction and biodiversity in ecosystem. Biodiversity and Conservation 2: 233-241.

Mitra, B. and Banerjee, D. 2007. Fly pollination: assessing their value in biodiversity and food security. Rec. Zool. Sur. India 107 (pt. 1): 33-48.

Polis, G.A., M.E. Power and G.R. Huxel, (eds.) 2004. Food webs at the landscape level. The University of Chicago Press, Chicago, Illinois.

Raman, A., Schafer, C.W. and Withers, T.M. 2005. Biology, Ecology and Evolution of gall inducing arthropods, Volume 1 & 2. *Science Publishers Inc, Enfield*, USA.

Thompson, R.M. 2006. Embedding food webs in ecological land capes. Global ecology and Biogeography 15: 109.

Villani, M.G., Albe L.L., Diaz, A. and Robins, P.S. 1999 Adaptive strategies of edaphic arthropods. Ann. Rev. Entomol. 44: 233-256.

Wheeler, A.G. (Jr.) and Hoebeke, E.R. 2009. In: Footit, R. and Adler (eds.): Adventive (non-native) insects: importance to security of society, 1st ed. Blackwell Publishing House: pages 476-521.

Dynamics of Insect Behaviour
T.N. Ananthakrishnan & K.G. Sivaramakrishnan
Scientific Publishers (India), Jodhpur, 2012, pp. 28-33

Chapter - 4

BEHAVIOURAL DIVERSITIES IN RELATION TO ECOLOGICAL SERVICES RENDERED BY INSECTS

T.N. Ananthakrishnan

Insects modify, maintain and create habitat playing a positive role in their formation, modification and maintenance. By so doing they influence community composition and directly or indirectly create or modulate the availability of resources by causing changes in their physical state. Ecosystem engineers, as they are usually designated modulate the supply of resources. Many single species of trees are good examples, since they modulate the flow of diverse resources to forest inhabitants. All ecosystems are controlled and organized by a set of key plant and animal processes that nurture the landscape at different scales (Holing, 1992). Different species of plants produce different quantitative and qualitative amounts of living and dead fuel, profoundly affecting the supply of resources to many species. Besides, dead leaves or twigs accumulating as litter tend to alter the microenvironment of the soil, resulting in decreased drainage. They also act as physical barriers for seeds and seedlings, ultimately affecting the structure and composition of communities (Facelli and Pickett, 1991). Our understanding of diverse ecological interactions is enhanced by the dimensions of engineering activities of insects, as observable in the new range of habitat niches exploited by a wide variety of specialists in these habitats, significantly contributing to species diversity. Insects also provide essential services such as weed suppression, and plant and animal decomposition as well as soil improvement facilitated by exotic herbivores. The rare occurrence of species such as that of *Leeuwenia ananthakrishnani* from oak plantation of Manipur and Nagaland at altitudes ranging from 900 – 1200 M colonizing tender leaves of oak, *Quercus sirrata,* the gravid females laying 66 ± 5 eggs, with adults protecting them till hatching (Fig. 1). Predatory coccinellids such as *Coccinella septumpunctata* and *Harmonia demidiata* feed on larval form. A colony will have an average of 65 egg/ leaf and emerging young and adults feeding gregariously, make the foliage completely devoid of chlorophyll (Courtesy Raju). Ants farm fungi, raise aphids as livestock, use chemical sprays to alarm and confuse enemies and have a massive environmental impact. Other ants and termite species are specialized to cultivate fungi underground, planting the spores, weeding gardens and fertilize them with compost mulched by chewed up leaves (Dawkins, 1996). The common house ant *Monomorium* performs the antennal tapping behaviour on aphids, stimulating them to excrete honey dew droplets, which the ants consume (Fig. 2). Ants modify their behaviour from tending aphids to acting as predator. In short, *antogenic engineers* change the environment via their physical structures, while *allogenic engineers* change the environment by transforming living or non-living material from one physical state to another (Jones *et al.,* 1993). Changing the environment by transforming living or non-living materials from one physical state to another is equally common (Jones *et al.,* 1993). Needless to indicate that insects being the most successful organisms contribute significantly to vital ecological functions such as pollination, gall formation, pest control and decomposition. An assessment of the value of these function become important to establish the stability of their function.

Pollination is the best form of ecosystem service performed by insects especially that by managed insects like honey bees. Native bees tend to provide all the pollination needs of some crops, with 90% of insect pollination of water melon being by honey bees (Kremen *et al.,* 2002, 2004). While honey bees do perform this behaviour, bumble bees tend to sonicate the flowers, resulting in cross pollination (Green leaf and Kremer, 2006). On an average, the amount of seed set is doubled by honey bees alone. In short, pollination modulates the supply of resources.

The size and growth form and variety of above ground parts have broad effects on plant architecture and herbivore diversity. Such studies reveal the effect of habitat diversity on species diversity. The resource diversity hypothesis, viz., plants with greater variety of resource types, viz., feeding resting, overwintering and hiding places support more herbivorous insects (Lawton, 1983).

Feeding and oviposition behaviour of gall inducing insects, host plant and habitat selection are key factors in the biology and ecology of gall insects. Galling has developed into a separate discipline, cecidology, involving the intimate interaction between the insect and the plants. Gall is an expression of a suite of adaptations achieved by the host plant for the inducer, including provisions of food and shelter (Raman, 2006). Gall diversity is unique, being formed in diverse plant parts. Feeding and site selection behaviour of the inducing insect determines the nature and shape of the gall. Galls are highly host specific and gall inducers belong to different natural orders display varied feeding behaviour. Gall is an expression of the host plant to isolate the insects in space and time, altering the host plant to its advantage. Both feeding and oviposition stimulate gall initiation. Gall inducing insects have the ability to discriminate and choose specific host plants to its advantage. They have the ability to discriminate and choose specific host plants and feed on particular plant parts of its specific host. The biology and ecology of the gallers is governed by life history performance, feeding, oviposition behaviour and host plant selection. Plant chemistry appears to control host specificity patterns among gallers (Anathakrishnan and Raman, 1977).

While innumerable instances of gall diversity are known from several host plants, induced by several groups of insects, some of the mere striking instances relate to the rosette galls on *Acacia leucophloea* by *Thilakothrips babuli* pouch gall induced by *Anstrothrips cochinchinensis* on *Getomia floribunda*, *Gnorimoschema gallaesolidaginis*, a lepidopteran stem galler on *Solidago* sp., increasing stem biomass, interestingly gall inducers on seeds are evident in seeds of Eucalyptus sp. induced by *Quadrastichodella*. Echinate galls are induced on flowers and galls of sweet pepper and brinjal, by *Ceratoneura indi*. Needless to indicate that the diversity of galls formed by insects is considerable, with the more common gall forms as leaf folds, leaf rolls, leaf curls, bud and inflorescence galls, enabling insects to survive in protected environment. In short, gall is an expression of a suite of adaptations achieved by the host plant for the inducer including food and shelter (Raman, 2006).

Among the diverse ecological services, soil fertilization by dung beetles of the sub family Scarabaeinae, play a significant role in soil nutrient cycling, soil fertilization, bioturbation or the displacement and mixing of sediment particles by animals and plants and seed dispersal, besides the removal and burial of dung from the surface to the soil (Nicholas *et al.,* 2008). The benefits that dung beetles provide to the soil environment from forest and pasture environments are considerable. Dung beetles exhibit a variety of dung processing strategies, with some species rolling dung balls away from the dung pile and bury them, as in dung rollers. Others deposit dung in burrow tunnels with branching brood chambers directly beneath the dung pile as in burrowers or tunnelers. Some stay within the dung pile as in endocoprids or dwellers. The most common dung removal and processing behaviour among dung beetles, is tunneling. Tunnelling species are numerous, very beneficial to soil health, enhancing the soil by increasing soil aeration and percolation, inducing organic matter with the soil (Nichols *et al.*, 2008). Dung beetles provide a good example of engineering in that a large proportion of dung deposited in the ground is processed by the dung beetles which promote decomposition of dung into labile form of nitrogen that can be assimilated by plants, thereby function as a fertilizer when dung is buried (Losey and Vaughan, 2006).

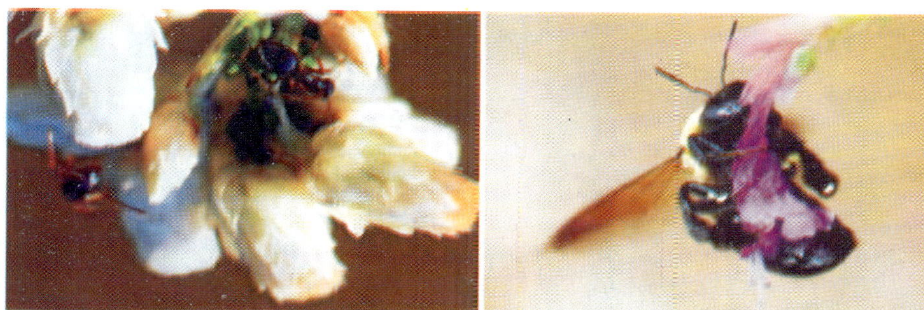

Fig. 1. Ant tending aphids Bee pollination (Courtesy: A. Ramdas)

Fig. 2. Butterfly in pollinating act Bee pollination (Courtesy: A. Ramdas)

Onthophagus dama

Coccobius meridionalis

L. ananthakrishnani (♀)

Colony comprising different life stages of *L. ananthakrishnani* on oak leaf

Coccinella septempunteta

Larvae of *Harmonia demidiata* feeding on thrips

Yellow pan water trap to collect thrips from Oak plantation.

With regard to phytophagous species plant size, design and structural complexity influence insect diversity. Plants with diverse varieties of resources tend to support more species of insects, than those with limited resources. Insects need oviposition sites, shelter, overwintering places and in all these aspects trees and bushes provide more opportunities. Needless to emphasize that twigs, bark and buds are efficient overwintering and oviposition sites, not to mention the occurrence of some species in the dead wood or lichen covered bark gold trees (Banerjee, 1981).

Needless to emphasize that the phenomenon of role of insects as ecosystem engineers suggest that diverse groups of insects supported by diverse habitats are influenced by their interactions.

REFERENCES

Ananthakrishnan, T.N. and Raman, A. 1977. In: T.N. Ananthakrishnan and A. Raman (eds.): Aspects of host specificity in some cecidogenon insects: Macmillan, India: pages 51-56.

Banerjee, B. 1981. An analysis of the effect of latitude age and area on the number of arthropod pest species on Tea. Journal of Applied Entomology 18: 339-342.

Dawkins, R. 1996. Climbing Mount Impossible, Oxford University Press, New York, London.

Facelli, J.M. and Pickett S.T.A 1991. Plant litter: Its dynamics and effects on plant community and structure. Science 185: 1058-1060.

Holling, C.S. 1992. Cross-scale morphology, geometry and dynamics of ecosystems Ecological Monitoring. 62: 447-502.

Jones, G., Lawton, J.H. and Shachak, M. 1993. Organisms as ecological engineers. Oikos 69: 373-386.

Kremen, C., Williams, N.M. and Thorp, R.W. 2002. Crop pollination from native bees at risk from agricultural intensification. *P Natl Acad Sci USA* 99: 16812–16.

Kremen, C., N.M. Williams, R.L. Bugg, J.P. Fay and R.W. Thorp. 2004. The area requirements of an ecosystem service: crop pollination by native bee communities in California. Ecology Letters 7:1109–1119.

Lawton, J.H. 1983. Plant architecture and the diversity of phytophagous insects. Annual Review of Entomology 28: 23-39.

Losey, J.E. and Vaughan, M. 2006. The economic value of ecological services provided by insects. Bioscience 56(4): 311-323.

Nichols, E., Spector, S., Lonzada, J., Larsen, T., Amezquita, S. and Favila, M.E. 2008. Ecological functions and ecosystem services provided by Scarabaeniae dung beetles. *Biological Conservation* 141(6): 1461-1474.

Raman, T.R.S. 2006. Effects of habitat structure and adjacent habitats on birds in tropical rainforest fragments and shaded plantations in the Western Ghats, India. *Biodiversity and Conservation* 15:1577-1607.

Dynamics of Insect Behaviour
T.N. Ananthakrishnan & K.G. Sivaramakrishnan
Scientific Publishers (India), Jodhpur, 2012, pp. 34-70

Chapter - 5

GALL-INDUCING BEHAVIOUR IN HEMIPTEROID INSECTS*

Anantanarayanan Raman

INTRODUCTION

Among phytophagous insects, certain species interact intimately with plants and induce specialized enclosing structures — the galls. Galls represent highly regulated growth manifestations on plants within which the inducing insects and their progeny live. Although these are usually absent in normal plants, nearly 90% of known galls develop as geometrical forms. Galls are the best examples for modified, natural-plant structures that arise solely because of messages from the inducing insects. These structures develop as an extension of the host-plant phenotype (Raman, 2011). That the morphology and physiology of the plant influence the biology, ecology, and evolution of a plant-feeding insect is established: interactions between diverse species of flowering plants (e.g., *Quercus*, *Rosa*) and different species of gall-inducing Cynipidae (Hymenoptera) demonstrate that plant architecture influences the oviposition behaviour of the Cynipidae (Abrahamson *et al.,* 1998, Shorthouse *et al.,* 2005). By inducing a gall, the insect ensures nutrition

---

* To Warren Gene Abrahamson III (Bucknell University, Lewisburg, Pennsylvania, USA), who blazed new trails in interpreting the evolutionary ecology of galls and inducing insects. Memories of my stay in his laboratory in 1990–1991 are pleasant.

and shelter for either shorter or longer periods of its life. Because of their usually concealed habit ('pit' galls induced by some Sternorrhyncha excepted), compared with non-gall inducing insects, the gall-inducing insects generally show novel traits in their nutritional physiology and population dynamics (Raman *et al.*, 2005a). Gall-inducing habit has evolved indepe-ndently and multiple times among insects (e.g., Cook and Gullan, 2004). Gall-inducing capability could be a characteristic of a subgroup: e.g., Ophelimini (Hymenoptera: Eulophidae) (LaSalle, 2005).

Ability to induce galls does not occur in all orders of Insecta. Within plant-feeding groups, gall-inducing capability principally occurs among insects belonging to Thysanoptera, Hemiptera, Diptera, and Hymenoptera, and in a lesser frequency among those of Lepidoptera and Coleoptera. Even in these orders, gall-inducing habit does not present any logical pattern. For example, the gall-inducing capability in the Psylloidea is high in the Triozidae, but low among the Psyllidae, Phacopteronidae, and Calophyidae (D Burckhardt, Basel, Switzerland, personal communication, e-mail 5 November, 2010). A similar pattern occurs in the Thysanoptera, with a majority of gall-inducing taxa restricted to the Tubulifera (viz., Phlaeothripidae), whereas less than five taxa are known, as of today, as gall inducers in the Terebrantia (Ananthakrishnan and Raman, 1989, Mound and Morris, 2005). Considering gall-inducing capability at the Order level, most of the gall-inducing Hemiptera remain in the Steronorrhyncha. However, the reports of *Aphelaenus nigripectus* (Hemiptera: Aphrophoridae) as roll-gall inducers on the leaves of *Prunus speciosa*, *P. yedoensis*, and *P. pendula* cv. *pendula* (Rosaceae) in Kyoto (Japan) (Sugiura and Yamazaki, 2003), *Cicadulina bipunctata* (Hemiptera: Cicadellidae) on different species of Poaceae (e.g., *Zea mays*, *Oryza sativa*, and species of *Triticum*) (Matsukura *et al.,* 2010), and a few scattered historical references to galls induced by species of Cercopoidea and Membracoidea (Meyer 1987, Raman *et al.,* 2005a) need to be recognized. In Western Kazakhstan (northeastern Kyzylkum desert near Baltakul, 43°12'N, 67°77'E), *Scenergates viridis* (= *Papyrina viridis*) (Hemiptera: Cicadellidae) induces closed leaf galls on a species of *Alhagi* (Fabaceae). Each gall on *Alhagi* would include one cicadellid, either an adult or an immature stage. Gall development commences with the feeding action of a young nymph on the midrib of a young leaf, which develops as a leaf fold concealing the insect (Mitjaev, 1968). *Copium* and *Paracopium* (Hemiptera: Tingidae) are the two taxa (plus three others: see Werner, 2000) claimed as gall inducers on different species of *Teucrium* (Lamiaceae) and *Clerodendrum* (Verbenaceae) (Schaefer, 2005), Four species of *Copium* remain tied to European *Teucrium*, whereas *C. japonicum* in the Eastern Palaearctic infests *Kesukea japonica*, also a taxon of Lamiaceae; about 40 species of *Paracopium* are hosted by species of *Clerodendrum* and *Scaevola* (Goodeniaceae) distributed in the Pantropics and Pan-Pacifics (Werner, 2000). The most preferred host plant of *Copium clavicorne* is *Teucrium chamaedrys*. Yet a few taxa from section *Teucrium chamaedrys* (*T. canum*, *T. marum*, and *T. flavum*) also serve as hosts of *C. clavicorne*. Host plants of *C. teucrii* occur on taxa of section *Teucrium polium* (*T. montanum, T. polium, T. capitatum, T. aureum*), and also on *T. radicans* (Zalat *et al.,* 2000, Lambinon and Schneider, 2004). These auchenorrhynchous taxa induce 'low' level alterations (viz., curls and crinkles) on leaves and flowers, and a few of them are specific to their hosts (Murphy, 1989). However, these Auchenorrhyncha are generally polyphagous, whereas the other gall-inducing taxa in the Cynipidae, Cecidomyiidae, Psylloidea, and Phlaeothripidae are highly monophagous. The debatable point is whether the Auchenorryncha induce 'true' galls as the Sternorrhyncha do[1] . Alternatively, do the 'galls' induced by non-sternorrhychous Hemiptera offer opportunities to explain the evolution of gall-inducing behaviour among the hemipteroids?

Regional lists of arthropod-induced galls (e.g., Trotter and Cecconi 1902, Docters van Leeuwen–Reijnvaan and Docters van Leeuwen 1926, Felt 1965, Mani 2000) indicate that nearly 90% of the galls occur on dicotyledons, few on monocotyledons, and still fewer on grasses (Mani 1964). Established as a few examples of galls on grasses are those induced by *Orseolia oryzae* (Diptera: Cecidomyiidae) in Asia (Gagné 2004, Krishnaiah 2004) and *O. oryzivora* in Africa (Harris and Gagné 1982) on different varieties of *Oryza sativa* and *O. perennis*

---

1 Daniel Burckhardt [email, 5 November 2010], remarked: "I wonder whether the high percentage of gall-inducing taxa in the Sternorrhyncha compared with Auchenorrhyncha is a direct consequence of their feeding habits." Roman Rakitov remarked [email, 10 November 2010]: "A majority of Auchenorrhyncha feed on phloem, similar to many of the Sternorrhyncha, but those of Cercopoidea and Cicadoidea feed on xylem. A majority of Cicadellidae feed on phloem, but the Cicadellinae and related subfamilies feed on xylem. Species of a single but diverse Typhlocybinae (Cicadellidae) feed from individual cells and they are unique in displaying such behaviour within Auchenorrhyncha".

(Poaceae), at least four species of gall midges that induce galls on *Phragmites australis* (= *P. communis*) (Poaceae) in Central Europe (Skuhravá and Skuhravý 1992), and a few Northern European species of Chloropidae (Diptera) inducing galls on different species of Poaceae (De Bruyn 2005). In such a context, the puzzling questions are 'why galls on monocotyledons are so few?' and 'are those claimed as galls on monocotyledons true galls?' The latter question gains in significance because the structure of 'galls' induced by *Giraudiella inclusa* (Diptera: Cecidomyiidae) on *P. australis* (Rohfritsch 1975) defies established cecidogenetic patterns (Rohfritsch 1992). Gall induction reported on monocotyledons, on grasses in particular, needs to be examined closely to arrive at generic conclusions.

Specific differentiation processes control growth and development of galls. Among the different insect-induced galls, distinct variations in differentiation and in final gall shapes occur. Insects attack plant tissues resulting in the alteration of the subcellular environment and placing those tissues in a state of chemical shock. This shock entails osmotic changes in the cells attacked, which establishes the earliest stage in gall induction. To neutralize the stress consequent to osmotic changes, aggravated by wounding, the plant responds by developing usually one, sometimes 2–3, metaplasied cell(s). Localized metabolic changes diffuse from the metaplasied cell(s), but not throughout either the involved plant organ (e.g., leaves) or the plant (Rohfritsch, 1978, 1980, 1992). When the shock is of low intensity, the plant responds with certain chemical–molecular factors that disperse from the metaplasied cell(s) triggering gall development. When the shock factor is of high intensity the cells under the influence of the insect (e.g., *Dasineura marginem-torquens*, Diptera: Cecidomyiidae; *Daktulosphaira vitifoliae*, Hemiptera: Phylloxeridae) die rejecting the inducing insect and thus defending the plant tissue (Ollerstam *et al.,* 2002; Raman *et al.,* 2009a).

Insects feed on gall tissue continuously for a specific period, throughout their developmental phase, and this beahviour usually synchronizes with their life history. Osmotic-change related stress prevails during that period of time, which in turn stimulates a sequence of plant-mediated changes including alterations in gas exchange and synthesis of growth promotors. Osmotic stress affects electrical properties of the plasma membrane and impacts on IAA activity, which in turn, also alters H^+–transport systems. During the physical action of insect feeding, the host-cell wall breaks down, and the degenerated wall materials act as elicitors. Gall-hosting plants employ varied strategies to mitigate and neutralize stress consequent to gall induction. Although the neutralizing strategies exist in the genetic constitution of the host plants, they are mediated by complex molecular interactions. Plants generally use a flexible short-term strategy to respond to stress; organisms that can modify gene expression reversibly have an advantage in evolutionary terms, since they can avoid rearrangements and species diversification. Mechanisms of DNA methylation and histone modifications possibly regulate inheritance of stress 'memories'. Inherited genetic traits also play a role in gall morphogenesis, followed by roles played by correlating morphogenetic factors.

The sequence of steps in the process of gall development are: osmotic-shock related stress → establishment of either one or a group of metaplasied cells → growth promotor-mediated cell expansion → commitment of the metaplasied cell(s) starting the 'novel' cell-cycle patterns, cell multiplication, and programmed differentiation. Most critically, galls are controlled manifestations that are, usually, symmetrical — either radial or bilateral — when mature. Each gall is unique because the inducing insect species controls its shape (Raman, 2011).

Gall initiation and development

Key phases in gall development are induction and growth; the latter includes increment in mass and qualitative differentiation. The earliest recognizable event is the isolation and insulation of either one or a few cells on the attacked plant organ (e.g., leaf) from their normal course of differentiation organizing a few metaplasied cells that are irreversibly changed by the stimulus from the inducing insect (Rohfritsch, 1978, 1980, 1992, Meyer and Maresquelle, 1983, Raman, 1991, 1996, Raman, *et al.,* 2005b). A chemical factor obviously stimulates division activity in the metaplasied cell(s) triggering the gall (Bronner, 1973, Rohfritsch, 1992, Rey 1992). The chemical factors variously implicated are high-molecular weight proteins (Carango *et al.,* 1988), bruchins (Doss *et al.,* 2000), and mitogenic lipids (Farmer, 2000). This stimulus activates and regulates growth by triggering novel patterns of differentiation, which ensue at the metaplasied site (Harper *et al.,* 2004).

During early stages, gall-inducing insects evoke stress responses at their attack sites, such as accumulation of active peroxisomes (Corpas *et al.,* 2004) and glyoxysomes (Kim *et al.,* 2004), The key role played by oxysomal microbodies in stress neutralization mediated by the production of reactive-oxygen species is established (Pastori and Rio, 1994). Stress responses evoke gene expression from cells in the vicinity, which in turn, activate new growth and differentiation, resulting in a gall offering nutrition and shelter to the inducing insect (Raman, 2010, 2011). Continuing stimulus from the insect is absolutely essential to control gall shape and structure (Rohfritsch, 1971). Gall development, in principle, involves two counteracting events: the insect stresses the host organ; the host counters it with new physiological activities supplemented by newly differentiated tissues. In the galls induced by *Phacopteron lentiginosum* (Hemiptera: Phacopteronidae) on the leaves of *Garuga pinnata* (Burseraceae) (Raman, 1987a, Burckhardt, 2005), a sac-like gall develops from within the leaf, the most complex gall type among those induced by different species of Phacopteronidae (Malenovský *et al.,* 2007). A vegetative mersitem differentiates within the primordial mesophyll, which grows into the saccular gall housing developing stages of *P. lentiginosum*. *Phacopteron lentiginosum*'s action has limited impact on the adaxial and abaxial epidermal cells of the host leaf. The gall, which shoots through the leaf, expresses itself differently on either side of the lamina by differentiating its own epidermises with varying structures: cells at the gall summit are similar to those of the host leaf, whereas those along the mid and lower regions, differentiated through intercalary meristematic activity, vary in their morphologies and even generate multicellular trichomes that are absent on normal leaves (Raman, 1993).

Ninety per cent of gall-inducing insects are specific to their hosts (Raman *et al.,* 2005b). Their levels of host fidelity are remarkable compared with those of related, but non-gall-inducing, plant-feeding taxa (Raman 1996). Fidelity of gall-inducing species of North-American Cynipidae (Hymenoptera) to *Quercus* (Fagaceae) demonstrates a high degree of monophagism (Abrahamson *et al.,* 1998); many similar examples from other groups of gall-inducing insects are available (Raman 1996, Gagné 2004, see various chapters in Raman *et al.,* 2005a). This trait is conservatively preserved among gall-inducing insects. For example, out of the nearly 250 insect species known to utilize *Mangifera indica* (Anacardiaceae), the 25 established taxa are gall-inducing Cecidomyiidae (Diptera), most of which are species of *Procontarinia* (Cecidomyiidae: Cecidomyiini) (Gagné 2004). Adaptive radiation of *Procontarinia* on *M. indica* has progressed conservatively among the diverse *M. indica*-infesting gall-inducing Cecidomyiidae, with a majority of them continuing to utilize the leaf. Species of *Procontarinia* on *M. indica* exhibit minimal variation in their overall gall shapes, in patterns of tissue differentiation, in the nature of dehiscence at maturity, on their location on leaves, and in the numbers of larval chambers per gall (Raman *et al.,* 2009b).

Although concrete evidences explaining how gall-inducing insects recognize their specific host plants and organs are not available, subtle evidences point to that host-plant surface waxes and lectin-recognizable carbohydrates (Smith 2005) act facilitate recognition. Chitosan (a polymer of ß–1,4–linked glucosamine) with a strong affinity for DNA was localized in the nuclei of leaf cells of *Solanum dulcamara* (Solanaceae) punctured by *Aceria cladophthirus* (Acarina: Eriophyidae) and also at the feeding sites on leaf-cell walls (Bronner *et al.,* 1989). Chitosan transmitted by *A. cladophthirus* induced changes in the DNA in cells under the direct feeding impact; accumulation of chitosan within nuclei and at feeding points on the walls could be the trigger factor (or, one of many factors?) for division activity in cells in the neighbourhood of the laminae of *S. dulcamara* eventuating in the gall. Polygalacturonase activity in the saliva of *Aceria caulobia* on *Suaeda fruticosa* (Chenopodiaceae) (Monfreda and Spagnuolo 2004, de Lillo and Monfreda 2004) has also been shown to be a stimulus-recognition factor eliciting response signals. Subcellular incompatible responses have been demonstrated in the *A. cladophthirus*–*S. dulcamara* system (Westphal 1980, Westphal *et al.,* 1981). Vacuolar alkalinization in the epidermal cells punctured by feeding *A. cladophthirus* within 24 min of attack has been shown as a key resistance response (Westphal 1982). During incompatible reactions, callose accumulates in unattacked cells and autofluorescent, phytolaexin-like materials accumulate in cells around the lesion; pathogenesis-related proteins also build up in resistant tissues (Westphal 1992). Building up of pathogenesis-related proteins (e.g., osmotines) in alkaline vacuoles indicate accelerated senescence similar to host-plant resistance to microbial-plant pathogens (Memelink *et al.,* 1990), but no more is known in the context of gall-inducing arthropods at early stages of their interactions with plants.

Whereas we know much of cell structure and physiology of plant galls, the nature and functions of signal receptors and steps involved in the signal-transduction pathway have not been convincingly explained. Genes analogous to the 'Nod' factors[2] involved in the signal transduction in the development of root nodules induced by *Rhizobium* (Lerouge *et al.*, 1990) have been characterized in galls induced by *Diplolepis spinosa* (Hymenoptera: Cynipidae) on *Rosa spinosa* (Rosaceae) (Schonrögge *et al.*, 1998) suggesting that these genes act as signal molecules in triggering gall induction. The Nod factors interrupt cytokinin-signal transduction pathway, by introducing concurrent changes in auxin-related responses during root-nodule initiation, in high likelihood, through increased levels and action of flavonoids (Mathesius *et al.*, 1998). Flavonoids occur at 'low — negligible' levels in insect-induced gall systems (Nyman and Julkunen-Tiitto 2000, Motta *et al.*, 2005), although soluble phenolic materials and tannins are usually high. Isolation of castasterone-4 (a brassinosteroid) from insect-induced galls of *Castanea crenata* (Fabaceae) (Yokota *et al.*, 1982) prompts the exploration of brassinosteroids in triggering signal-transduction pathways in galls.

SCOPE OF THIS CHAPTER

Because considerable literature on the gall-inducing behaviour of the specialist gall-inducing insects belonging to the Diptera (Cecidomyiidae: Skuhravá *et al.*, 1984, Yukawa and Rohfritsch 2005) and Hymenoptera (Cynipidae: Askew 1984, Csóka *et al.*, 2005) exists, in this chapter, I would restrict to a précis of gall-inducing behaviour among Thysanoptera and sternorrhynchous Hemiptera. Hemipteroid insects induce galls through feeding action by either juvenile stages or adults, sometimes by both, which in principle, is similar to the gall-

---

2. Nodules (free-nitrogen fixing sites) develop in the roots of legumes consequent to an association between the plant and symbiotic soil bacteria, the rhizobia. During early phases of association between the legumes and rhizobia, certain molecular-signalling molecules (MSM) are synthesized by the rhizobia. These MSM-s, now known as the 'Nod' factors — characterized as lipochito-oligosaccharides — are the key for initiating nodulation in legumes. Structures of these oligosaccharides vary with species and contribute to host specificity (Dénarié et al. 1996). For example, in the nodules of *Medicago truncatulata* (Fabaceae) the Nod factors of *Rhizobium meliloti* ('NodRm') possess a SO_4 moiety attached to the reducing end of the chitin tetrasaccharide backbone, further to an acetyl group and an unsaturated lipid chain attached to the non-reducing sugar (Crespi and Gálvez 2000).

inducing behaviour in the Cecidomyiidae (Diptera). The Hymenoptera (Tenthredinidae, Braconidae, Chalcidoidea, and Cynipidae), on the contrary, induce galls by inserting their eggs into plant tissue. For an understanding the role of oviposition and eggs in triggering gall development among the Hymenoptera, the reader is referred to Bronner (1973) and Rey (1976), and several citations in them.

THYSANOPTERA

Thysanoptera (thrips)-induced galls at maturity house several individuals in different developmental stages. Thrips-induced galls occur on leaves in far higher frequency than those induced on vegetative buds and flowers (Ananthakrishnan and Raman 1989). An extensive variation in final gall shapes usually manifests: from simple leaf distortions and puckering to leaf folds and rolls, and of complex shapes. Extreme gall shapes include rosettes (e.g., *Thilakothrips babuli* [Phlaeothripidae]—*Acacia leucophloea* [Fabaceae: Mimosoideae]), pouches (e.g., *Austrothrips cochinchinensis* [Phlaeothripidae]—*Calycopteris floribunda* [Combretaceae]), and horn-like growths (e.g., *Liothrips ramakrishnae* [Phlaeothripidae]—*Schefflera racemosa* [Araliaceae]). Numbers in thrips populations are vital for the final gall form (Krishnamurthy *et al.,* 1975, Raman *et al.,* 1978, Raman 2003); a staggering number of over 100,000 individuals of *A. cochinchinensis* is known in a single pouch gall on *C. floribunda* (Ananthakrishnan and Raman 1989). Several studies made in the last decade confirm eusocial behaviour among gall-inducing thrips (see Mound and Morris 2005). Greater levels of humidity benefit thrips (Crespi *et al.,* 1997). Thrips usually attack a tender plant organ (usually a developing leaf) that has the morphogenetic potential to redirect its course of development and grow into a gall. An exception is the multilocular woody stem gall-inducing *Iotatubothrips crozieri* (Phlaeothripidae) on *Casuarina cristata* (Casuarinaceae) in Australia (Mound and Crespi 1992).

Several species of tropical and subtropical Tubulifera induce galls; on the contrary, only five species of Terebrantia, throughout the world, are the known gall-inducing taxa (Raman 2003). Gall-inducing thrips, as known from subtropical southern India, have an almost uniform life cycle. Terebrantia insert their colourless and smooth-shelled eggs singly into plant tissue with their saw-like ovipositor (Childers and Achor 1995). Tubuliferans, on the contrary, lay their eggs exposed on plant surface in clusters. Life-cycle periods vary with species, temperature, photoperiod, and food availability; the shortest known is 10–15 d; in subtropical peninsular India, thrips show a life-cycle time of 30-40 d and occur throughout the year with 8–10 generations (Ananthakrishnan and Raman 1989).

Mouth parts of thrips form a cone that occurs on the underside of the head. Much of the cone occurs beneath of the first thoracic segment in a resting thrips. When extended, it is directed downwards and backwards. Mouth parts are characteristically asymmetrical because of the vestigial right mandible. In species of Phlaeothripidae (Tubulifera), maxillary stylets are usually longer and more deeply retracted into the head than they are in either Thripidae or Aeolothripidae (Terebrantia) (Mound 1971). Tubuliferan cones are usually longer than those of the Terebrantia, and this endowment, in high likelihood, enables the Tubulifera to explore host tissues with greater efficacy (Mound *et al.,* 1980, Childers and Achor 1995). Musculature of the head and pharynx enable ingestion of cell sap with their contractile and pumping action (Moritz 1995). Leaf-feeding thrips rasp epidermal cells with their left-mandibular stylet and then insert the paired maxillary stylets. They empty the host-cell contents through maxillary stylets and use them to probe and feed on deeper-lying leaf tissues. When a thrips commences feeding, it presses the tip of its mouth cone against the host-plant surface and the labral pad, thus maintaining the closest contact with the host surface. Thrips feed through rapid protraction and retraction of the mandibular stylet by shortening the mouth cone [the mouth cone is similar in function to that of an accordion bellows — with a circular outline — in a traditional photographic camera,]. The mandibles, however, can move only within certain limits because of the nature of their orientation, musculature, and articulation within the head capsule (Heming 1978, Moritz 1982). For a detailed description of thrips mouth parts, read Kirk (1997).

In *Franklienella occidentalis* (Terebrantia: Thripidae), a non-gall inducing taxon on *Phaseolus vulgaris* cv. Prelude (Fabaceae), the stylets can utmost extend to about 30 μm, just reaching the first and second cell layers immediately below the epidermis (Kindt *et al.,* (2003). However, in *P. vulgaris* cv. Prelude infested leaves the further-lying tissues suffer subcellular changes ('damage') although not by wounding. The damage could be due to either the chemicals in the saliva of *F. occidentalis* (Kindt *et al.,* 2003) or due to desiccation (Wardle and Simpson 1927), or both acting concurrently. Similar observations are available for *F. bispinosa*-fed *Citrus* flower tissues (Childers and Achor 1991). However, what needs to be recognized here is that a subtle difference exists between the non-gall-inducing, plant-feeding Thysanoptera (e.g., species of *Frankliniella*, Thripidae) and gall-inducing Phlaeothripidae (e.g., species of *Gynaikothrips*). Gall-inducing thrips elicit a defined response, which includes the organization of a specialized tissue of nutrition with pumped-up levels nutrients (Raman and Ananthakrishnan 1983a, Ananthakrishnan 1992), whereas the non-gall-inducing taxa never elicit a similar response in their host tissues. For example, the feeding action of *Crotonothrips* (Tubulifera: Phlaeothripidae) that induces galls on the leaves of *Memecylon* (Melastomaceae) evokes a specific response in the host, by establishing specialized transfer-cell-like cells, which increase surface areas of nutritive cells, thus enabling short-distance nutrient transport (Raman 1987b).

A gravid-female (the gall founder) selects and occupies the potential gall site. Her progeny feeds at that site contributing to gall development. Salivary glands of the gravid females of *Arrhenothrips ramakrishnae* (Phlaeothripidae) that induce fold galls on the leaves of *Mimusops elengi* (Sapotaceae) include greater quantities of hydrolyzing enzymes (e..g., proteases, amylases, and lipases) than those in either males or pupae (Raman *et al.,* 1991). Maximum levels of amylases occur in the first and second larvae. The pupae have shrunken (non-functional?) salivary glands; except for insignificant lipase values, neither proteases nor amylases occur. Compared with any other development stages, the males and pupal stages, the saliva of the gravid female includes a wide range of soluble proteins. In high likelihood, salivary proteases and soluble proteins in the gravid female alter the host-tissue metabolism in enabling them to develop into galls. Higher concentrations of hydrolyzing enzymes in the saliva of *A. ramakrishnae* show them to be a better adapted taxon for mesophyll feeding.

Final gall shape, be it either a simple fold or roll or a complex pouch, materializes by the collective feeding action of several individuals — the progeny from the gall-founding female. During early stages of gall development, 4–5 individuals occur at a gall site. From the time of unfurling and expansion of the lamina, host leaves remain susceptible for a specific period of time as evident in some of the studied examples in subtropical southern India: *Pavetta hispidula* (Rubiaceae), the host of *Teuchothrips longus* (Phlaeothripidae) — 5 d; *Ventilago maderasapatana* (Rhamnaceae), the host of *Schedothrips orientalis* (Phlaeothripidae) — 7 d. If thrips invade after the susceptible period, chances of galls developing are unlikely. Usually in 48–72 h, the invaded tissue responds with dramatic changes leading to gall development. The period of existence of thrips in old galls and the time necessary for the completion of gall growth generally coincide with the pre-oviposition period of the inducing thrips (Varadarasan 1979).

The developmental stage of the host organ and the size of thrips population together regulate variations in gall shapes. For example, when ±20 adults and 10 larvae of *Gynaikothrips flaviantennatus* inhabit, partial galls develop on the leaves of in *Casearia tomentosa* (Flacourtiaceae); when ±80 adults 90 larvae inhabit, complete galls develop (Raman *et al.,* 1978, Raman 2003). Similar effects have been confirmed in the horn galls on the foliage of *Schefflera racemosa* (Araliaceae) induced by *Liothrips ramakrishnae* (Phaleothripidae), where smaller populations inhabit galls made of 7–12 layers of parenchyma, whereas larger populations inhabit galls made of 25—35 layers (Krishnamurthy *et al.,* 1975). For remarks on the unique 'company' galls induced by thrips, see Ananthakrishnan (1992).

Thilakothrips babuli inducing rosette galls on *Acacia leucophloea* shows a unique behaviour. It not only induces galls on the axillary vegetative meristems, but also on the florets. In an annual cycle, peak populations of *T. babuli* exist for *c.* 90 d: *c.* 60 d in galls induced on vegetative meristems, *c.* 30 d in galls on florets. When galls on meristems dry, thrips populations decline, but a few apterous adults survive diapausing. With the onset of flowering, these apterous adults move to the flower head and initiate galls on the florets. The floret-gall inhabiting *T. babuli* populations exhibit wing polymorphism, whereas the meristem-gall inhabiting populations of *T. babuli* do not (Ananthakrishnan and Raman 1989). Such a capability of *T. babuli* inducing galls both on vegetative meristems and florets is striking because even among the sophisticated gall-inducing insects (e.g., Cecidomyiidae, Cynipidae) such a behaviour seldom occurs (Raman 2007): two of the rare examples are *Asphondylia* (Diptera: Cecidomyiidae) on *Larrea tridentata* (Zygophyllaceae) and *Andricus* (Hymenoptera: Cynipidae) on species of *Quercus* and *Cerris* (Fagaceae), where radiation of the inducing insect taxon has occurred between organs in the same host (Cook *et al.*, 2002, Joy and Crespi 2007). However, in the evolution of insect phytophagy, fidelity to feeding-site is more critical for speciation than fidelity to host plants (Favret and Voegtlin 2004).

STERNORRHYNCHA

The Aphidoidea, Psylloidea, Coccoidea, and to a limited extent the Aleyrodoidea include gall-inducing taxa. For an understanding of the range of galls they induce, the reader is referred to chapters by Wool (2005), Burckhardt (2005), Gullan *et al.*, (2005), and Byrne (2005). The number of insects in mature galls would vary with groups: several in galls induced by the Aphidoidea and Coccoidea, whereas one (rarely two) in galls induced by the Psylloidea, although exceptions occur. Hemipteran mouth parts are of piercing-and-sucking nature, adapted for an exclusively liquid diet. The liquid diet could be either the plant sap from vascular elements or the fluids from living non-vascular cells. Social behaviour and incidence of a special class of gall-inducing Aphididae (the *samurai* aphids) have been reported by Aoki and Aoki *et al.*, in Japan between 1977 and 2000 (see Wool 2005 for references).

Several papers have appeared in recent times explaining the morphology of mouth parts of plant-feeding Hemiptera. For the sake of contextual clarity, a brief note is provided here. Mouth parts of plant-feeding Hemiptera (for a comprehensive review of the nutritional physiology of plant-feeding Heteroptera, see Hori 2000) include the labrum, labium, and a sclerotized stylet bundle that includes two mandibular and two maxillary stylets. The mouth-part complex is essentially tubular and devoid of either labial or maxillary palpi. The labral cone, usually endowed with sensilla, is attached proximally to the clypeus and occurs overlying the labial groove.

The included stylets are sharp tipped and are elaborately sculptured both at the tips and along the edges. The first maxillae are closely pressed to each other so that the oppositely lying grooves along their interfaces are arranged in such a manner that they bear two superposed capillary tubes. Through one of the tubes the hemipteran flushes its saliva and through the other sucks plant sap. Endowed with a variety of sensilla, the distal tip of the labium guides the stylet into the host organ.

The second maxilla fused into a labium constitutes the rostrum, with a groove in which the distal parts of the stylets slide. Each stylet is manipulated by two sets of retractor and protractor muscles. Large muscles attached to the ceiling of the cibarium enable a suction force, which helps in either drawing or injection through the food and salivary canals that lie between the maxillary stylets. The two maxillary stylets interlock with each other along their full length, thus constituting a smooth hollow tube that bears an armature of denticles at the tips (Hori 2000). The articulation on the opposite side of the stylet bears the salivary canal, which opens terminally between the denticles and the extreme end of the stylets. Although each maxillum is similar in shape and dimension, length of stylets changes as the insect develops: first nymphal instars: 300–600 µm; adults: 1000–1400 µm.

Because stylet bundles become longer with each successive moult of the nymphal instar, developing nymphs shift their feeding sites from superficial cells to deeper-lying tissues as they mature. Gall-founding aphid females change their feeding sites several times during gall development; they usually generate and introduce new stylets at different sites (Rohfritsch and Anthony 1992).

During feeding the labium does not pierce the plant tissue, but is positioned perpendicular on the surface to insert the stylets into the host. Although a majority of the Sternorrhyncha feed passively on the contents of phloem, several studies on gall-inducing Sternorrhyncha, especially on the nymphal instars indicate them to be non-vascular tissue feeders (e.g., parenchyma) (Raman 1991, Rohfritsch and Anthony 1992, Raman *et al.*, 1996). Because we know more about the saliva of non-gall-inducing aphids, we will use that as a model here. Aphids produce two types of saliva. The first is dense and proteinaceous, which jels around the stylets thus forming stylet sheaths, isolating plant tissues from the mouth parts and preventing any possible adverse plant reactions (Felton and Eichenseer 1999). On reaching the phloem, aphids secrete the second type of saliva, which is less dense and is injected directly into the vascular system of the plant. The watery saliva contains diverse digestive and lytic enzymes. The aphid's feeding action inflicts a 'fine' wound, but the salivary proteins interact with Ca^{++} of host-plant tissues (Will *et al.*, 2007) preventing possible wound-healing effort by the plant. In general, wounding does not induce cell necrosis. Non-gall inducing aphids impress as more sophisticated taxa than their allies, because they avoid plant-based allelochemicals and indigestible compounds by their elegant feeding mechanism (Schoonhoven *et al.,* 2007). In *Bemisia argentifolii* (Aleyrodidae), the adult stylet bundle enters the labial groove of the labium between the first and second segments and is included within the labium except during feeding; stylet length is equal to the combined length of 2, 3, and 4 labial segments; the physical force necessary for adult stylet penetration is derived from changes in the position of the *B. argentifolii* head during feeding; the head is bent over the labium, which is attached to the leaf surface, forcing the stylet bundle down the labial groove and into the host tissue (Freeman *et al.,* 2001).

Miles (1999) provides the state-of-the-art knowledge on aphid saliva; a few key points are summarized here: (i) Several amino acids occur in the watery saliva, but their concentrations vary with species. In high probability, the amino acids that occur are the same that occur in the sieve-tube sap, and they represent 'unutilized dietary products absorbed into the haemolymph and excreted *via* the salivary glands' (Srivastava 1987). (ii) Watery saliva has reducing and surfactant properties to lubricate stylet activity. (iii) Watery saliva is of pH 8–9, similar to that of many Pentatomomorpha (Taylor and Miles 1994). (iv) Pectin-hydrolyzing enzymes such as pectinestrases and polygalacturonases occur. (v) Cellulase, arabinogalactan breaking down 1, 3-glucosidases and exopolygalacturonase also occur. (vi) In the non-aphidid aphids, α–amylases exist; in the aphidid taxa, α–amylases are not from the saliva, but are from the gut tissues that houses a bacterium. (vii) Glucose-breaking down α–glucosidases occur in both saliva and gut tissues. (viii) The sheath material and watery saliva of *Rhopalosiphum padi* release *p*–nitrophenol from both *p*–nitrophenyl–α–d–glucopyranoside and *p*–nitrophenyl–ß–d–glucopyranoside, indicating a salivary α–glucosidase and ß–glucosidase, respectively (Urbanska and Leszczynski 1997). (ix) The watery saliva and the stylet sheath contain polyphenol oxidase originating from the accessory glands occurring near the salivary glands; the difficult-to-demonstrate peroxidases have been shown in *Aphis pomi* (Aphididae) (Leszczynski *et al.,* 1996). (x) Salivary soluble proteins vary considerably between species (Miles and Harrewijn 1991). The saliva of *Therioaphis trifolii* (Aphididae) includes soluble proteins of molecular weights 6--200 kDa, and the three that showing oxidase activity had molecular weights of

20, 90, and 200 kDa (Madhusudhan and Miles 1998); the saliva of a non-gall-inducing species *Acyrthosiphon pisum* (Aphididae) includes two minor and two major soluble protein fractions (120–300 kDa), and the three that showed oxidase activity with molecular weights 120, 180, and 300 kDa.

Stylet tracks (the proteinaceous sheath) left in the host tissues by gall-inducing and non-gall-inducing Sternorrhyncha after the withdrawal of the stylets accept colouring by cationic dyes (e.g., methylene blue, bismark brown) and can be easily detected in a light microscope. In some species the track is straight (e.g., *Eriosoma lanigerum* [Pemphigidae] on *Malus domestica* [Rosaceae]), whereas in other species it is meandering and branched (e.g., *Adelges abietes* [Adelgidae] on *Picea excelsa* [Pinaceae]). Sternorrhyncha extensively probe the plant surface before they commence feeding (e.g., *Hormaphis hamamelidis* [Aphididae] on *Hamamelis virginiana* (Hamamelidaeceae); Lewis and Walton 1958), whereas others do not (e.g., *Daktulosphaeria vitifoliae* [Phylloxeridae] on *Vitis vinifera* [Vitaceae]; Raman et al., 2009a). In a majority of instances the stylet path travels intercellularly dissolving the middle lamella, principally made of pectic compounds, which cements adjacent cells (Rohfritsch 1976, 1988). Pectinase activity in aphid saliva is known from the 1960s (Auclair 1963, Ma *et al.,* 1998, Miles 1999).

One key characteristic of gall-inducing Sternorrhyncha is that their first-instar nymphs (the crawlers) are mobile. Settling at the most-suitable feeding site by these nymphs is vital, because movement over greater distances after they moult into subsequent stages is generally not possible. Byrne and Bellows (1991) and Gullan and Kosztarab (1997) have shown that the crawlers disperse over short distances before settling under favourable conditions. Crawlers of the Coccoidea show different kinds of movements. For example in a species of *Matsucoccus* (Margarodidae), the crawler movement combines thigmotactic behaviour and negative phototaxis over a short distance away from the mother (Gullan *et al.,* 2005). Crawlers of *Anonidiella aurantii* (Diaspididae) move upwards in light, downwards in dark, and settle beneath either in leaf sheaths or at stem angles. Crawlers of *Aulacaspis tegalensis* (Diaspididae) respond phototactically and geotactically, generally clustering at leaf tips, enabling their wind dispersal (Greathead, 1990). Once at the periphery of the plant, the first-instar nymphs can orientate themselves towards wind currents and get easily dispersed over long distances (Washburn and Washburn 1984). The first-instar nymphs of Aleyrodidae also move either short distances on the same plant or long distances between plants. For example, those of *Aleurotrachelus jelinekii* (Aleyrodidae) crawl for 1–2 days on the abaxial leaf surface before they settled (Southwood and Reader 1976). Young nymphs of *Bemisia argentifolii* move for about 50 mm on the same plant and travel as far as 200 mm during interplant movement; 80–100% of *B. argentifolii* crawlers settle on the abaxial surface, but the behavioural cues for site selection are undetermined as of today (Summers *et al.,* 1996).

The first-instar nymphs of *Trioza jambolanae* (Triozidae, Raman 1991) and *Diaphorina truncata* (Psyllidae; Balakrishna and Raman 1992) settle on the abaxial side of their host leaves to initiate a gall. Neonate nymphs of *T. jambolanae* crawl to venal angles along the abaxial leaf surface. The choice of settling site is dictated by the availability of stomata. After settling, these nymphs feed on the primordial mesophyll parenchyma cells lying immediately below the abaxial epidermis by inserting their stylets through a stoma. The potential cues, which the nymphs of *T. eugeniae* use when discriminating between the two morphological sides of a leaf, include the differences in external cuticular chemistry, differences in internal leaf chemistry, presence of stomata or other structural features, gravity, and light (Luft and Paine 1997). Walker (1987) determined that adults of *Parabemisia myricae* (Aleyrodidae) discriminate between young and mature lemon leaves from leaf-surface cues picked up by the sensilla on the rostral tip. Probing by first-instar nymphs could be limited by physical and/or chemical factors in leaf cuticle (Walker, 1985). For instance, sucrose, in high likelihood, is a phagostimulant for inducing the settling response (Walker and Bednar, 1986). Actively growing tissues rich in nutrients could be another factor for sternorrhynchan nymphal preference for young leaves (Walker and Aitken, 1985). Consequently, settling is a response to the presence and absence of both inhibitory and stimulatory behavioural cues. Direct-contact chemoreception could be playing a reduced role in discrimination. Adaxial and abaxial faces of leaves could be producing a different quantity and quality of volatiles that nymphs use in discrimination. Discrimination of host quality is critical for the fitness of insects with less mobile further stages during nymphal growth.

Sea-urchin shaped galls induced by *Mangalorea hopeae* (Coccoidea: Beesoniidae) arise on the shoots of *Hopea ponga* (Dipterocarpaceae) along the western coastal plains of peninsular India. The gall consists of a 1–1.5 cm wide head-like structure (the columella) on which several, compactly arranged sharp lance-like structures arise. The neonate wingless female nymphs of *M. hopeae* invade axillary angles of vegetative meristems of *H. ponga*, exploiting the space made available by their extra-axillary positions on the stem, shortly after monsoon rains (Raman and Takagi 1992). How they invade new sites remains a supposition. The conjecture is that the older winged brothers transport female nymphs enabling them to occupy new sites, similar to what occurs in the gall-inducing Australian Eriococcidae (Gullan and Cockburn, 1986). Immediately after settling, the female nymph of *M. hopeae* starts feeding on the cortical tissues of the vegetative bud. A gall develops arching over the nymph, enclosing it nearly completely. The trapped female nymph grows into an adult fitting snugly in that space. In between the lance-like appendages that develop through the modification of the multicellular trichomes along the outer edges, male nymphs – arising from the trapped adult female – live feeding on the columellar tissue. When the gall dries, the lance-like appendages separate enabling the males, which would have grown into adults by that time to escape. Galls induced by *M. hopeae* stand distinctly from the remainder of the known galls induced by Coccoidea (Gullan *et al.,* 2005) because of the following characteristics: (i) two life stages of *M. hopeae* contribute to gall development, (ii) a first-nymphal instar initiates the gall on the vegetative shoot bud, exploiting the axillary gap, and (iii) several male nymphs that emerge from the entrapped female after its maturation and mating, occupy the spaces between the lance-like appendages. Although occupation of a maternal gall by several males is usual among galls induced by Coccoidea (Gullan *et al.,* 2005), the feeding behaviours of the inducing female and the male progeny are strikingly different: the inducing female *M. hopeae* feeds on cortical parenchyma, whereas the *M. hopeae* males feed on phloem elements. An adult male from outside mates with the female trapped within the gall, through an aperture that occurs close to the stalk of the gall. This aperture not only enables mating, but also the discharge of nymphs that develop viviparously within the females' saccular body.

Verification of oviposition preference patterns of *Trioza apicalis* (Triozidae) that lives on *Daucus carota* as the preferred host plant, on different plants of Apiaceae, and its alternate host *Picea abies* (Pinaceae) provides the following results (Valterová *et al.,* 1997): the number of eggs laid by *T. apicalis* per day varies from 18 on *Daucus carota* subsp. *sativus* to zero on *Aegopodium podagraria*; individual *T. apicalis* occupy quickly and oviposit earlier on *D. carota* subsp. *sativus* compared with *Anethum graveolens*; *T. apicalis* survived on *P. abies* for more than 30 weeks, but died on *Phleum pratense* and *Brachypodium sylvaticum* in 15 d. Day length was a factor that regulated hibernation; at 20°C, light regimes shorter than 17 h induced hibernation, while longer light periods led to reproduction. The Apiaceae taxa and *P. abies* released monoterpene hydrocarbons, but with distinct differences among tested species. *Daucus carota*, the most preferred host plants released (+)– and (–)–c~-pinene and (+)-sabinene, whereas the other lesser preferred Apiaceae released either (+)– or (–)–limonene. *P. abies* included (–)–c~–pinene, (–)–J3–pinene, and (–)–limonene as the principal volatiles.

Among the different Indian Anacardiaceae-infesting psylloids, *Apsylla cistellata* (Calophyidae) infesting *Mangifera indica* induces fir-cone shaped galls. *Apsylla cistellata* is restricted to the Indo-Gangetic Plains and the lower valleys of the Himalaya (20–34°N; 77–85°E; Singh 2003). Gravid females lay 75–150 eggs on leaves, burying them partly into leaf tissue. Eggs hatch approximately in 200 d and adult development occurs in 140 d. The first-instar nymphs feed on the leaves on which the eggs were laid and their feeding action 'triggers' gall development in the vegetative axillary buds that occur approximately 10 cm away. The feeding effect of many neonate nymphs induces the modification of nearby vegetative buds into galls in about 30 d. Stimulation of gall development at a site farther away, viz., the vegetative axillary bud, is implicated by the translocation of a 'stimulus' over a distance of 10–15 cm. Such a behaviour is unusual among the gall-inducing psylloids (Burckhardt 2005b), but is analogous to the behaviour of *Adelges cooleyi* (Adelgidae) that induces galls on the vegetative buds of the hybrid interior spruce (*Picea glauca* × *P. engelmannii*) in North America, where a dose-dependent chemical stimulus is transported over long distances from the point where the gall-founding female settled (Sopow *et al.,* 2003). In spite of an apparent similarity, in the bud galls of *M. indica*, the first-instar nymphs of *A. cistellata* are the purported gall inducers, whereas in the bud galls of *Picea glauca* × *P. engelmannii*, adult females of *A. cooleyi* induce galls. Burying eggs – either partly or fully – into host tissue by the gravid female is unusual among the Psylloidea. Among the Triozidae such behaviour is unknown (Boujou and Nguyen 1974, Ratsimialarivo-Ravoahangiariosoa 1983, Raman 1991, Raman *et al.,* 1996). However, the gall-inducing females of *Phacopteron lentiginosum* (Phacopteronidae) bury their eggs in the leaves of *Garuga pinnata* (Burseraceae) in southern India. Leaf tissues of *G. pinnata* respond to the buried eggs with metabolic changes (e.g., accumulation of phenolics) with features of reaction tissue, similar to that shown in the leaves of *Lactuca sativa* (Asteraceae) consequent to mechanical wounding (Choi *et al.,* 2005). The emerging nymphal instars of *P. lentiginosum*, living within leaf tissues, initiate galls by their feeding action. The gall assumes its complex shape with the further development of *P. lentiginosum* (Raman 1987a)

Injection of aphid saliva alters the hormonal balance in the host, thus leading to gall development. However, apart from three triacylglycerides containing (E,E,E)–octa–2,4,6–trienoic acid isolated from the gall induced by *Colopha moriokaensis* (Pemphigidae) on the leaves of *Zelkova serrata* (Ulmaceae) responsible for cell hypertrophy, no potential gall-inducing compound has been determined so far (Otha *et al.,* 2000). Soluble proteins in the saliva of the nymphs of *Trioza jambolanae* have been shown to be a critical factor for gall development in the leaves of *Syzygium cumini* (Rajadurai *et al.,* 1990). In the saliva of *Trioza apicalis*, an undetermined amine exists in addition to plant-derived sugars (Markkula *et al.,* 1976). Gall-inducing Hemiptera vigorously take up oxygen from gall tissue (several examples in Miles 1999), along with a stimulation of auxin activity. Use of oxygen in the tissues under arthropod attack might be so great that the IAA-oxidase activity that regulates the concentration of IAA might be deprived of oxygen and therefore inhibited. Such a deprivation of oxygen (Florentine *et al.,* 2002) results in the concentration of IAA increasing disproportionately at feeding sites with a consequential hypertrophy of meristematic plant tissues. Although the specific agent in the hemipteran saliva that induces galls has not been determined, salivary oxidases should be playing a role in the disruption of IAA-oxidase pathway.

CONCLUDING REMARKS

Rohfristch (1992) indicates that the pattern evident in the feeding and gall-induction behaviours of Thysanoptera and Sternorrhyncha is, in principle, similar to that evident in the feeding and gall-induction behaviours of Cecidomyiidae (Diptera). Her argument is based on her study of the gall-inducing behaviour of *Hartigiola annulipes* (Cecidomyiidae) on leaves of *Fagus sylvatica* (Fagaceae) and that of different Sternorrhyncha (Rohfritsch 1971, 1976, 1988). Neonate larvae of *H. annulipes* (and many other Cecidomyiidae, for that matter) initiate galls by sucking cell sap using their sharp, but short mandibles (Rohfritsch 1978, 1980) eliciting a covering growth. She has also argued that in a majority of the Cecidomyiidae-induced galls, covering growth is the common pattern, which is also the basic pattern in the galls induced by Thysanoptera and Sternorrhyncha (Rohfritsch 1992).

A comparison of gall-inducing behaviours in the Thysanoptera and Sternorrhyncha presents at least two similarities: (i) gall induction occurs by the feeding action of a single female; (ii) gall-founding females disperse over short distances seeking young host-plant organs (e.g., differentiating leaves). More importantly, thrips saliva is considered 'similar' to that of the Sternorrhyncha in several ways (Kirk 1997). Differences, however, exist in gall-inducing behaviours. Among the Thysanoptera, a gravid female founds the gall. Collective feeding effort of her progeny is critical in achieving the final-gall shape. In a majority of galls induced by thrips, a marked level of wounding occurs because of their feeding action with asymmetrical mouth parts and their probing behaviour (which includes rasping) followed by piercing and sucking. Cell necrosis occurs to a limited extent at the wounded sites of host plants. Among gall-inducing Aphidoidea (e.g., Adelgidae) and Coccoidea (e.g., Eriococcidae, Beesoniidae), a first-instar female nymph initiates the gall. Among Psylloidea (e.g., Triozidae), a first-instar nymph initiates the gall by settling on the stomatal aperture on the abaxial side of the leaf; however, in the Triozidae, whether it is a female nymph, similar to what is known among gall-inducing Aphidoidea and Coccoidea, or whether it is a male nymph is uncertain presently, although chances of a male inducing a gall are highly unlikely. Studied examples from the Triozidae indicate that their galls house a single individual (Raman 1991, Raman *et al.,* 1996). The trait of occupation of a gall as a colony is shared between the Thysanoptera on the one hand and the Aphidoidea and Coccoidea on the other. In the Psyllidae, e.g., *Diaphorina truncata* that induce galls on the leaves of *Strychnos nux-vomica* (Loganiaceae), gravid females colonize young leaves and by feeding on their abaxial surfaces they initiate galls, which require the feeding-action support from her progeny (Balakr-ishna and Raman 1992). Such behaviour is similar to those in a majority of leaf-roll, leaf-fold gall-inducing Phlaeothripidae (e.g., *Arrhenothrips ramakrsihnae*, *Gynaikothrips flaviantennatus*), which require at least 8–15 individuals to reinforce their mother's feeding initiative.

Among the Psylloidea, gall-inducing behaviours of *Apsylla cistellata* (Calophyidae) and *Phacopteron lentiginosum* (Phacopteronidae) are different compared with those of the Triozidae and Psyllidae. In gall-inducing Triozidae, gravid females deposit their eggs on the potential site where galls would develop, and only the egg stalks remain buried in the leaf tissue. However, *P. lentiginosum* buries its eggs entirely in the host-leaf tissue, although gall initiation is triggered only by the feeding action of the neonate nymph. Young leaves of *Garuga pinnata* respond to oviposition by generating a reaction tissue. On the contrary, in *Apsylla cistellata*, Singh (2003) reports that the egg is 'partly buried' on the leaves of *Mangifera indica* and the emergent nymphs feeding on the leaf initiate galls on the axillary vegetative buds, at least 10 cm away. Samui and Jha (2009) describe *A. cistellata*'s oviposition behaviour as follows: eggs are laid singly in slits dug by ovipositor, eggs remain embedded in midrib tissues along the under sides of new leaves, eggs are inserted alternatively by puncturing the tissue along both sides of the on the dorsal face of the midrib, the intensity of egg laying depends on the availability of new flush of tender leaves and number of adults emerging at a particular time in an orchard, and if many females had only a few leaves for egg laying, then they lay eggs along both sides of lateral veins along the under sides of *M. indica* leaves. Burying eggs in host tissue, therefore, emerges as a non-Triozidae character in the Psylloidea, shared only in species of Phacopteronidae and Calophyidae.

Claims of gall induction by the Tingidae and Cicadellidae need to be discussed. In terms of their general biology, the Tingidae prefer to feed on the abaxial-leaf sides seeking humid microenvironments — a trait shared by a majority of gall-inducing Sternorrhyncha. Nonetheless, among the purported gall-inducing Tingidae, their gregarious occurrence on host organs (e.g., leaves, flowers) stands out (Schaefer, 2005). Their preference for flowers and capability to induce floral galls impress as specialized traits among the gall-inducing hemipteroids, because floral galls induced by Sternorrhyncha and Thysanoptera are rare. The singular near match is the southern-Indian phlaeothripid *Thilakothrips babuli* that induces galls on the florets of *Acacia leucophloea* (Raman and Ananthakrishnan, 1983). Gall-bearing *Teucrium polium* (Sinai desert, Egypt; 29°30'N, 33°50'E) include leaves and floral axes reduced in size compared with the uninfested, although the petals in galled flowers are 'enlarged' ('hypertrophied?) (Zalat *et al.*, 2000). The other key behaviour that distinguishes gall-inducing Tingidae (e.g., *Copium*) from that of the gall-inducing Sternorrhyncha is that they insert their eggs in host plants to trigger gall development (Behr, 1952, Monod and Carayon, 1958). Evidences of a differently structured internal reproductive system in *Copium* show it to be better adapted for such specific behaviour (Schaefer, 2005). Nonetheless, without arguing how relevant the following could be, the oviposition behaviour of gall-inducing Tingidae indicates a situation parallel to what occurs in the few gall-inducing Terebrantia (Thysanoptera), which insert their eggs into plant tissues triggering gall development. Among the reported instances of gall-induction in the Cicadellidae (Mitjaev, 1968, Matsukura *et al.*, 2010), a common behaviour is that both nymphs and adults are gall inducers, which is strikingly different to the behaviour in the gall-inducing Sternorrhyncha; however, the purported 'gall'-inducing behaviour of the Cicadellidae reminds us of the behaviour known in gall-inducing Tubulifera in that both immature stages and adults participate in gall induction. Notwithstanding that nymphs and adults of *Cicadulina bipunctata* induce galls not at the locations they feed, but on distant leaves through dose-dependent stimulation (Matsukura *et al.,* 2009) reminds us the gall-inducing behaviours of *Apsylla cistellata* and *Adelges cooleyi*. Nevertheless, the polyphagous behaviour of *Copium clavicorne* and *C. teucrii* (Drake and Rohoff, 1965) infesting different species of southern Palaearctic and Middle-Eastern *Teucrium* and *Cicadulina bipunctata* on different Poaceae in south-east Asia raises the question 'could these be considered true gall-inducing taxa because of the low level of tissue complexity in their so-called galls and their polyphagous nature?'[3]

A trait shared in all of the gall-inducing Tubulifera (Phlaeothripidae) (*Iotatubothrips crozieri* excepted), the few gall-inducing Terebrantia, and a majority of the gall-inducing Sternorrhyncha is that they induce galls on soft plant organs (e.g., leaves, vegetative shoot buds, and flowers) of tree species, and that capability is considered an evolved trait. Molecular investigations (Morris and Mound, 2003) and evolutionary reconstructions based on fossil evidences (Grimaldi *et al.*, 2004) suggest Phlaeothripidae as the most derived family in Tubulifera and this family includes the highly diverse set of taxa including a majority of gall-inducing species. Gall-inducing Phlaeothripidae radiated in the Cenozoic (or at the latest, in the Cretaceous) in conjunction with the evolution of flowering plants (Grimaldi and Engel, 2005). Preference to tree species by these specialist phytophages could be interpreted as reliance on a 'predictable resource' (*sensu* Raffa 1989). However, their strong level of fidelity to respective plant species, and even plant organs, does not impress as an evolutionary advantage. Considering species of *Procontarinia* and *Mangifera indica* interactions, Raman *et al.*, (2009b) propose that the gall midges have progressed conservatively specializing along the host-plant lines to match with the plant phenology and to minimize competitive interactions with the parasitic, predatory, and inquilinous arthropods. Established schemes outlining the phylogeny of plant-feeding Sternorrhyncha indicate changes in feeding preferences enabled their radiation historically: for instance, the Cretaceous coccoids fed on conifers (Grimaldi *et al.*, 2000); host switching from conifers to angiosperms that occurred in the Cretaceous triggered radiation of aphids, and in high likelihood, in the Tertiary (c. 40 MYA) (von Dohlan and Moran, 1995). At the other end of the spectrum, the gall-inducing arthropods have limited neurological capacity to process different hosts, and therefore different diets, which has restricted them to specific plants and organs. We need to factor here Labandeira and Phillips's (1996) explanation that in the later Middle Pennsylvanian coal-swamp forests of Euramerica

3. Carl Schaefer, [email, 9 November 2010], remarked: "Why heteropterans and the auchenorrhynchans do not induce galls? Do they penetrate too far into the plant and so avoid leaf structures?" Dietrich Werner (email, 13 November 2010) clarified that a few Tingidae induce leaf rolls ('galls').

the dominant arborescent plants possessed vascular tissues unavailable to insects, because they were either deeply embedded in a thick layer of cortical tissues or protected by outer indurated, peridermal tissues. Subsequent tree-fern forests of the Late Pennsylvanian provided accessible vascular and other tissues to surface-dwelling insects with stylate mouthparts — a condition which continued into the Permian and propelled the radiation in the Hemiptera. By Middle Pennsylvanian–Late Pennsylvanian period, a rapid expansion of phytophagy occurred, from simple foliage feeding to complex gall-inducing capability (Labandeira, 2006).

The more I explore, new questions emerge; I am left nonplussed.

ACKNOWLEDGEMENTS

Daniel Burckhardt (Naturhistorishes Museum, Basel, Switzerland), Roman Rakitov (Paleontological Institute, The Russian Academy of Sciences, Moscow, Russia), Carl Schaefer (University of Connecticut, Storrs, Connecticut, USA), and Dietrich Werner (Universität zu Köln, Köln, Germany) reviewed the pre-final draft. I am grateful to Carl, Daniel, Dietrich, Murray and Roman. Rakitov kindly supplied I.D. Mitjaev's 1968 paper, enabled with an English summary of Mitjaev's findings. Werner supplied me his papers on gall-inducing Tingidae. Both Roman and Dietrich were admirably rapid in their responses to my requests; I thank them for that help.

REFERENCES

Abrahamson, W.G., Melika, G., Scrafford R. and Csóka, G. 1998. Gall-inducing insects provide insights into plant systematic relationships. American Journal of Botany 85: 1159–1165.

Ananthakrishnan, T.N. 1992. Unique aspects in the biology of thrips-induced galls. In *Biology of insect-induced galls* edited by J.D. Shorthouse and O. Rohfritsch, Oxford University Press, New York, USA, pages 185–195.

Ananthakrishnan, T.N. and Raman, A. 1989. *Thrips and gall dynamics,* Oxford & IBH Publishing Company, New Delhi, India.

Askew, R.R. 1984. The biology of gall wasps. In *Biology of gall insects* edited by T.N. Ananthakrishnan, Oxford & IBH Publishing Company, New Delhi, India, pages 223–271.

Auclair, J.L. 1963. Aphid feeding and nutrition. Annual Review of Entomology 8: 439–490.

Balakrishna, P. and Raman, A. 1992. Cecidogenesis of leaf galls of *Strychnos nux-vomica* (Loganiaceae) induced by the jumping plant louse species *Diaphorina truncata* (Homoptera: Psylloidea: Psyllidae). Entomologia Generalis 17: 285–292.

Behr, L. 1952. Über die Blütengalle des *Teucrium chamaedrys* L. Berichte der Deutschen Botanischen Gessellschaft 65: 326–330.

Boujou, B. and Nguyen, T.X. 1974. Observations sur la morphogénèse et la structure de la galle de *Trioza alacris* Flor. (Homoptera—Psyllidae) sur *Laurus nobilis* L. Marcellia 33: 237–254.

Bronner, R. 1973. Propriétiés lytiques des œufs de *Biorrihza pallida* Ol. Comptes Rendus de l'Académie des Sciences, Paris 276: 189–192.

Bronner, R., Westphal, E. and Dreger, F. 1989. Chitosan, a component of the compatible interaction between *Solanum dulcamara* L. and the gall mite *Eriophyes cladophthirus* Nal. Physiological and Molecular Plant Pathology 34: 117–130.

Burckhardt, D. 2005. Biology, ecology, and evolution of gall-inducing psyllids (Hemiptera: Psylloidea). In *Biology, ecology, and evolution of gall-inducing arthropods* edited by A Raman, CW Schaefer, and TM Withers, Science Publishers, New Hampshire, USA, pages 143–158.

Byrne, D.N. 2005. Gall-inducing whiteflies (Hemiptera: Aleyrodidae). In *Biology, ecology, and evolution of gall-inducing arthropods* edited by A. Raman, C.W. Schaefer and T.M. Withers, Science Publishers, New Hampshire, USA, pages 133–142.

Byrne, D.N. and Bellows, T.S. Jr. 1991. Whitefly biology. Annual Review of Entomology 36: 431–457.

Carango, P., McCrea, K.D., Abrahamson, W.G. and Chernin, M.I. 1988. Induction of a 58,000 Dalton protein during goldenrod gall formation. Biochemical and Biophysical Research Communications 152: 1348–1358.

Corpas, F.J., Barroso, J.B., Carreras, A., Quirós, M., León, A.M., Romero-Puertas, M.C., Esteban, F.J., Valderrama, R., Palma, J.M., Sandalio, L.M., Gómez, M. and del Río, L.A. 2004. Cellular and subcellular localization of endogenous nitric oxide in young and senescent pea plants. Plant Physiology 136: 2722–2733.

Childers, C.C. and Achor, D.S. 1991. Feeding and oviposition injury to flowers and developing floral buds of 'Navel' orange by *Frankliniella bispinosa* (Thysanoptera: Thripidae) in Florida. Annals Entomological Society of America 84: 272–282.

Childers, C.C. and Achor, D.S. 1995. Thrips feeding and oviposition injuries to economic plants, subsequent damage and host responses to infestation. In *Thrips biology and management* edited by BL Parker, M Skinner, and T Lewis, Plenum Press, New York, USA, pages 31–51.

Choi, Y.J., Tomás-Barberán, F.A. and Saltveit, M.E. 2005. Wound-induced phenolic accumulation and browning in lettuce (*Lactuca sativa* L.) leaf tissue is reduced by exposure to *n*-alcohols. Postharvest Biology and Technology 37: 47–55.

Cook, L.G. and Gullan, P.J. 2004. The gall-inducing habit has evolved multiple times among the eriococcid scale insects (Sternorrhyncha: Coccoidea: Eriococcidae). Biological Journal of the Linnean Society 83: 441–452.

Cook, J.M., Rokas, A., Pagel, M. and Stone, G.N. 2002. Evolutionary shifts between host oak species and host plant organs in *Andricus* gall wasps. Evolution 56: 1821-1830.

Crespi, B.J., Carmean, D.A. and Chapman, J.W. 1997. Ecology and evolution of galling thrips and their allies. Annual Review of Entomology 42: 51–71.

Crespi, M. and Gálvez, S. 2000. Molecular mechanisms in root nodule development. Journal of Plant Growth Regulation 19: 155–166

Csóka, G., Stone, G.N. and Melika, G. 2005. Biology, ecology, and evolution of gall-inducing Cynipidae. In: *Biology, ecology, and evolution of gall-inducing arthropods* edited by A. Raman, C.W. Schaefer and T.M. Withers, Science Publishers, New Hampshire, USA, pages 573–642.

De Bruyn, L. 2005. The biology, ecology, and evolution of shoot flies (Diptera: Chloropidae). In *Biology, ecology, and evolution of gall-inducing arthropods* edited by A. Raman, C.W. Schaefer, and T.M. Withers, Science Publishers, New Hampshire, USA, pages 373–405.

de Lillo, E. and Monfreda, R. 2004. 'Salivary secretions' of eriophyoids (Acari: Eriophyoidea): first results of an experimental model. Experiemtnal and Applied Acarology 34: 291–306.

Dénarié, J., Debellé, F. and Promé, J.C. 1996. *Rhizobium* lipochitooligosaccharide nodulation factors: signalling molecules mediating recognition and morphogenesis. Annual Review of Biochemistry 65: 503–535.

Docters van Leeuwen–Reijnvaan, J. and Docters van Leeuwen, W.M. 1926. *The zoocecidia of the Netherlands East Indies*. Drukkerij de Unie, Batavia, Indonesia.

Doss, R.P., Oliver, J.E., Proebsting, W.M., Potter, S.W., Kuy, S.R., Clement, S.L., Williamson, R.T., Carney, J.R. and DeVilbiss, E.D. 2000 Bruchins: insect-derived plant regulators that stimulate neoplasm formation. Proceedings of the National Academy of Sciences USA 97, 6218–6223.

Drake, C.J. and Ruhoff, F.A. 1965 Lacebugs of the world: a catalog (Hemiptera: Tingidae). U S National Museum Bulletin 243: 1–634.

Farmer, E.E. 2000 Potent mitogenic lipids from gall-inducing insects. Trends in Plant Sciences 5: 359–360.

Favret, C. and Voegtlin, D.J. 2004 Speciation by host-switching in pinyon *Cinara* (Insecta: Hemiptera: Aphididae). Molecular Phylogeny & Evolution 32: 139–151.

Felt, E.P. 1965 *Plant galls and gall makers* (Reprint of 1940 edition). Hafner Publishing Co Inc., New York, USA.

Felton, G.W. and Eichenseer, H. 1999 Herbivore saliva and its effects on plant defense against herbivores and pathogens. In: *Induced plant defenses against pathogens and herbivores* edited by A.A. Agrawal, S. Tuzun and E. Bent. APS Press, St Paul, USA, pages 19–36.

Florentine, S.K., Raman, A. and Dhileepan, K. 2002 Responses of the weed *Parthenium hysterophorus* (Asteraceae) to the stem gall-inducing weevil *Conotrachelus albocinereus* (Coleoptera: Curculionidae). Entomologia Generalis 26: 171–182.

Freeman, T.P., Buckner, J.S., Nelson, D.R., Chu, C.C. and Henneberry, T.J. 2001. Stylet penetration by *Bemisia argentifolii* (Homoptera: Aleyrodidae) into host leaf tissue. Annals of the Entomological Society of America 94: 761–768.

Gagné, R.J. 2004. A catalog of the Cecidomyiidae (Diptera) of the world. Memoirs of the Entomological Society of Washington 25: 1–408.

Greathead, D.J. 1990. Crawler behavior and dispersal, In *World crop pests* edited by D, Rosen, Vol. 4A Armored scale insects: their biology, natural enemies and control. Elsevier, Amsterdam, The Netherlands, pages 305–308.

Grimaldi D. and Engel, M.S. 2005. *Evolution of the Insects*, Cambridge University Press, Cambridge, UK.

Grimaldi, D., Shedrinksy, A. and Wampler, T.P. 2000. A remarkable deposit of fossiliferous amber from the Upper Cretaceous (Turonian) of New Jersey. In *Studies on fossils in amber, with particular reference to the Cretaceous of New Jersey* edited by D Grimaldi, Backhuys Publishers, Leiden, The Netherlands, pages 1–76.

Grimaldi, D., Shmakov, A. and Fraser, N. 2004. Mesozoic thrips and early evolution of the order Thysanoptera (Insecta). Journal of Paleontology 78: 941–952.

Gullan, P.J. and Cockburn, A. 1986. Sexual dichronism and intersexual phoresy in gall-forming coccoids. Oecologia 68: 632–634.

Gullan, P.J. and Kosztarab, M. 1997. Adaptations in scale insects. Annual Review of Entomology 42: 23–50.

Gullan, P.J., Miller, D.R. and Cook, L.G. 2005. Gall-inducing scale insects (Hemiptera: Sternorrhyncha: Coccoidea). In *Biology, ecology, and evolution of gall-inducing arthropods* edited by A. Raman, C.W. Schaefer, and T.M. Withers, Science Publishers, New Hampshire, USA, pages 159–230.

Harper, L.J., Schonrögge, K., Lim K.Y., Francis, P. and Lichtenstein, C.P. 2004. Cynipid galls: insect-induced modifications of plant development create novel plant organs. Plant, Cell, Environment 27: 327–335.

Harris, K.M. and Gagné, R.J. 1982. Description of the African rice gall midge *Orseolia oryzivora* with comparative notes on the Asian rice gall midge *Orseolia oryzae* Wood-Mason (Diptera: Cecidomyiidae). Bulletin of Entomological Research 72: 467–472.

Heming, B.S. 1978. Structure and functions of mouthparts of *Haplothrips verbasci* (Osborn.) (Thysanoptera, Tubulifera, Phaleothripidae). Journal of Morphology 158: 1–58.

Hori, K. 2000. Possible causes of disease symptoms resulting from the feeding of phytophagous Heteroptera. In *Heteroptera of economic importance* edited by CW Schaefer and AR Panizzi, CRC Press, Boca Raton, Florida, USA, pages 11–35.

Joy, J.B. and Crespi, B.J. 2007. Adaptive radiation of gall-inducing insects within a single host-plant species. Evolution 61: 784-795.

Kim, K.W., Park, E.W. and Kim, K.S. 2004. Glyoxysomal nature of microbodies complexed with lipid globules in *Botryosphaeria dothidea*. Phytopathology 94: 970–977.

Kindt, F., Joosten, N.N., Peters, D. and Tjallingii, W.F. 2003. Characterisation of the feeding behaviour of western flower thrips in terms of electrical penetration graph (EPG) waveforms. Journal of Insect Physiology 49: 183–191.

Kirk, W.D.J. 1997. Feeding. In *Thrips as crop pests* edited by T Lewis, CAB International. Wallingford, Oxford, UK, pages 119–174.

Krishnaiah, K. 2004. Rice gall midge, *Orseolia oryzae* – an overview. In *New approaches to gall midge resistance in rice* edited by J. Bennett, J.S. Bentur, I.C. Pasalu and K. Krishnaiah, International Rice Research Institute, Los Baños, The Philippines, pages 1–5.

Krishnamurthy, K.V., Raman, A. and Ananthakrishnan, T.N. 1975. On the morphology of the ceratonean thrips galls of *Schefflera racemosa* Harms. Marcellia 38: 179–184.

Labandeira, C.C. 2006. Silurian to Triassic plant and hexapod clades and their associations: new data, a review, and interpretations. Arthropod Systematics & Phylogeny 64: 53–94.

Labandeira, C.C. and Phillips, T.L. 1996. Insect fluid-feeding on Upper Pennsylvanian tree ferns (Palaeodictyoptera, Marattiales) and the early history of the piercing-and-sucking functional feeding group. Annals of the Entomological Society of America 89: 157–183.

Lambinon, J. and Schneider, N. 2004. Contribution à la connaissance des galles d'hémiptères (Insecta, Hemiptera) du Luxembourg. Bullétin de la Société des Naturalistes Luxembourgeois 105: 109–132.

LaSalle, J. 2005. Biology of gall inducers and evolution of gall induction in Chalcidoidea (Hymenoptera: Eulophidae, Eurytomidae, Pteromalidae, Tanaostigmatidae, Torymidae). In *Biology, ecology, and evolution of gall-inducing arthropods* edited by A Raman, CW Schaefer, and TM Withers, Science Publishers, New Hampshire, USA, pages 507–538.

Lerouge, P., Roche, P., Faucher, C., Maillet, F., Truchet, G., Promé, J.C. and Dénarié, J. 1990. Symbiotic host specificity of *Rhizobium meliloti* is determined by a sulphated and acylated glucosamine oligosaccharide. Nature 344: 781–784.

Leszczynski, B., Urbanska, A., Matok, H., Leskowska, I., Jozwiak, B., Wozniak, M. and Niraz, S. 1996. Metabolism of cereal xenobiotics by grain aphid. Experimental Toxic Pathology (Suppl. II) 48: 238–243.

Lewis, I.F. and Walton, L. 1958. Gall formation on *Hamamelis virginiana* resulting from material injected by the aphid *Hormaphis hamamelidis*. Transactions of the American Microscopic Society 77:146–200.

Luft, P.A. and Paine, T.D. 1997. Evaluation of environmental and plant-associated cues for nymphal settling preference by *Trioza eugeniae*. Entomologia Experimentalis et Applicata 85: 105–111.

Ma R., Reese J., Black W.C. and Bramel-Cox P. 1998. Chlorophyll loss in a greenbug-susceptible sorghum due to pectinases and pectin fragments. Journal of the Kansas Entomological Society 71: 51–60.

Madhusudhan, V.V. and Miles, P.W. 1998. Mobility of salivary components as a possible reason for differences in the responses of alfalfa to the spotted alfalfa aphid and pea aphid. Entomologia Experimentalis et Applicata 86: 25–39.

Malenovský, I., Burckhardt, D. and Tamesse, J.L. 2007 Jumping plant-lice of the family Phacopteronidae (Hemiptera: Psylloidea) from Cameroon. Journal of Natural History 41: 1875–1927.

Mani, M.S. 1964 *The ecology of plant galls*, W. Junk Publishers, The Hague, The Netherlands.

Mani, M.S. 2000 *Plant galls of India* (second edition), Science Publishers, New Hampshire, USA.

Markkula, M., Laurema, S. and Tiittanen, K. 1976. Systemic damage caused by *Trioza apicalis* on carrot. Symposia Biologica Hungarica 16: 153–155.

Mathesius, U., Schlaman, H.R.M., Spaink, H.P., Sautter, C., Rolfe, B. and Djordjevic, M.A. 1998. Auxin transport inhibition precedes root nodule formation in white clover roots and is regulated by flavonoids and derivatives of chitin oligosaccharides. Plant Journal 14: 23–34.

Matsukura, K., Matsumura, M. and Tokuda, M. 2009. Host manipulation by the orange leafhopper *Cicadulina bipunctata*: gall induction on distant leaves by dose-dependent stimulation. Naturwissenschaften 96:1059–1066.

Matsukura, K., Matsumura, M. and Tokuda, M. 2010. Both nymphs and adults of the maize orange leafhopper induce galls on their host plant. Communicative & Integrative Biology 3: 388–389.

Memelink, J., Linthorst, H.J.M., Schilperoort, R.A., and Hoge, H.C. 1990. Tobacco genes encoding acidic and basic isoforms of pathogenesis-related proteins display different expression patterns. Plant Molecular Biology 14: 119–126.

Meyer, J. 1987. *Plant galls and gall inducers*. Gebrüder Bornträger, Stuttgart, Germany.

Meyer, J. and Maresquelle, H.J. 1983. Anatomie des galles. Handbuch der Pflanzenanatomie XIII, 1–662.

Miles, P.W. 1999. Aphid saliva. Biological Reviews 71: 44–85.

Miles, P.W. and Harrewijn, P. 1991. Discharge by aphids of soluble secretions into dietary sources. Entomologia Experimentalis et Applicata 59: 1123–1134.

Mitjaev, I.D. 1968. A gall-forming leafhopper. Trudy Instituta Zoologii Akademii Nauk Kazakhskoi SSR (Transactions of the Zoological Institute of the Academy of Sciences of Kazakh SSR) 30: 205–206 [In Russian].

Monfreda, R. and Spagnuolo, M. 2004. Enzyme activity of an eriophyoid 'salivary' secretion: preliminary report of polygalacturonase. Phytophaga 14: 611–614.

Monod, Th. and Carayon, J. 1958. Observations sur les *Copium* (Hemipt., Tingidae) et leur action cécidogène sur les fleurs *Teucrium* (Labiacées). Archives de Zoologie Éxperimentale et Génerale 95: 1–31.

Moritz, G. 1982. Zur Morphologie und Anatomie des Franzenflüglers *Aeolothrips intermedius* Bagnall 1934. I Mitteilung: Der Kopf. Zoologisher Jahrbuch, Anatomie 107: 557–608.

Moritz, G. 1995. Morphogenetic development of some species of the Order Thysanoptera (Insecta). In *Thrips biology and management* edited by BL Parker, M Skinner, and T Lewis. Plenum Press, New York, USA, pages 489–504.

Morris, D.C. and Mound, L. 2003. Thysanoptera phylogeny: the molecular future. Entomologische Abhandlungen 61: 153—155.

Motta, L.B., Kraus, J.E., Salatino, A. and Salatino, M.L.F. 2005 Distribution of metabolites in galled and non-galled foliar tissues of *Tibouchina pulchra*. Biochemical Systematics and Ecology 33: 971–981.

Mound, L.A. 1971. The feeding apparatus of thrips. Bulletin of Entomological Research 60: 547–548.

Mound, L.A., Heming, B.S. and Palmer, J.M. 1980. Phylogenetic relationships between the families of recent Thysanoptera (Insecta). Zoological Journal of the Linnean Society 69: 111–141.

Mound, L.A. & Crespi, B.J. 1992. The complex of phlaeothripine thrips (Insecta, Thysanoptera) in woody stem galls of *Casuarina* in Australia. Journal of Natural History 26: 395–406.

Mound, L.A. and Morris, D.C. 2005. Gall-inducing thrips: an evolutionary perspective. In *Biology, ecology, and evolution of gall-inducing arthropods* edited by A Raman, CW Schaefer, and TM Withers, Science Publishers, New Hampshire, USA, pages 59–72.

Murphy, D.H. 1989. A *Paracopium* (Hemiptera: Tingidae) from *Clerodendrum inerme* (Verbenaceae) which is not gall forming, with a note on other herbivores. Raffles Bulletin of Zoology 37: 168–170.

Nyman, T. and Juljkunen–Tiitto, R. 2000. Manipulation of the phenolic chemistry of willows by gall-inducing sawflies. Proceedings of the National Academy of Sciences USA 97: 13134–13187.

Ollerstam, O., Rohfritsch, O., Höglund, S. and Larsson, S. 2002. A rapid hypersensitive response associated with resistance in the willow *Salix viminalis* against the gall midge *Dasineura marginemtorquens*. Entomologia Experimentalis et Applicata 102: 153–162.

Otha, S., Kajino N., Hashimoto H., and Hirata T. 2000. Isolation and identification of cell hypertrophy-inducing substances in the gall forming aphid *Colopha moriokaensis*. Insect Biochemistry and Molecular Biology 30: 947–952.

Pastori, G.M. and Río, L.A. 1994. An activated-oxygen-mediated role for peroxisomes in the mechanism of senescence of *Pisum sativum* L. leaves. Planta 193: 385–391.

Raffa, K. 1989. Genetic engineering of trees to enhance resistance to insects. BioScience 39: 524–534.

Rajadurai, S., Mani T., Balakrishna P. & Raman A. 1990. On the digestive enzymes and soluble proteins of the nymphal salivary glands of *Trioza jambolanae* Crawford (Triozinae: Psyllidae: Homoptera), the gall maker of the leaves of *Syzygium cumini* (L.) Skeels (Myrtaceae). Phytophaga 3: 47–53.

Raman, A. 1987a. On the cecidogenesis sand nutritive tissues of the leaf galls of *Garuga pinnata* Roxburgh (Burseraceae) induced by *Phacopteron lentiginosum* (Pauropsyllinae, Psyllidae, Homoptera), Phytophaga 1: 141–160.

Raman, A. 1987b. On the transfer cell-like nutritive cells of the galls induced by thrips (Thysanoptera: Insecta). Current Science 56: 737–738.

Raman, A. 1991. Cecidogenesis of leaf galls on *Syzygium cumini* (L.) Skeels (Myrtaceae) induced by *Trioza jambolanae* Crawford (Homoptera: Psylloidea). Journal of Natural History 25: 653—663.

Raman, A. 1993. Architecture and behavioural dynamics of the epidermal cells of the leaf galls of *Garuga pinnata* Roxburgh (Burseraceae) induced by *Phacopteron lentiginosum* Buckton (Insecta: Homoptera: Psylloidea). Annals of Forestry 1: 74–81.

Raman, A. 1996. Nutritional diversity in gall-inducing insects and their evolutionary relationships with flowering plants. International Journal of Ecology and Environmental Sciences 22: 150–160.

Raman, A. 2003. Cecidogenetic behavior of some gall-inducing thrips, psyllids, coccids, and gall midges, and morphogenesis of their galls. Oriental Insects 37: 359–413.

Raman, A. 2007. Biogeographical implications in species richness, biological diversity, and evolution of gall-inducing insects of the Orient and the Eastern Palearctic. Oriental Insects 41: 9–25.

Raman, A. 2010. Insect–plant interactions: the gall factor. In *All flesh is grass: plant—animal interrelationships* edited by J Seckbach and Z Dubinsky, Springer, Heidelberg, Germany, pages 121–146.

Raman, A. 2011. Morphogenesis of insect-induced plant galls: facts and questions. Flora 206: 000–000.

Raman, A. and Ananthakrishnan, T.N. 1983a. Studies on some thrips (Thysanoptera, Insecta) galls. Fine structure of the nutritive zone. Proceedings of the Indian National Academy of Sciences B49: 525–561.

Raman, A. and Ananthakrishnan, T.N. 1983b. On the developmental morphology of the rosette galls of *Acacia leucophloea* Willd. (Mimosaceae) induced by *Thilakothrips babuli* Ramk. (Thysanoptera: Insecta). Proceedings of the Indian Academy of Sciences (Animal Sciences) 92: 342–350.

Raman, A. and Takagi, S. 1992. Galls induced on *Hopea ponga* (Dipterocarpaceae) in southern India and their gall-maker belonging to the Beesoniidae. Insecta Matsumurana (New Series) 47: 1–32.

Raman, A., Ananthakrishnan, T.N. and Swaminathan, S. 1978. On the simple leaf galls of *Casearia tomentosa* Roxb. (Samydaceae) induced by *Gynaikothrips flaviantennatus* Moulton (Tubulifera, Thysanoptera, Insecta). Proceedings of the Indian Academy of Sciences 87: 231–242.

Raman, A., Rajadurai, S., Mani, T. and Balakrishna, P. 1991. On the salivary enzymes and soluble proteins of *Mimusops* gall thrips *Arrhenothrips ramakrishnae*. Zeitschrift für Angewandte Zoölogie 78: 131–136.

Raman, A., Singh, R.N. & Maryanska–Nadachowska, A. 1996. Biology and karyology of a cecidogenous psylloid, *Trioza fletcheri minor* (Homoptera: Psylloidea) and morphogenesis of galls on the leaves of *Terminalia tomentosa* and *Terminalia arjuna*. Insecta Matsumurana (New Series) 53: 117-134.

Raman, A., Schaefer, C.W. and Withers, T.M. [Editors] 2005a. *Biology, ecology, and evolution of gall-inducing arthropods*, Science Publishers, New Hampshire, USA.

Raman, A., Schaeffer, C.W. and Withers, T.M. 2005b. Galls and gall-inducing arthropods: an overview of their biology, ecology, and evolution. In *Biology, ecology, and evolution of gall-inducing arthropods*, Science Publishers, New Hampshire, USA. pp. 1–33.

Raman, A., Beiderbeck, R. and Herth, W. 2009a. Early subcellular responses of susceptible and resistant *Vitis* taxa to feeding by grape phylloxera *Daktulosphaira vitifoliae*. Botanica Helvetica 119: 31–39.

Raman, A., Burckhardt, D. and Harris, K.M. 2009b. Biology and adaptive radiation in the gall-inducing Cecidomyiidae (Insecta Diptera) and Calophyidae (Insecta Hemiptera) on *Mangifera indica* (Anacardiaceae) in the Indian subcontinent. Tropical Zoology 22: 27–56.

Ratsimialarivo-Ravoahangiariosoa, A. 1983. Observations on a psyllid leaf gall on *Cunonia bogeri* Seem. caused by a new species of *Trioza*. Cecidologia Internationale 4: 1–7.

Rey, L. 1976. Recherches comparées sur trios types d'hymenopterocécidies (Cynipidae et Tenthredinidae). Étude morphologique, histologique et ultrastructurale du développement et de la differenciation, essais éxperimentaux. Thèse (Docteur es Sciences), University of Nantes, Nantes, France.

Rey, L. 1992. Developmental morphology of two types of hymenopterous galls. In *Biology of insect-induced galls* edited by J.D. Shorthouse and O. Rohfritsch, Oxford University Press, New York, pages 87–101.

Rohfritsch, O. 1971. Développement cécidien et rôle du parasite dans quelques galles d'arthropodes. Marcellia 37: 233–339.

Rohfritsch, O. 1975. Étude du développement de la galle de *Giraudiella inclusa* Fr. sur *Phragmites communis* Trin. Marcellia 38: 251–270.

Rohfritsch, O. 1976. Trâces de succion de deux Chermisidae *Chermes abietes* L. et *Chermes strobilobius* Kalt. dans les bourgeons de *Picea excelsa* L. Marcellia 39: 69–84.

Rohfritsch, O. 1978. Premières manifestations de l'action parasitaire du *Hartigiola annulipes* Hartig sur la feuille du hêtre. 103e Congrès Societé des Savantes de Nancy, Science 3: 311–322.

Rohfritsch, O. 1980. Relations hôte–parasite au début de la cécidognèse du *Hartigiola annulipes* Hartig sur le hêtre. Bulletin de la Societé Botanique France, Actualités Botanique 127: 199–208.

Rohfritsch, O. 1988. A resistance response of *Picea excelsa* to the aphid, *Adelges abietes* (Homoptera: Aphidoidea). In *Mechanisms of woody plant defenses against insects: search for pattern* edited by W.J. Mattson, J. Levieux and C. Bernard-Dagan, Springer, New York, USA, pages 253–266.

Rohfritsch, O. 1992. Patterns in gall development. In *Biology of insect-induced galls* edited by J.D. Shorthouse and O. Rohfritsch, Oxford University Press, New York, USA, pages 60–86.

Rohfritsch, O. and Anthony, M. 1992. Strategies in gall induction by two groups of homopterans. In *Biology of insect-induced galls* edited by J.D. Shorthouse and O. Rohfritsch, Oxford University Press, New York, USA, pages 102–117.

Samui, G. and Jha, S. 2009. Biology, seasonal incidence and management of *Apsylla cistellata* Buckton on mango in West Bengal. The Journal of Plant Protection Sciences 1: 16–20.

Schaefer, C.W. 2005. Gall-inducing heteropterans (Hemiptera). In *Biology, ecology, and evolution of gall-inducing arthropods* edited by A. Raman, C.W. Schaefer and T.M. Withers, Science Publishers, New Hampshire, USA, pages 231–238.

Schoonhoven, L.M., van Loon, J.J.A. and Dicke, M. 2007. *Insect–plant biology*. Oxford University Press, Oxford, UK.

Schonrögge, K., Harper, L.J., Brooks, S.E., Shorthouse, J.D. and Lichtenstein, C.P. 1998. Reprogramming plant development: two approaches to study the molecular mechanism of gall formation. In *The biology of gall-inducing arthropods* edited by G. Csóka, W.J. Mattson,

G.N. Stone and P.W. Price, General Technical Report NC 199, USDA Forest Service, St Paul, Minnesota, pages 153–160.

Shorthouse, J.D., Leggo, J.J., Sliva, M.D. and Lalonde, R.G. 2005. Has egg location influenced the radiation of *Diplolepis* (Hymenoptera: Cynipidae) gall wasps on wild roses? Basic and Applied Ecology 6: 423–434.

Singh, G. 2003. *Mango shoot gall: its causal organism and control measures* Indian Council of Agricultural Research, New Delhi, India.

Skuhravá, M. and Skuhravý, V. 1992 Biology of gall midges on common reed in Czechoslovakia. In *Biology of insect induced galls* edited by J.D. Shorthouse and O. Rohfritsch, Oxford University Press, New York, USA, pages 196–207.

Skuhravá, M., Skuhravý, V. and Brewer, J.W. 1984 Biology of gall midges. In *Biology of gall insects* edited by T.N. Ananthakrishnan, Oxford–IBH Publishing Company, New Delhi, India, pages 169–222.

Smith, C.M. 2005 Plant resistance to arthropods: molecular and conventional approaches. Springer, Heidelberg, Germany.

Sopow, S.L., Shorthouse, J.D., Strong, W. and Quiring, D.T. 2003. Evidence for long distance, chemical gall induction by an insect. *Ecology Letters* 6: 102–105.

Southwood, T.R.E. and Reader, P.M. 1976. Population census data and key factor analysis for the viburnum whitefly, *Aleurotrachelus jelinekii* (Frauenf.) on three bushes. Journal of Animal Ecology 45: 313–325.

Srivastava, P.N. 1987. Nutritional physiology. In *Aphids: their biology, natural enemies and control*, vol. 2A, edited by A.K. Minks and P. Harrewijn, Elsevier, Amsterdam, The Netherlands, pages 99–121.

Summers, C.G., Newton, A.S. Jr. and Estrada, D. 1996. Intraplant and interplant movement of *Bemisia argentifolii* (Homoptera: Aleyrodidae) crawlers. Environmental Entomology 25:1360–1364.

Taylor, G.S. and Miles, P.W. 1994. Composition and variability of the saliva of coreids in relation to phytotoxicoses and other aspects of the salivary physiology of phytophagous Heteroptera. Entomologia Experimentalis et Applicata 73: 265–277.

Trotter A. and Cecconi G. 1902. *Cecidotheca Italica* : o raccolta di galle Italiane determinate, preparate ed illustrate. Fascicules XVI–XVIII (No: 376–450), Padua, Italy

Urbanska, A. and Leszczynski, B. 1997. Enzymatic adaptations of cereal aphids to plant glycosides. *Communication of Plants with the Environment*, Phytochemical Society of North American and Phytochemical Society of Europe, Noordwijkerhout, The Netherlands, page 25.

Valterová, I., Nehlin, G. and Borg-Karlson, A.K. 1997. Host plant chemistry and preferences in egg-laying *Trioza apicalis* (Homoptera, Psylloidea). Biochemical Systematics and Ecology, 25: 477–491.

Varadarasan, S. 1979. *Bioecological investigations on some gall-inhabiting Thysanoptera.* Thesis (Doctor of Philosophy), University of Madras, Madras, India.

von Dohlan, C.D. and Moran, N.A. 1995. Molecular phylogeny of the Homoptera: a paraphyletic taxon. Journal of Molecular Evolution 41: 211–223.

Walker, G.P. 1985. Stylet penetration by the bayberry whitefly, as affected by leaf age in lemon, *Citrus limon.* Entomologia Experimentalis et Applicata 39: 115–121.

Walker, G.P. 1987. Probing and oviposition behavior of the bayberry whitefly (Homoptera: Aleyrodidae) on young and mature lemon leaves. Annals of the Entomological Society of America 80: 524–529.

Walker, GP and Aitken, DCG 1985. Oviposition and survival of bayberry whitefly, *Parabemisia myricae* (Homoptera: Aleyrodidae) on lemons as a function of leaf age. Environmental Entomology 14: 254–257.

Walker, G.P. and Bednar, L.F. 1986. Behavioral responses of California red scale crawlers (Homoptera: Diaspididae) to solutions behind artificial membranes. Annals of the Entomological Society of America 79: 549–553.

Wardle, R.A. and Simpson, R. 1927. The biology of Thysanoptera with reference to the cotton plant. 3. Relation between feeding habits and plant lesions. Annual Applied Biology 14: 513–528.

Washburn, J.O. and Washburn, L. 1984. Active aerial dispersal of minute wingless arthropods: exploitation of boundary-layer velocity gradients. Science 223: 1088–1089.

Werner, D.J. 2000. Gallwanzen und Wanzengallen (Heteroptera: Tingidae). Verhandlungen Westdeutscher Entomologentag (Löbbecke-Museum, Düsseldorf) XX: 211–228.

Westphal, E. 1980. Response of some Solanaceae to attack by gall mite *Eriophyes cladophthirus*. Plant Disease 64: 406–409.

Westphal, E. 1982. Modification du pH vacuolaire des cellules épidermiques foliaires de *Solanum dulcamara* L. soumises à l'action d'un acarien cécidogéne. Canadian Journal of Botany 60: 2882–2888.

Westphal, E. 1992. Cecidogenesis and resistance phenomena in mite-induced galls. In *Biology of insect-induced galls* edited by JD Shorthouse and O Rohfritsch, Oxford University Press, New York, USA, pages 141–156.

Westphal, E., Bronner, R. and Le Ret, M. 1981. Changes in leaves of susceptible and resistant *Solanum dulcamara* infested by the gall mite *Eriophyes cladophthirus* (Acarina, Eriophyoidea). Canadian Journal of Botany 59: 875–882.

Will, T., Tjallingii, W.F., Thonnessen, A., and van Bel, A.J.E. 2007. Molecular sabotage of plant defence by aphid saliva. Proceedings of the National Academy of Sciences USA104: 10536–10541.

Wool, D. 2005. Gall-inducing aphids: biology, ecology, and evolution. In *Biology, ecology, and evolution of gall-inducing arthropods* edited by A.

Raman, CW Schaefer, and T.M. Withers, Science Publishers, New Hampshire, USA, pages 73–132.

Yokota, T., Arima, M. and Takahashi, N. 1982. Castasterone, a new phytosterol with plant hormone potency, from chestnut insect gall. Tetrahedron Letters 32: 1275–1278.

Yukawa, J. and Rohfritsch, O. 2005. Biology and ecology of gall-inducing Cecidomyiidae (Diptera). In *Biology, ecology, and evolution of gall-inducing arthropods* edited by A. Raman, C.W. Schaefer and T.M. Withers, Science Publishers, New Hampshire, USA, pages 273–304.

Zalat, S., el-Akkad, S., Henediq, S., Gadalla, S. and Gilbert, F. 2000. An insect–plant interaction in the Sinai desert ecosystem. Egyptian Journal of Biology 2: 8–14.

Dynamics of Insect Behaviour
T.N. Ananthakrishnan & K.G. Sivaramakrishnan
Scientific Publishers (India), Jodhpur, 2012, pp. 71-80

Chapter - 6

BEHAVIOURAL DYNAMICS OF ACRIDIDS

K.P. Sanjayan

Grasshoppers and locusts have evinced considerable interest in terms of their abundance diversity and the behavioural intricacies. They occupy a wide range of habitats, with a tendency to migrate in large number enabling survival in adverse condition. In their feeding, locomotion, and reproductive behaviour they evince dynamism stereotyped and instructive behavioural patterns serve them remarkably well, in particular in adjusting to the wide range of fluctuation of physical factors. They effectively exploit resources of their habitat and their characteristic rapid, jumping and flying responses, many a time evading several natural enemies are key characteristic. Climbing a small shrub or a tall grass, they rest at various heights within the canopy. Under favourable conditions of temperature and humidity, they tend to be actively feeding at night. Warming by barking in the sun taken place through turning aside perpendicular to the rays or they tend to jump to the ground to bark, after which they repeat to their customary shelters. A common orientation is to turn aside perpendicular to the rays of the sun and lower the hind legs, exposing the abdomen. They usually bark for one to two hours on sunny days in a prelude to becoming active. Their character jumping and flying responses make them to escape several enemies which parasitise or feed on them. Different species have diverse ways of attacking and feeding on their host plants, geoplan or ground-dwelling species feeding on seeds and leaves, often performing young green leaves. They evince dynamism not only in their feeding, ranging from grasses, leafy plants and bushes, but also in their locomotion, reproduction and developmental behaviour. Their ability to jump repeatedly several times their body length, projecting through the air with an explosive trick, they are able to propel themselves in the air for short distances. A common orientation is to turn aside perpendicular to the rays of the sun, exposing the abdomen.

GENERAL BEHAVIOUR

Steriotyped and instinctive behaviour patterns serve grasshoppers remarkably well in making adjustments to wide fluctuations of physical factors that otherwise might be fatal. Grasshoppers effectively exploit the resources of their habitat and at the same time are able to tolerate or evade the extremes of physical factors. Their characteristic rapid jumping and flying responses help them escape numerous enemies that parasitize or feed upon them. Grasshopper species show certain behaviour patterns which are common. They occasionally stir, preen, turn around to expose the opposite side and sometimes crawl to a more favourable basking location. After basking for one to two hours on sunny days, grasshoppers become active, seeking mates, or feed. During warm sunny weather of the late spring and summer, grasshoppers take advantage of two foraging periods, one in the morning and one in the afternoon. Different species of grasshoppers have different way of attacking and feeding on their host plants. Individuals of certain species climb the host plant to feed on leaves, petals, buds, or soft seeds while others cut and fell a grass leaf and feed on it while sitting on the ground. Geophilous (ground-dwelling) species regularly search the ground for food, picking up and feeding on seeds, dead arthropods, and leaves felled by other grasshoppers. In grasslands with considerable bare ground, very little litter is produced by grasshoppers. Several leaves, dropped by the grasshoppers that feed on the host plant, are soon found and devoured by the ground foragers. In habitats infested by dense populations of grasshoppers, pellets of their excrement, rather than litter, accumulate in small conspicuous piles. Only in habitats of tall grass do grasshoppers produce leafy ground litter that goes uneaten and 'wasted'. This is because the grasshopper species in these habitats are phytophilous and feed resting on the host plant.

Grasshoppers may even make choices among the leaves of a single host plant. They prefer young green leaves and discriminate against old yellowing ones. Nevertheless, individuals of ground-dwelling species, often feed in short bouts on old plant litter lying on the ground as well as on dry animal dung. This feeding may be a means of restoring water balance – either losing or gaining moisture (Gangwere *et al.,* 1989). Amazingly, grasshoppers are able to communicate visually and acoustically among themselves. They produce sounds with structural adaptations on hind legs and wings and receive these signals with auditory organs (ears) located in the first abdominal segment. Using their colourful wings and hind legs, they also flash visual messages and receive these with their compound eyes. Intra-specific communication that occurs between members of the same species is used to attract and recognize mates, to ward off natural enemies.

LOCOMOTOR BEHAVIOUR

Grasshoppers are known for their characteristic jumping and flying behaviours apparently to escape capture. The stimuli initiating these responses may be of many kinds but circumstantial observation indicates that any image movement across the compound eye is usually the primary stimulus. Noise appears to seldom have an effect in eliciting the locomotor response of the grasshopper. Also other insects, such as large darkling beetles or large grasshoppers, may crawl close and elicit no response until they touch a resting grasshopper. Phytophilous species of grasshoppers may not jump or fly to escape intruders, but may shift their position away from the intruder to the opposite side of a stem, retreat deeper into the canopy, or drop to the ground. During most daylight hours grasshoppers are largely quiescent. Considerably less time is devoted to feeding, locomotion, mating, and oviposition. The adults of most species of grasshoppers possess long, functional wings that they use effectively to disperse, migrate, and evade predators. Adults regularly fly out from deteriorating habitats caused by drought or depletion of forage, but they may also leave a site for other reasons. Some individuals of a population instinctively fly and disperse while others remain in the habitat. The most notorious migrating grasshoppers are old world locusts, *Schistocerca gregaria* and *Locusta migratoria*.

Many insects have evolved elaborate mechanisms for displacing their body rapidly away from potential predators or as a means of increasing their forward speed of locomotion. The forces underlying these movements can be generated by different parts of the body in orthopteran insects the legs have become the main propulsive mechanism, with the hind pair of legs showing the greatest specializations, most obviously in their large size relative to the others. The main thrust in these insects is provided by a rapid, coordinated extension of the two hind tibiae. Furthermore, rapid and powerful movements of an individual hind leg can also be used in defensive kicking.

FEEDING BEHAVIOUR IN GRASSHOPPERS

Grasshoppers are often categorized as grass-feeders, forb-feeders, or mixed-feeders. Grasshoppers can also be thought of as generalists (polyphagous) or specialists (monophagous), depending on whether they feed on many different items or focus on a few specific items. Grasshoppers mostly eat vegetation, but some also scavenge on grasshopper feces and even on dead grasshoppers of the same species. Most grasshoppers are polyphagous, as they feed on more than a few species of plants, although some are monophagous (Lockwood et al., 1996; Chapman, 1990). For polyphagous grasshoppers, growth rate is generally increased by mixed feeding (Bernays et al., 1997). This diet mixing helps to compensate a diet that would otherwise be lacking in certain nutrients (Bernays and Simpson, 1990). Distance between food sources affects grasshopper feeding, possibly limiting a generalist to a specialist diet. Beyond a certain distance, grasshoppers would not leave one plant to feed on another, possibly because the increased time and distance would have resulted in more energy used than nutrients gained. Monophagous grasshoppers on the other hand may receive certain benefits that outweigh the costs of slower growth.

Leaf pubescence, hardness and size can deter grasshoppers from feeding (Chapman, 1990). Abiotic factors in the environment, competition, and predation also shape the feeding behaviour of grasshoppers. Grasshoppers modify their behaviour accordingly in order to minimize costs and maximize benefits associated with foraging and feeding. If food plants are abundant, grasshoppers may spend less time foraging and level shorter distances between foods, thus reducing the total energy spent foraging (Cambers et al., 1996; Chapman, 1990).

The feeding of individual grasshopper is a result of several behavioural stimuli. From a distance, grasshoppers are guided by visual and olfactory cues from a food source (Chapman, 1990) and once in contact with a plant, may be stimulated to make a test bite based on inputs from several sensillae present on the antennal tip, maxillary and labial palp and tarsal sensillae. Upon biting the plant, nutrient chemicals as well as secondary chemicals are sensed by the sensilla on the labrum and labium and provide crucial information pertaining to the continuance of feeding (Ananthakrishnan *et al.,* 1986). Based on the nutrient quality of the plant, the water content of the plant, the secondary chemicals within the plant, the physical properties of the plant (hairiness, hardness), and the physiological state of the grasshopper, the grasshopper will make a decision whether and for how long to continue feeding (Sanjayan and Ananthakrishnan, 1987; Sanjayan and Murugan, 1987). Feeding in grasshoppers is governed by the need for nutrients, minerals and water (Chapman, 1990); Bernays and Simpson, 1990) and factors such as food availability, environmental conditions, competition and predation are important (Bernays *et al.,* 1997).

Nutritional quality, size and time since the previous meal are primary factors determining feeding. Stretch receptors in the gut, nutrient feedbacks and blood osmolality aid in monitoring feeding. Short term endogenous cycles also govern the foraging and feeding behaviours of grasshoppers regardless of the previous meal. Many of these behaviours, such as locomotion, may occur without the grasshopper actually initiating feeding.

BEHAVIOURAL CHANGES IN RELATION TO LOCUST PHASE CHANGE

The desert locust, Schistocerca gregaria (Forskal) provides an excellent system for investigating this issue since the same individual can exists in two forms (phases), depending on population density (Uvarov, 1977; Pener, 1991; Simpson *et al.,* 1999). Solitarious phase nymphs are cryptically coloured (green or brown) and relatively inactive. They also behave cryptically, being repelled by other locusts and freezing when visually disturbed. Although they are potentially highly polyphagous, solitarious locusts live in low-density populations in desert environments that are usually dominated by a small number of host plant species. In contrast, gregarious phase locusts are conspicuously coloured as nymphs (yellow, black and orange), highly mobile and adapted to living at high population densities. When in crowds, they form marching bands that can travel considerable distances (up to ca. 1000 m per day as final instar nymps) (Ellis and Ashall, 1957). Their conspicuous coloration may confer protection against vertebrate predators, particularly during the intial stages of an outbreak. Being less mobile and more cryptic in behaviour and coloration, it might be predicted that solitarious phase should show greater fidelity to a feeding site than gregarious nymphs, and recent work indicates that they are better able to tolerate nutritional imbalances that would result from limited movements between foods.

Schistocerca gregaria Forskal shows phase polyphenism, which is a form of density-dependent phenotypic plasticity in which the same genotype may express graded changes between two extreme forms, the cryptic phase *solitaria* and the aggregating, highly mobile phase *gregaria* (Pener, 1991). The desert locust has the genetic potential to exist in either phase, chaning between them within a lifetime and epigenetically across generations through a water soluble factor added to the egg foam by the mother in response to recent crowding (Bouaichi *et al.,* 1995). Phase change involves many morphological, physiological and behavioural changes. Individual phase change is stimulated by mechanical contact with other locusts. A clumped resource distribution promotes change to the gregarious phase by increasing crowding and contact between individuals. The close contact between solitary individuals triggers a rapid behavioural phase change, which results in beginning to actively aggregate than being repelled by each other.

REPRODUCTIVE BEHAVIOUR

Grasshoppers have an amazing ability to identify their mates. Each species has its individual song, produced by rubbing or flicking the lower back legs on the fore wings to create either a chirping or a clicking sound. This behaviour is known as stridulation. Females sing more softly than males, facilitating differentiation between both sex and species. Species that make no sound rely on sight and scent to find a mate. Males emit pheromones, external hormones which attract females, while other species use their excellent eyesight to enable identification by colour.

Mating occurs when the male alights on the female's back and may last anywhere from 45 minutes to well over a day. In some species, a female mates with several males. Most of the sperm in her genital tract from the first suitor is replaced by the sperm of her next mate. Males therefore mate many times with the same partner and other females to gain the maximum opportunity to pass on their genes. Males of some species die shortly after mating. The females die after egg-laying, which may last until cold weather begins.

OVIPOSITION BEHAVIOUR

During oviposition behaviour several different sequences of rhythmical movements occur. The first part of the behaviour consists of digging a deep hole in the ground. To accomplish this, the animal begins by pressing the abdomen tip into the ground while making sweeping opening and closing valve movements. Once the ovipositor has become engaged in the substrate, the female simply stands on the surface while the ovipositor burrows beneath her. The result is that the abdomen becomes enormously extended into the ground. To allow this, the soft cuticle between some abdominal segments and the interseg-mental muscles are modified permitting about 10-fold length changes without damage. After the hole has been dug, the abdomen tip retracts slightly from the bottom, and pauses. Soon, very rapid opening and closing movements of the valves ensue, and then the first egg slowly emerges during a sustained gaping posture assumed by the valves. Once the egg has emerged, the valves close, whereupon the egg drops to the bottom of the hole. The closing of the valves reverses the initial orientation of the egg (50100 eggs in all), which assures that when the future embryo develops the head will be positioned upwards. Following the deposition of each egg, rapid opening and closing movements are repeated. These movements are accompanied by the secretion of a small amount of white frothy substance from the accessory glands. The egg laying process continues until all eggs are laid forming an 'egg-pod' while abdomen slowly retracts from the bottom of the hole. The final portion of oviposition behaviour is the capping of the egg-pod with a large amount of the frothy substance. The froth then hardens and darkens. The occurrence of oviposition behaviour is related to the cycle of maturation of the eggs, and is expressed once each 7-10 days, but females can postpone egg-laying for several days if a suitable site is not available. Capulation and fertilization are not necessary requisites for the behaviour because quite successful oviposition and hatching can occur parthenogenically in these animals.

Grasshoppers produced acoustical signals by rubbing the hind legs against the tegmina or the sides of the abdomen. They may also communicate by rapidly flexing or snapping their hind wings in flight, behaviour called cryptation. Each species apparently produces its own unique sound. The songs of grasshoppers relate to various behaviours including aggressive behaviour, mating behaviour, calling songs and are unique for each species making them a non-invasive way of biodiversity enumeration.

Oviposition behaviour differs among species of grasshoppers. Most species lay their eggs in the ground. Females of some species choose bare ground, while others choose to lay among the roots of grasses or forbs. A female ready to lay often probes the soil several times before finally depositing a clutch of eggs. Experimental evidence indicates that probing is a mean by which a female obtain sensory information on the physical and chemical properties of the soil. The ovipositor is supplied with a variety of sensilla that allow a female to monitor soil conditions. Temperature of the soil must be favorable, water content must be in a suitable range, and acidity and salt content of soil must be tolerable. Females of some species select loose soils such as sand, some rocky soils, while the majority of females prefer compact loamy soils for oviposition. The pods of different species vary in size, shape, and depth in the soil. The latter condition profoundly affects incubation temperature, hatching, and egg survival. A female actively depositing eggs is often attended by one or more males. Depending at least partly on clutch size, females take from 35 to 9- minutes to lay a full complement of eggs. After withdrawing her ovipositor a female will take a minute or two to cover the aperture of the hole with particles of soil and ground litter. The female uses either her ovipositor or her hind tarsi, depending on species, to do this. The act appears to be instinctive maternal care that provides some protection for the eggs from predation by birds, rodents, and insects. In certain species, as soon as the female retracts her ovipositor, the attendant male mates with her; females and attending males of other species merely walk away from each other.

CONCLUSION

Preservations of grasshoppers in their natural habitat have revealed various behavioural responses among species. Each appears to have its won way of reacting to a battery of environmental factors. Because a relatively few species have been investigated, much remains to be discovered. The whole subject of grasshopper behaviour provides a fertile area for research both in the field and in the laboratory. Acquiring sufficient information on the various aspects of grasshopper behaviour will not only serve to improve integrated management of pest species and the protection of beneficial ones, but will also advance the science of animal behaviour.

REFERENCES

Ananthakrishnan, T.N., Sanjayan, K.P. and Suresh Kumar, N. 1986 Host plant preference in *Cyrtacanthacris ranacea* Stoll in some malvaceous hosts in terms of food utilization. Proceedings of Indian National Science Academy B51(3): 352-357.

Bernays, E.A., Angel, J.E. and Augner, M. 1997 Foraging by a generalist grasshopper: the distance bweteen food resources influences diet mixing and growth rate (OrthopteraL Acrididae). Journal of Insect Behaviour 10: 829-840.

Bernays, E.A. and Simpson, S.J. 1990. Nutrition. In *Biology of Grasshoppers* (Eds. R.F. Chapman and A. Joern), John Wiley and Sons, New York., pages 105-127.

Bouaichi, A., Roessingh, P. and Simpson, S.J. 1995 An analysis of the behavioural effects of crowding and re-isolation on solitary reared adult desert locusts (*Schistocerca gregaria*) and their offspring. Physiology Entomology 20: 199-208.

Chambers, P., Sword, G., Angel, J.E., Behmer, S. and Bernays, E.A. 1996 Foraging by generalist grasshoppers: two different strategies. Animal Behaviour 52: 155-165.

Chapman, R.F. 1990 Food selection. In *Biology of Grasshoppers* Eds. R.F. Chapman and A. Joern), John Wiley and Sons, New York., pages 39-72.

Ellis, P. and Ashall, C. 1957 Field studies on diurnal behaviour, movement and aggregation in the desert locust (*Schistocerca gregaria* Forskal*)*. Anti-Locust Bulletin*,* 25. Anti-locust Research Centre.

Gangwere, S.K., Muralirangan, M.C. and Meera Muralirangan, (eds.) 1989. Food selection and feeding behaviour in acridoids: A review. Contributions of American Entomological Institute 25(5): pages i + 55.

Lockwood, J.A., Struttmann, J.M. and Miller, C.J. 1996 Temporal patterns in feeding of grasshoppers (Orthoptera: Acrididae); Importance of nocturnal feeding. Environmental Entomology 25: 570-581.

Pener, M.P. 1991 Locust phase polymorphism and its endocrine relations. Advances in Insect Physiology 23: 1-79.

Sanjayan, K.P. and Ananthakrishnan, T.N. 1987 Host preference of some acridids (Insecta: Orthoptera) in relation to some biochemical parameters. Proceedings of Indian Academy of Sciences 96(1): 15-21.

Sanjayan, K.P. and Murugan, K. 1987 Nutritional influence on the growth and reproduction in two species of acridids (orthoptera: Insecta). Proceedings of Indian Academy of Sciences 96(1): 229-237.

Simpson, S.J., McCaffery, A.R. and Hagele, B. 1999. A behavioural analysis of phase change in the desert locus. Biological Review 74: 461-480.

Uvarov, B.P. 1977 *Grasshoppers and Locusts. A handbook of general Acridology*. Vol. 2. Centre for overseas Pest Research, London.

Dynamics of Insect Behaviour
T.N. Ananthakrishnan & K.G. Sivaramakrishnan
Scientific Publishers (India), Jodhpur, 2012, pp. 81-91

Chapter - 7

AGGREGATION AND DORMANCY BEHAVIOUR OF *LUPROPS TRISTIS*, THE RUBBER BEETLE INVADING DOMESTIC ENVIRONMENTS

Sabu K. Thomas

INTRODUCTION

Luprops tristis Fabricius, 1801 (Coleoptera: Tenebrionidae), is an inconspicuous detritivore beetle abundant in the rubber plantation litter in the Western Ghats. Their high abundance in the range of 200 per m^2 of litter or about 20 million per hectare in rubber plantations makes *L. tristis* as the most prominent litter dwelling beetle in the agriculture belts along the western slopes of the Western Ghats. With their litter feeding habits, harmless nature, nocturnal surface activity and diurnal passivity in lower litter layers they would have remained inconspicuous facilitators of litter decomposition and nutrient cycling in monoculture rubber plantation forests in the region. But massive seasonal invasion of huge aggregations of *L. tristis*, following summer showers into residential buildings, continued presence of clusters of thousands of nocturnally active beetles crawling inside buildings and their prolonged stay in a state of dormancy is a frustrating nuisance of unimaginable magnitude for the farming community (Figures 1-3; Sabu *et al.*, 2007, 2008). The continued presence and attraction of these beetles towards light, following overnight invasion into buildings is a frustrating nuisance for local people. Clusters of several hundred to thousands crawl inside the living rooms and fall off into beds and food from ceilings. They do not sting or bite, but when disturbed, such as picking them off the walls or when they are squashed or pressed against while sleeping; they release an irritating odoriferous phenolic secretion that causes a burn to the skin. Subsequently they congregate in dark, undisturbed areas such as attics and wall voids and remain dormant for several months. Its amazingly synchronous massive infestation, abundance and selection of specific buildings as shelter for more than 20 consecutive years, remains as a curious ecological phenomenon to biologists whereas it is a matter causing agony and concern to the affected people in rubber plantation belts.

Figure 1: Massive aggregation of *Luprops tristis* (Fabricius, 1801), in a residential building in the rubber tracts of Kerala state, India.

Figure 2: Nocturnal dispersal of *Luprops tristis* during the early phase of aggregation.

Figure 3: Killed aggregated beetles in a residential building

This beetle previously identified as *Lyprops curticollis* Fairmaire, 1896 *probably* following the terminology used in Beeson (1941) and has been referred to by that name in many publications (Narendran, 1998; Thomas and Jacob, 2005). Recent works (sabu *et al.,* 2007) found out that it is *Luprops tristis* and not *Luprops curticollis* prevalent in the north west of the country. Its familiar presence below coconut sheaths in thatched sheds and on woodworks supporting earthen tile roofing in traditional buildings has led to the popular names "Ola prani" and "Oadu vandu" (Ola=coconut fronds; Oadu=clay tile; prani & vandu= beetle in the regional language) in north Kerala. In central and south Kerala, they are known as "Mupli vandu", in reference to the first reporting of such huge aggregations in the rubber estates at Muplium in central Kerala during the 1970s. *L. tristis* beetles breed in the litter stands of rubber plantations and all feeding stages have a strong preference towards tender leaves. Larval, pupal and teneral adult stages were observed in the field only after annual leaf fall and biology studies indicated that their life cycle is linked to the phenology of leaf shedding of rubber plants (Sabu *et al.*, 2008; Joby, 2006). Adult beetles live for 11 months and about 8-9 months of the adult phase is spending in aggregation and dormancy. They enter into aggregation sites before entering into reproductive facets of their life cycle.

The most striking behavioural characteristic of *Luprops* is their mass invasion and formation of overwintering aggregations prior to monsoon rains in to residential buildings and below boulders and continued selection of same sites as aggregation spots. It is a phenomenon common in the rubber plantation belts in the moist western slopes of the Western Ghats but not in the dry eastern slopes of the Western Ghats, The beetles in these residential building aggregations may be so numerous that 5 Lakhs -15 Lakhs living individuals may be collected from each building on a single day. Beetles may remain in these aggregations for extended periods up to 10 months. Beetles from overwintering aggregations are characterized by the presence of a well-developed fat body, an increase in body weight as compared to the reproductively active post dormancy adults present in litter stands. The extremely high fat deposit in the body cavity in early spring aggregations corresponds to the advance preparation of the beetles towards dormancy. Since no such fat deposition is present in post dormancy adults, huge deposition of fat in the early phases prior to dormancy suggest that the fat reserves were accumulated to tide over the anticipated dormancy period. Absence of mating pairs in field conditions and larval instars in the field prior to dormancy or pre-monsoon period in rubber belts indicates lack of reproductive activity prior to aggregation in the field. *L. tristis* could enter into reproductive activity and produce eggs in simulated laboratory conditions within a very short span of time at any stage of the dormancy, hence hypertrophy of reproductive organs reported in other overwintering coleoptera (Roubik and Skelly, 2001; Hodek, 1973) is not occurring in the group.

AGGREGATION

Reasons for the home invasion and aggregation are easy to find when its habitat requirements are understood. *Luprops* can not survive in the wet litter conditions during the rainy reason that persists for months in the region. Rainfall is the most important factor driving them to dormancy and making it a very crucial event determining the life cycle of *L. tristis* in the region (Vinod & Sabu, 2010). It leads them away from the rainfall mediated wet litter to warm and dry residential buildings and boulders. Since Tenebrionidae are dry habitat preferring groups (Endrödy-Younga & Tschinkel, 1993; Watt, 1992) and most majority of litter and soil dwelling Tenebrionidae are nocturnal detritivores, no other limiting factor would have lead to such behaviour. Non-record of Tenebrionidae from the cold wet cloud forests and very low presence in evergreen forest is additional proof to the importance of dry conditions in deciding the habitat selection and distribution of Tenebrionidae and also towards the generally low abundance of Tenebrionidae in the moist Western slopes of the Western Ghats. Short duration home invasion and aggregation of *Luprops* in the dry rain shadow belts in the eastern slopes of the Western Ghats indicates the importance of rainfall mediated wetness in evoking home invasion and aggregation. During the short refractory period, beetles are readily attracted to tender leaves and dry litter conditions and could come out from dormancy (Sabu *et al.*, 2008; Sabu & Vinod, 2009). However, in natural conditions, although leaf resources and favorable habitat conditions prevailed for a considerable span of time prior to the onset of monsoon rains, home-invaded beetles in the refractory phase of dormancy do not return to field. Voracious feeding behaviour and initiation of reserve food storage long before the summer showers by teneral adult stages and non-return to natural habitats (= rubber plantation litter stands) even after the return of favorable habitat conditions (Sabu *et al.*, 2008; Sabu & Vinod, 2009), indicate that *L. tristis* have evolved some anticipatory mechanism to predict that short-duration summer showers will precede adverse conditions in the natural habitat (Sabu & Vinod, 2009). It is obvious that the beetles aggregate to overcome the cold wet environmental/habitat conditions prevailing in the region during the 8-9 months wet period, and to utilize the short duration (3-4 months) seasonal leaf resource availability and to establish themselves successfully for the short period of favourable conditions prior to the monsoon period in the region.

AGGREGATION SITES

Aggregation of *L. tristis* are found in the attics of tile roofed buildings, traditional tile roofed wooden buildings and thatched sheds, hay stacks in cow sheds, hollow wooden blocks, coconut fronds and coconut husks in firewood stocks, and crevices below boulders in rubber plantations. All selected spots are dark and closed places with open space around and located within 1 km vicinity of rubber plantations or in the middle of rubber plantations.

Probably the most curious aspect of the aggregation behaviour of *L. tristis* is their return and aggregation in specific buildings and thatched sheds successively for years even after application of both organic and inorganic repellents and insecticides by the affected people. What enables them to identify the same spots and aggregate in specific buildings and thatched sheds successively for years even after application of both organic and inorganic repellents and insecticides? Adults are univoltine and the longevity of *L. tristis* ends 5–7 weeks following their return to the field from dormancy. And it is the new generation beetles that enter the homes every year (Sabu *et al.*, 2008). Since generations are not overlapping and the same buildings are regularly infested leaving neighboring buildings uninfested, it raises the question how do the newly emerged adults with no previous experience select the same buildings used by the previous generation for aggregation? These questions remains unanswered its likely that either the presence of specific semiochemicals/aggregation pheromones left behind by the parental generation in the aggregation sites might be providing them the cue to return to the aggregation sites of the parental generation located about 0.5 to 1 km away.

BEHAVIOUR DURING AGGREGATION

Grouping: Aggregating beetles were 3–4 layers deep and occasionally honey comb-like aggregations hung precariously from the wooden supports. They clung to one another very strongly and separating the individual members from the outer layers is not easy. Dormant beetles showed no response towards light or towards leaf materials in the aggregation sites. The activity of *L. tristis* in aggregations appears to be temperature related. Except for short duration (2–3 hours) massive diurnal dispersal within the buildings on intermittent rainless, hot, and humid days of rain break and sunshine dormant beetles in aggregation do not disperse. Disturbing the beetles on such occasions resulted in the beetles moving around in the aggregation sites. They will move about rapidly or take flight whereas on wet rainy days they showed no response towards disturbances.

During the initial phase of aggregation (refractory period), beetles were nocturnally surface active for 8–11 days, with a strong attraction towards light and fresh leaf materials. Diurnally, they remained inactive in the corners of the shelter in small groups. Subsequently they clustered and entered into dormancy for 9 months. Last of phase of dormancy, was indicated by the nocturnal dispersal within the shelters lasted for 15–17 days, which is initiated by a few individuals and is followed by the rest within 2–3 days. No beetles from the shelters returned to the litter at this stage.

Dormancy: Dormancy in *L. tristis* can be broadly categorized as wet season dormancy. The long duration dormancy shown by *L. tristis* involving adults of various emergence classes does not fall under either quiescence or diapause. The term quiescence is used situations with a shorter period and no induction phase and with active feeding throughout the stage (Mansingh, 1971; Danks, 1987). *L. tristis* does not diapause either, as diapause is preceded by a long preparatory phase during which energy reserves are accumulated at a specific stage in life cycle. Preparation for diapause by accumulation of energy reserves prior to active dormancy is not mandatory in *L. tristis*. In classical dormancy, the preparatory phase succeeds a typically long refractive phase and the insects cannot terminate diapause and continue morphological development even if exposed to optimal environmental conditions during this stage (Mansingh, 1971; Tauber & Tauber, 1976; Denlinger, 1986, 2002; Danks, 1987). However, *Luprops* beetles are readily attracted to food resources and water during the short refractory phase. Adults of all emergence classes irrespective of the active feeding duration available entered into dormancy without a preparatory phase nor an induction period in response to an unexpected environmental token stimulus, the short mid summer rainfall that mediated wetness in the rubber plantation litter floor. During the short refractory period, beetles are readily attracted to tender leaves and dry litter conditions in experimental conditions. These observations weighed in favor of categorization of dormancy in *L. tristis* as oligopause and not as diapause, which is intermediate between quiescence and diapause (Mansingh, 1971; Danks, 1987; Leather *et al.*, 1995).

Recognizing the environmental stimuli leading to and arousal from dormancy is a difficult task for many tropical insects (Denlinger, 1986, 1994). However, synchrony of diapause arousal and subsequent field return of beetles with the beginning of annual leaf fall in the rubber plantations in the region points makes it easier for *Luprops*. Return of dry conditions following the prolonged monsoon season, and the related variation in moisture, photoperiod and temperature in the region are the environmental cues leading to the arousal of *L. tristis* from oligopause. Seasonal availability of a thick layer of freshly fallen litter in rubber plantations both as a habitat and food resources facilitates the return of post-dormant beetles to the field. Generally, mortality during dormancy is very low in natural populations (Wolda & Denlinger, 1984; Danks, 1987). High mortality during dormancy in *Luprops* is arising from the presence of higher number of beetles in shorter emergence class (shorter pre-dormancy period) with low feeding duration.

Flight capacity: Their flight muscles did not degenerate, as was evidenced by the mass flight of beetles in response to experimental wetting of the tiles and during the post dormancy return to the field shows that flight muscles did not degenerate and *Luprops* did not lose their capacity for flight during the dormancy period. This is contrary to the reports in *S. rotundus* where the aggregating beetles lost their flight capacity during aggregation and dormancy as mechanism to conserve energy (Wolda & Denlinger, 1984).

Advantages of aggregation

For some species insects, aggregation has been theorized to be beneficial and in others it has been theorized to be disadvantageous. Aggregation is beneficial to *Luprops* as it provides protection from adverse habitat conditions and secondarily as a strategy to synchronize its breeding stages with phenology of rubber plants. Aggregation of aposematically colored insects aggregate and these insects have some form of chemical defense against predation. These insects being brightly colored and living in aggregations are very easy to detect. Aggregation of aposematically colored insects helps to emphasize the insects' warning coloration and also to dilute predation as predators learn to avoid these insects after experiencing their unpleasant defense chemicals. (Honek *et al.,* 2007; Gagliardo & Guilford, 1993; Sillen-Tullberg & Leimar, 1988). Cryptic coloration is assumed to have evolved to minimize conspicuousness to decrease the risk of detection (Edmunds, 1974; Evans & Schmidt, 1990; Endler, 1991) and it makes *Luprops* inconspicuous in the litter habitat as well as in the aggregation sites.

How ever, what is the need for *Luprops* with cryptic coloration and chemical defense to aggregate? Defensive gland secretions and the repelling odour prevailing in the aggregation sites must be a signal to the potential predators that may stray into the aggregation sites or attempted to feed upon *Luprops* during the refractory phase or at any other occasion.

Further, aggregation and dormancy during the rainy season, when litter habitat is less conducive for beetles, being wet and in the advanced phases of faster decomposition enables escape from adverse habitat conditions (Jobi, 2006). Subsequently it allows synchronization of the life cycle of *L. tristis* and synchronized development of post dormancy adults irrespective of age class variations in the following dry season when the leaf litter habitat becomes suitable by the influx of fresh litter and the availability of tender leaves as a food resource. The simulta-neous return of post dormancy adults to the litter habitat facilitates synchronized feeding and preparation for breeding activities through temporal and spatial coincidence of mates. Dormancy in *L. tristis* becomes a classical example of how seasonal patterns of development optimize reproductive effici-ency by synchronising mating (Denlinger, 1986) and food availability. Synchronized development of feeding stages with fresh fallen leaves enables efficient utilization of the tender leaf resources that are transient in nature, and the brief period of dry litter habitat provide conditions for larval, pupal and teneral adults before the onset of the long wet season.

Synchronized emergence of physically fit adult reproductive population in a subsequent season, optimization of the time of reproduction, synchronization of the larval and adult feeding stages with food resource availability (annual leaf shedding of host plants) in rubber plantations, prevention of the arrival of vulnerable early developmental stages during the monsoon periods, and synchronization of the feeding stages and reprodu-ctive stages. It is clear that aggregation in a state of dormancy during the seasonally inimical wet seasons, arousal from dorma-ncy synchronized with the return of dry seasons and annual litter fall in the rubber plantations, dispersal of aggregation and completion of sensitive stages in the life history before return of inimical conditions and signaling effect of obnoxious odour prevailing in the aggregation sites contributed to the enormous success of *L. tristis* in the region.

REFERENCES

Beeson, C.F.C. 1941. The Ecology and Control of the Forest Insects of India and the Neighboring Countries. Government of India publications, New Delhi, 767 pp.

Danks, H.V. 1987. *Insect Dormancy: An Ecological Perspective.* Biological Survey of Canada, Ottawa. 439 pp.

Denlinger, D.L. 1986. Dormancy in tropical insects. *Annual Review of Entomology* 31: 239-4.

Denlinger, D.L. 1994. The beetle tree. *American Entomologist* 40: 168-171.

Denlinger, D.L. 2002. Regulation of diapause. *Annual Review of Entomology* 47: 93-122

Edmunds, M. 1974. Defense in animals: a survey of anti-predator defenses. *Longman*, Harlow, UK, 357 pp.

Endler, J.A. 1991. Variation in the appearance of guppy color patterns to guppies and their predators under different visual conditions. *Vision Research.* 31, 587-608.

Endrödy-Younga, S. and Tschinkel, W. 1993. Estimation of population size and dispersal *Entomologique de Belgique,* 40, 6–62.

Evans, D.L. and Schmidt, J.O. 1990. *Insect Defenses: Adaptive Mechanisms and Strategies of Prey and Predators*, State University of New York Press, Albany

Fairmaire, L. 1896. Hétéromères de l'Inde recueillis par M. Andrewes. *Annales de la Société Entomologique de Belgique*, 40, 6–62.

Gagliardo, A. and Guilford, T. 1993. Why do warning-colored prey live gregariously? *Proceedings of the Royal Society of London B-Biological Sciences* 251: 69-74

Hodek, I. 1973. Biology of Coccinellidae. *Academia,* Prague and W. Junk, The Hague, *260 pp.*

Honek, A., Martinkova, Z. and Pekar, S. 2007. Aggregation characteristics of three species of Coccinellidae (Coleoptera) at hibernation sites. *Europian Journal of Entomology* 104: 51–56, 2007

Jobi, M.C. 2006. Studies on the biology of Luprops curticollis (Coleoptera: Tenebrionidae) and rubber litter insect dynamics in relation to the rubber litter fall patterns. Ph. D. thesis. University of Calicut – Kerala.

Leather, S.R., Walters, K.F.A. and Bale, J.S. 1995. *The ecology of insect over wintering.* Cambridge University Press.

Mansingh, A. 1971. Physiological classification of dormancies in insects. *Canadian Entomologist* 103: 983-1009.

Narendran, T.C. 1998. Incidence of *Lyprops curticollis* as a household pest in Kerala. *Insect Environment* 3(4): 115.

Roubik, D.W. and Skelley, P.E. 2001. *Stenotarsus subtilis* Arrow, the aggregating fungus beetle, of barro Colorado island nature monument, panama. *The Coleopterists Bulletin*, 55(3): 249-263.

Sabu, T.K., Merkl, O. and Abhitha, P. 2007. A new *Luprops* species from Western Ghats with redescriptions of and identification key to the species of Peninsular India and Sri Lanka (Tenebrionidae: Lagriinae: Lupropini). *Zootaxa* 1636: 47-58.

Sabu, T.K., Vinod, K.V. and Jobi, M.C. 2008. Life history, aggregation and dormancy of the rubber plantation litter beetle, *Luprops tristis*, from the rubber plantations of moist south Western Ghats. *Journal of Insect Science* 8: 01, available online: insectscience.org/8.01. 17pp.

Sabu, T.K. and Vinod, K.V. 2009. Population dynamics of rubber plantation litter beetle *Luprops tristis*, in relation to annual cycle of foliage phenology of its host, the para rubber tree, *Hevea brasiliensis*. *Journal of Insect Science* 9: 56, available online insectscience.org/9

Sillen-Tullberg, B. and Leimar, O. 1988. The evolution of gregariousness in distasteful insects as a defense against predators. *The American Naturalist* 132: 723-734

Tauber, M.J. and Tauber, C.A. 1976. Insect seasonality: Diapause maintenance, termination and postdiapause development. *Annual Review of Entomology* 21: 81-107.

Thomas, B.B. and Jacob, M. 2005. Life history of the *Luprops curticollis* Fairmaire (Coleoptera: Tenebrionidae). *Entomon* 30(3): 269-272.

Vinod, K.V. and Sabu, K.T. 2010. Dormancy inducing factors of rubber litter beetle, *Luprops tristis* (Coleoptera: Tenebrionidae). *Insect Science* 17 (47-51).

Watt, J.C. 1992. Fauna of New Zealand, No. 26: Tenebrionidae, 7–11 DSIR, Auckland.

Wolda, H. and Denlinger, D.L. 1984. Diapause in a large aggregation of a tropical beetle. *Ecological Entomology* 9: 217-30.

Dynamics of Insect Behaviour
T.N. Ananthakrishnan & K.G. Sivaramakrishnan
Scientific Publishers (India), Jodhpur, 2012, pp. 92-116

Chapter - 8

BEHAVIOURAL ASPECTS OF SOME KEY PESTS OF PLANTATION FORESTRY IN INDIA

John Prasanth Jacob, N. Ramasubramanian and A. Naveenkumar

INTRODUCTION

Insect population behaviour influences the dynamics of forest ecosystems. It is necessary to understand the dynamics of specific herbivore populations and to develop general strategies for managing destructive herbivore outbreaks particularly in forest plantations. During the last two decades species like teak, poplar, eucalyptus and casuarina have been extensively planted commercially as well as in many other afforestration programmes. Short rotation multipurpose tree species are mainly grown for timber, pulp and fuel wood in several tropical countries including India. Some tree species like Casuarina play a major role as a windbreak and shelterbelt along coasts.

Insect pests, native as well as invasive, infesting these tree species leading to outbreaks in plantation areas have become a regular event. Insect pests like *Hyblaea puera* (Lepidoptera: Hyblaeidae), *Indarbela quadrinotata* (Lepidoptera: Metarbelidae) and *Leptocybe invasa* (Hymenoptera: Eulophidae) pose a serious problem in teak, casuarina and eucalypts respectively which are extensively used in plantation forestry (Fig. 1-3). Under forestry conditions, pest management techniques aimed at maintenance of pest populations at moderate levels have greater chance of success than conventional methods of pest control. Simple behavioural observations can sometimes be used to great advantage in the development of such methods. In recent years, the wide spread use of chemical pesticides aggravated insect pest problems by eliminating beneficial insects and development of resistance against synthetic pesticides by pests. Such environmental conditions has stimulated great interest in exploring the possibility of using behaviour modifying chemicals like pheromones for control of pests through mass trapping or disruption of the insect's normal communication systems.

Fig. 1. Larva of *H. puera* on teak; Fig. 2. Larva of *I quadrinotata* on Casuarina; Fig. 3. Galls on Eucalyptus. *Inset*- Adult *L. invasa* wasp

Information on volatile profiles which attract or deter insect pests of forest tree species is scanty. Barata *et al.* (2000) identified the host and non host semiochemicals of eucalypts wood borer *Phoracantha semipunctata* by gas chromatography and electroantennography. Host plant resistance, the inherited ability of a given plant to resist the damaging effects of herbivores to a greater extent than related plants when grown under the same conditions is a desirable feature. Plant characteristics which contribute to resistance include surface texture, leaf nutrient content, concentration and identity of chemical toxins including volatiles in plant tissues. Enhancing such plant characteristics would increase host plant resistance. In forestry, provenances or clones or different species of same genus is used extensively. In order to enhance the biomass productivity to meet the industrial needs for timber and pulp, provenances / land races are identified locally or introduced from their native homelands to establish trials and identify promising candidates through tree breeding programmes. Clonal forestry is now a well-recognized method for mass production of desired trees to get maximum economic benefits since it enables the capturing of genetic gain by skillful selection. Many clones or provenances of species like Teak, casuarina and eucalypts showed varying levels of tolerance/ resistance during natural incidence of key pests in field trials (Jacob *et al.,* 2002; Sasidaran *et al.,* 2005: Jacob and Ramesh Kumar, 2009) (Table 1). Finding volatile chemicals that affect the behaviour of pests associated with the plants indicate that the field of plant-insect chemical ecology is a useful area for further research.

Table 1. Details of species, clones and seed origin germplasm used in the study.

Species	Source	Status
	Seed	
Eucalyptus camaldulensis	Pudukkottai, Tamilnadu	Susceptible
	Panampalli, Kerala	Susceptible
	Sathyavedu-1, Andhra Pradesh	Tolarent/ Resistant
	Sathyavedu-2, Andhra Pradesh	Tolarent/ Resistant
	Kennedy River, Pudukkottai, Tamilnadu	Susceptible
	Ongole commander (white), Andhra Pradesh	Susceptible
	Ongole local (red), Andhra Pradesh	Susceptible
E. tereticornis	Rudrapur, Uttar Pradesh	Susceptible
	Clones	
	ITC -7	Tolarent/ Resistant
	ITC-10	Susceptible
	E-226	Tolarent/ Resistant

	IFGTB-7	Tolarent/ Resistant
	IFGTB-19	Tolarent/ Resistant
	IFGTB-30	Tolarent/ Resistant
	IFGTB-31	Susceptible
E. pellita	Seedlings	Tolarent/ Resistant
E. citriodora		Tolarent/ Resistant
Tectona grandis	**Clones**	
	MYSA-3	Tolarent/ Resistant
	APKKA-1	Tolarent/ Resistant
	APAKB-1	Tolarent/ Resistant
	APKKP-1	Tolarent/ Resistant
	APKZB-1	Tolarent/ Resistant
	APNBB-1	Tolarent/ Resistant
	APKKC-2	Tolarent/ Resistant
	TNT-16	Tolarent/ Resistant
	APNPL-6	Tolarent/ Resistant
	APKKR-4	Susceptible
Casuarina equisetifolia	**Clones**	
	Kenya, Sarawak	Tolarent/ Resistant
	Local germplasm (seedlings)	Susceptible
C. junghuhniana	Seedlings	Tolarent/ Resistant

Insect Pests of Plantation Forest Trees

Teak (*Tectona grandis* Linn. F.) is one of the most favoured timber species all over the world, used for a range of products and services. Ever increasing demand for teak timber has resulted in large scale plantations, both within and outside its range of natural distribution. There has been a steady expansion of teak plantation over the decades in order to meet the dwindling supply of wood from natural forests and due to an increase in demand for wood products. *H. puera,* a polyphagous pest feeds on many tree species, but teak is considered as the principal host. The first and second instars feed mainly on the leaf surface. From the third instar onwards the larva cuts out a leaf flap, folds it over and fastens it with silk and feeds from within. Incidence of heavy defoliation, particularly in young plantations results in losses averaging 44 percent of potential growth increment in four-to eight-years-old trees due to the teak defoliator (Nair *et al.*, 1985). Though the pest is found throughout the year in the forest areas, its population reaches peak during pre-monsoon showers. The moths when reach a critical density in a particular location migrate to facilitate exploitation of new food sources).

Casuarina equisetifolia Forst and Forst (Casuarinaceae) is an important multipurpose tree species grown for wind break, shelter belt along the sea coast, pulp wood and floor poles. *C. equisetifolia* is planted extensively during last three years in coastal areas of Tamil Nadu and Andaman group of Islands as a bio shield to reduce the impact of Tsunami in future. This species is highly susceptible to bark feeding caterpillar *I. quadrinotata*. The larvae tunnel into the wood through the axils of side shoots and come out to feed on the bark through a sleeve made of frass and excreta webbed together extending from the tunnel mouth. Several larvae may attack the same tree at different locations in heavy infestations and this may result in serious injury to the bark. Apart from this, the holes left on the trunk lead to infestation by pathogenic organisms and the trees suffer from die-back. Affected trees also break at the points of attack.

Eucalyptus, a native tree of Australia, is a highly favoured exotic species used in large scale plantations and in afforestation programmes all over the world, more especially in the Mediterranean region, Africa, and south Asia. Several species of Eucalyptus are planted for the production of not only electric and telegraph poles, but also pulp to meet the increasing demand from paper-production and related industries. *Eucalyptus camaldulensis* and *Eucalyptus tereticornis* (Myrtaceae) are two species raised extensively in semi-arid forests, because of their relatively fast growth rate, high productivity/unit area, and least attention needed in post-plantation plant-care and management. India has 39,42,600 ha area under eucalypts cultivation (Global Eucalypts Map, 2008). Invasive wasp *L. invasa* induces galls on shoot tips, petioles, and midribs in eucalyptus seedlings maintained in nurseries, and on coppiced shoots and in juvenile plants in plantations. The average yield of pulp wood is 70-100 mt/ha during a 5 year rotation which fetches about Rs. 1.4 lakhs per ha. The coppiced shoots are ideal breeding sites for *L. invasa*, which by building into massive populations on young shoots destroy them and thus force planters to abandon extensive areas of the crop (Jacob, 2009).

Insect behaviour in Relation to Germplasm variation

Highly productive provenances / land races of these tree species had been introduced from their home lands and through cost effective clonal propagation methods, superior-quality trees with high biomass production leading to maximum pulp yield based on fiber length and cellulose content have been raised and evaluated in India by government and various pulp- and paper-based industries to meet the industrial needs and to meet the needs of local livelihood. Field observations with reference to teak, casuarinas and eucalypts showed considerable variation in the patterns of incidence of its key pests on hosts raised from different seed sources or clonal route and provenances. Because characters such as tolerance / resistance to insects are critical in tree-improvement programmes, an understanding of the mechanisms involved in imparting tolerance to insects in plants is necessary to develop resistant–candidate germplasm through tree breeding. Behavioural response of *H. puera, I. quadrinotata*, and *L. invasa* with reference to feeding or oviposition on clonal germplasm of teak, casuarina and eucalypts will help in identifying allelochemical based variations in different clones which could be manipulated through breeding or biotechnological approaches in plants for insect pest management. Plants release a variety of secondary plant chemicals which attract or deter insect pests. Alleochemicals emanating from plants trigger a chain of behavioural response in insects like host finding, host plant selection for feeding or egg laying, survival and reproduction. These chemicals serve as easily detectable and distinctive chemicals cues not only for insect pests but also for the natural enemies attacking the pests. The volatiles, a blend of compounds like fatty acids, derived aldehydes and alcohols, terpenes and aromatic metabolites, vary in their relative and absolute concentration in plant genotypes like clones and provenances of trees. Lack of acceptable levels of resistance in clones/provenances to key insect pests tends to be due to the qualitative and quantitative variations in their volatile profiles interfering with the behavioural modulations in insects pests. An array of allelochemicals imparts resistance to insect pests. There are some volatile compounds that a species have in common as well as specific compounds that interfere significantly with the response of the insect pest. The blend of allelochemicals can radically change for every genotype of a plant species and have varying effects on the insects. Volatiles may not convey the same message even among clones/provenance of tree species. Enhancing such plant characteristics would increase host plant resistance. In forestry, provenances or clones or different species of same genus is used extensively. Many such clones or provenances of species like Teak, casuarina and eucalypts have been shown to be resistant to their key pests. Finding such Volatile chemicals that affect the behaviour of pests associated with the plants indicate that the field of plant-insect chemical ecology is a useful area for further research.

Heterogeneity in structure and chemical components of plants interferes with the exploitation of host plants by insects and notable instances of the varying degrees of resistance are shown by increasing numbers of plants. As a result of selection pressure exerted by insects feeding, host plants tend to diversify their biochemical pathways and production of newer compounds enables plants to overcome feeding pressures. In this way distinct chemotypes in plant populations tend to evolve (Ananthakrishnan, 1990). The implications of volatiles being released from different parts of the host plant can be far reaching (Ananthakrishnan and Senrayan, 1991). Majority of the volatiles detected in different cultivars of the plants significantly modulate the behaviour of the insects.

Plantation forestry is facing many of the same problems as modern agriculture with regard to pest management. Many of the tools that are commonly used for pest control in agriculture are not available in forestry or are not suitable for application in forestry. Even under conditions of plantation forestry large scale pesticide application (owing to the negative ecological impact) and rapid crop rotation on annual or semi annual cycles (owing to perennial biology of trees) are not an option in forestry. In spite of the potential benefits of genetic engineering of pest resistance in trees, many obstacles have prevented the deployment of transgenic trees. Therefore, sustainable management of insect pests in planted forests will rely on further development of knowledge of the natural defense and resistance mechanism of forest tree species and on the integration of such knowledge at molecular level and its application in tree breeding (Raffa *et al.,* 2005)

Natural defense or resistance in trees against insects may be caused by morphology and anatomy, chemical repellants and/or attractants produced and nutritional status of the host tree. Chemical repellants exert their effect in the volatile state causing an insect to avoid the tree completely. Genetic variation in the chemical properties within land races, provenances or clones of tree species is far less studied and should be exploited to develop resistant strains of trees. Modification or suppression of the chemicals through breeding could provide useful insect tolerance /resistance.

Keeping the importance of host plant chemicals in view, identification of the allelochemical profiles of insect pest resistant candidates of teak, casuarina and eucalyptus could be useful in genetic improvement for insect resistance. The intricacies of insect-plant interactions involve eco-physiological, biochemical and behavioural processes in which secondary metabolites or allelochemicals play a vital role. The diversity of behaviour modifying phyto-chemicals could be feeding, mating or ovipositional stimulants or deterrents. The nature of the primary metabolites and allelochemicals tends to change with the age of the plant, parts of the plants and from between plants (Raupp and Denno 1983). Painter (1966) points out that before attempting an investigation into the physiological basis of insect resistance, it is first necessary to find a resistant tree with demonstrable genetic variation in physiological traits so that resistant and susceptible individuals can be identified. Relatively few cases of such a rigorous demonstration of resistance to insects exist within tree species, but they are the prerequisite to physiological investigations in to tolerance / resistance mechanisms.

An undamaged plant maintains a baseline level of volatile metabolites that are released from plant tissues. These constitutive chemical reserves, which often include monoterpenes, sesquiterpenes, and aromatics, accumulate to high levels in specialized glands or trichomes (Pare and Tumlinson, 1997). Odors consisting of a blend of saturated and unsaturated six-carbon alcohols, aldehydes, and esters are produced by autolytic oxidative breakdown of membrane lipids and are released when leaves are mechanically damaged. This pattern of constitutive compounds has been analyzed in the field for perennials, including beech (Tollsten and Muller, 1996) and ash (Markovic *et al.*, 1996) trees. The presence of certain chemical in tree tissues can make them undesirable host and the insect is prevented from feeding, ovipositing or completing a phase of its development in that host. A large number of compounds like fatty acid derived aldehydes and alcohols (Croft *et al.,* 1993) terpenes derived from mevalonic acid and aromatic metabolites such as indole and methyl salicylate derived from Shikimic acid (Mann. 1987) have been identified as volatiles interfering with insect behaviour. The green odour or chlorophyll odour of plants, essentially due to the oxidative integration of fatty acids linoleic and linolenic acids results in host of 6 carbon compounds and insects make use of particular odour molecules like esters, alcohols, ketons, terpenoids as signals to detect their host plants.

Table 2. Percentage of gall incidence on *Eucalyptus* seedlings from different seed sources.

Percentage of gall infestation (Average no. of galled seedlings /30 seedlings)									
Species	*E camaldulensis*	*E camaldulensis*	*E camaldulensis*	*E. tereticornis*	*E camaldulensis*	*E camaldulensis*	*E camaldulensis*	*E. tereticornis*	*E camaldulensis*
Seed Source / Level	Sathyavedu_1#	Panampalli#	Sathyavedu-2#	Pudukkottai*	Pudukkottai #	Ongole Commander (white) #	Kennedy River Pudukkottai#	Rudrapur*	Ongole local (red) #
Four-month old seedlings									
Low	25	22	22	20	17	18	19	15	4
Medium	5	7	5	10	10	9	6	13	10
High	0	1	3	0	3	3	5	2	16
Six-month old seedlings									
Low	21	19	18	15	14	13	13	10	2
Medium	8	9	8	12	7	13	10	16	2
High	1	2	4	3	9	4	7	4	26
Average number of exit holes/30 seedlings									
Early July 2007	11	15	14	17	11	10	16	10	26
Mid-July 2007	8	7	8	10	9	15	12	16	12
Early August 2007	1	2	4	3	9	4	7	8	2

Defoliation by *H. puera* leads to epicromic branching in teak and bark feeding damage in casuarina results in stem girdling and breaking of the poles while gall development damages growing shoot terminals and leaves of *E. camaldulensis* and *E. tereticornis* inducing rapid abscission of leaves and drying of shoot tips. An intense infestation by *L. invasa* and *H. puera* results in loss of vigour and growth retardation in seedlings (Jacob and Rameshkumar, 2009). Capability of insects in distinguishing between plant hybrids and genotypes is well known (Raman, 1994, Floate and Whitham, 1995, Ollerstam *et al.,* 2002, Raman *et al.,* 2005). Variation in distribution of galls induced by *L. invasa* in different plant parts such as midrib, petiole and stem was evident in different seed sources of *Eucalyptus* (*camaldulensis* and *tereticornis*). Whereas maximal preference for gall induction was on stems, petioles and midribs of both high and moderately susceptible seed sources and clones of *E. camaldulensis* and *tereticornis*, viz., Ongole– red, Kennedy River, Rudrapur, Pudukkottai, ITC 7 and IFGTB 7, 19 and 30 (Table 2). Occurrence of galls was comparatively less in the stems of the less-susceptible seed source Sathyavedu 1 and clones like ITC 7 and IFGTB 7, 19 and 30 (*E. camaldulensis).* Although *L. invasa*-induced galls on the less-susceptible seed sources such as Sathyavedu 1 and clones like ITC 7 and IFGTB 7, 19 and 30, these galls remained isolated, sporadic, and occurred in limited numbers on the stem, petiole and midribs demonstrating traits of resistance to *L. invasa.* Insects respond to diverse stimuli when selecting their host plants for either feeding or oviposition (Ananthakrishnan *et al.,* 1985). Insects have the ability to choose the right host plant and the right host tissue for oviposition initially and feeding subsequently (Raman *et al.,* 2005).

Observations on the feeding and reproductive behaviour of teak defoliators in the lab revealed that the adults did exhibit ovipositional preference among the clones. However in the field, early larval instars and feeding damage has been observed which shows that eggs were also laid on resistant clones. The larvae hatched out cut small leaf flaps and fold it over and fed form inside. However no further feeding was observed in clones like MYSA-3 and APKKA-1. The larvae fed initially but died or subsequently left the leaf flaps. Oviposition behaviour of *H. puera* showed that the moths did not prefer to lay eggs on surfaces treated with extracts from resistant clones (Figure 4). Under choice situation the bark caterpillar did not settle on clones of *C. equisetifolia* like Chandipur, Kenya and *C. junghuhniana* for feeding. Larvae continued to remain settled on clone Malaysia and in all other clones only one larva resettled or left in between before the end of the experiments (Figure 6). Adults of *L. invasa* prefer to oviposit on young shoot tips, midribs and petioles of juvenile leaves of seedlings raised from different seed sources. In *E. camaldulensis* seedlings raised from Sathyavedu 1 seed sources and clones ITC 7 and IFGTB 7, 19 and 30, oviposition preference of *L. invasa* was more to midribs than either to the petiole or to the stem. Similarly in a Y tube olfactometer experiment, more *L. invasa* wasps oriented towards *E. camlderleusis* when given a choice with species like *E. pellita* and *E. citriodora* and *E. tereticornis*. Also the more number of wasps tend to fly towards the arm of Y tube where extracts of highly susceptible clones like ITC clone 10, and IFGTB 31 are maintained. By directing their movements not only towards preferred seed sources, but also towards specific sites and tissues within the seed sources, insects demonstrate adaptive tactics in the feeding selection and oviposition behaviour. Insects respond to various feeding stimuli when selecting their host plants. The plant surface is embedded with physical and chemical factors like trichomes, leaf surface chemicals, tissue toughness, nutrient deficiency and repellant and deterrent chemicals responsible for antixenosis to feeding insects (Southwood, 1986). Variation in the trichome diversity and density in teak clones (Jacob and Balu, 2007) and the thick cuticle of gall resistant eucalypt clones (Jacob and Rameshkumar, 2009) inhibit feeding and oviposition respectively by *H. puera* and *L. invasa*.

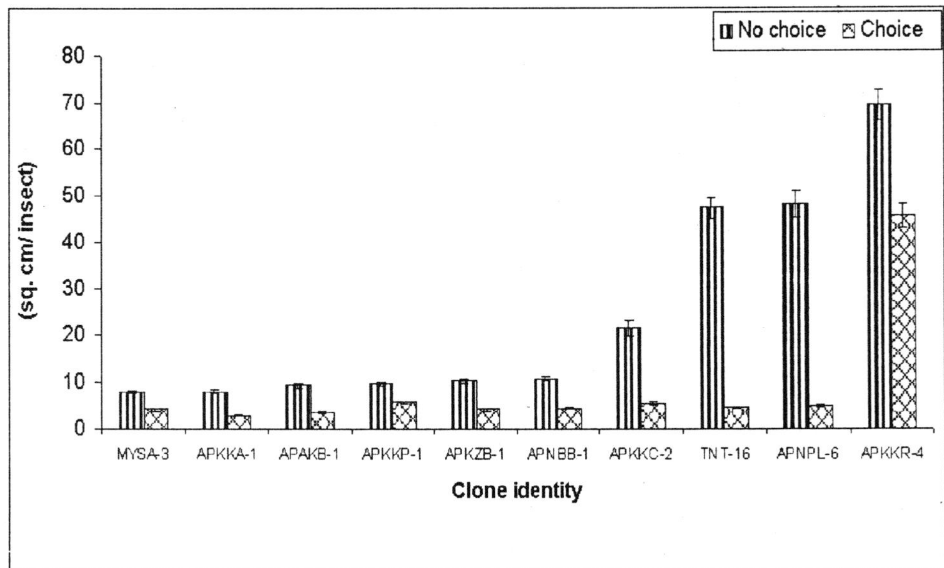

Fig. 4. Area fed by *H. puera* of teak clone leaves under choice and No-choice situation.

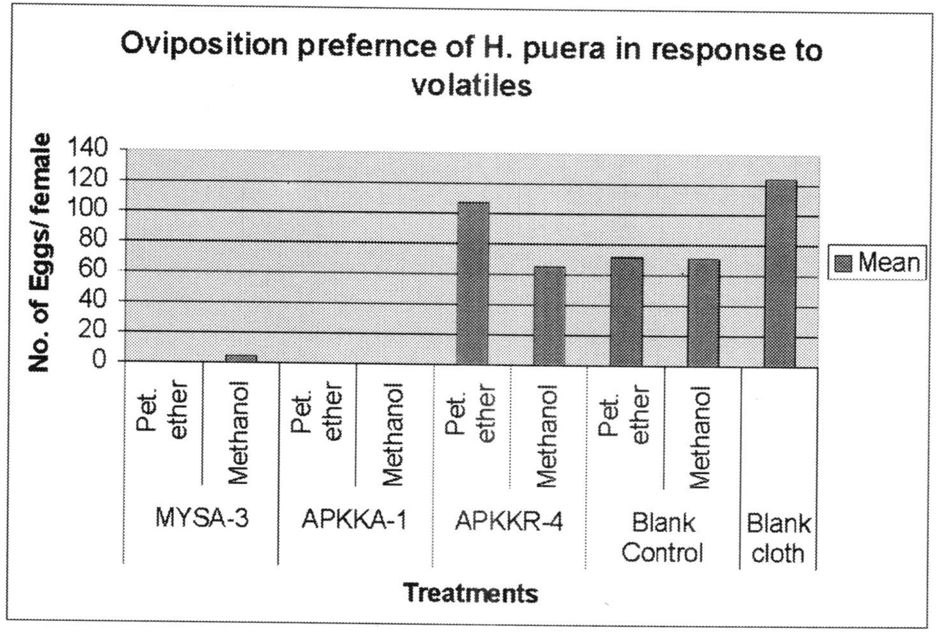

Fig. 5. Oviposition prefernce of *H. puera* in response to volatiles

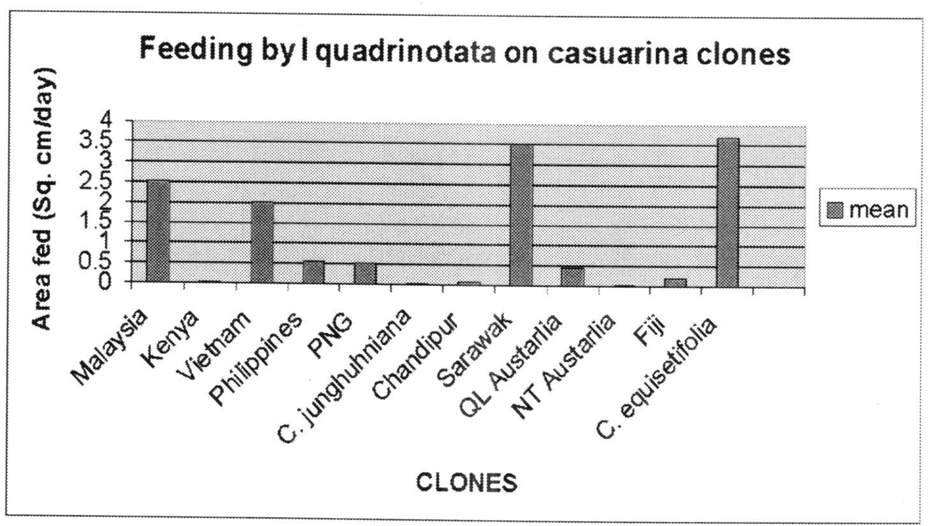

Fig. 6. Feeding by *I. quadrinotata* on Casuarina clones in a choice situation

Insect behaviour in Relation to Variation in Chemical Profiles of Germplasm

Secondary chemical constituents in plants play a predominant role in tree resistance to insects. Works of Feeny (1968)

demonstrated how tannins in oak leaves regulated the growth and development of winter moth. Defoliation induced changes in leaf biochemistry has been shown in many trees as basis of resistance against key pests. Defoliation by winter moth during spring increases tannins in oak leading to ceasation of feeding during later part of the year (Feeny, 1976). Infestation by *Epirrita autumnata* on birch *Betula pubescens* decreases host plant quality during the following year (Kaitaniemi *et al.,* 1999). Mutikainen *et al.* (2002) showed insect resistance through condensed tannin against *E. autumnata*. Tikkanen and Tiitto (2003) showed intraspecific variation in bud burst and accumulation of condensed tannin in *Quercus robur* against *Operophtera brumata*. Abundant sap flow in rain tree *Koelreuteria paniculata* and river birch *Betula nigra*, resist Cerambycid borer *Anolphophora glabripennis* attack (Morewood *et al.*, 2004). In a study on damage induced changes in woody plants and their effect on insect's performance. Nykanen and Koricheva (2004) showed that early season damage caused more pronounced changes in plants than late season damage which is in accordance with the assumption that vigorously growing foliage has a greater capacity response to damage. Natural as well as clonal variation in defoliator incidence on teak has been reported (Kedharnath and Singh, 1975; Jacob *et al.*, 2002). Attempts were made to evaluate teak clones for their natural variation in resistance to *E. macharalis,* by Misra (1992). Ahmad (1987) reported Clone TNT 6 of Tamilnadu as a comparatively resistant clone for *Hyblaea puera*. However Nair *et al*, (1989) showed that no genetic resistance to the pests exists in teak clones in Kerala. Roychoudhury *et al.* (1997) identified certain teak clones as resistant against *E. machaeralis* in terms of their Nitrogen, phenol and water contents. In eucalyptus incidence and intensity of attack of termites in nurseries and young plantation varies from species to species and from one locality to another (Thakur and Sen Sarma, 1982).

There tends to a relation between the chemical composition of a clone/ seed origin plant, germplasm and the insect behaviour. Kruel and Tencher (1958) demonstrated the racial and individual differences in the resistance of Douglas fir to the wolly aphid *Gilletteella cooleyi*. Zavarin and Sanjberk (1985) showed distinction of races on the basis of chemotaxonomic characteristics. In some cases casual relationships of certain chemicals like monoterpenes, singly of in combinations and resistance to insect pests could be demonstrated (Schuck, 1982).

While no significant correlation between resistance against autumn gum moth *Mnesmpula private* and total wax yield estimates in clonally replicated families of *E. globulus* was identified, significant differences were detected between resistant and susceptible classes in the relative quantities of aliphatic phenyl ethyl and benzyl wax esters within families (Jones *et al.,* 2002)

Behavioural response in terms of feeding and ovipositon in the key pests has clearly indicated the variations in their extent of damage on clones/seed sources/species of teak, eucalyptus and Casuarina. Resistance to insect damage is due to the secondary metabolites or allelochemicals including volatile (Elzen *et al.*, 1984). Defoliator resistant teak clones showed high phenol and tannin content where as in eucalyptus clones and seed origin plants, gall resistant plants showed low phenol and tanin levels. With reference to casuarina, bark feeder resistant clones showed high phenol and high tannin. Phenolic profiles of teak clones did not show much variation where in vanillic and protocatechuic acid were common fractions. Occurrence of Salicylic acid was also observed in few resistant clones. Distribution of phenolic acids was not consistent as far a gall resistant and susceptible clonal or seed origin plants are concerned. The same situation prevailed in Casuarina clones except *C. junghuhniana* where the phenolic acids profiles were much diverse. Free fatty acids on resistant candidates also showed diverse fractions compared to susceptible ones (Table 3 & 4). Genetically variable chemical trait acts as a strong antifeedant to insects in Eucalypts (Andrew *et al.,* 2007). The tendency of different species of eucalypts to hybridize easily (Morrow *et al.,* 1993) results in some level of uniformity in the biochemistry of eucalypts, which enable the feeding insects to either survive or oviposit on more than one species of eucalypts. However variations in the quality of nutrients (e.g., sugars, proteins, lipids) and non-nutrients (e.g., phenolic materials) between plant species or even within a plant influences food preference, growth rate, developmental time and hence the biotic potential of the insect (Slansky, 1982). In the present study quantitative variation in individual phenols, tannins, lipids, free fatty acids, phenolic acids and volatiles besides enzyme levels in clones of teak, casuarina and in different seed sources and clones of *E. camaldulensis* plays an active role in regulating either resistance or susceptibility of plants (Jacob, 2010).

Table 3. Phenolic acids in plant tissues

	PHENOLIC ACIDS
TEAK	
MYSA-3	Vanillic acid Protocatechuic
APKKA-1	Protocatechuic Vanillic
APAKB-1	Vanillic acid Protocatechuic
APKKP-1	Salicylic
APKZB-1	Vanillic acid Salicylic
APNBB-1	Vanillic acid Protocatechuic
APKKC-2	Vanillic acid Protocatechuic
TNT-16	Vanillic acid Protocatechuic
APNPL-6	Vanillic acid Protocatechuic
APKKR-4	Vanillic acid Protocatechuic
EUCALYPTS	
Species	
E. terticornis	Pyrogallol Salicylic acid
E. camaldulensis	2 methyl resorcinol
E. pelita	Salicylic acid Protocatechuic acid
E. citriodara	Protocatechuic Salicylic acid
Clone	
E.226	Hydroquinone Genetisic acid
IFGTB clone 19	Vanillic acid
IFGTB clone 7	Salicylic acid
IFGTB clone 30	2 methyl resorcinol
ITC clone 3	2 methyl resorcinol
ITC clone 7	Protocatechuic acid

ITC clone 10	Salicylic acid
IFGTB clone 5	Salicylic acid
IFGTB clone 31	2 methyl resorcinol
Seed origin	
Sathyaveedu 1	Salicylic acid
Ongole red	Salicylic acid
CASUARINA	
KENYA WOOD	Protocatechuic Vanillic acid
BARK	Salicylic acid
SARAWAK WOOD	Protocatechuic
BARK	Protocatechuic Vanillic
C .junghuhniana WOOD	---
BARK	Phloroglucinol Pyrogallol Salicylic acid Protocatechuic Genetisic
C. equisetifolia WOOD	---
BARK	Salicylic acid

Solvent I: Ethyl acetate: Benzene, 9:11
Solvent II: Benzene: Methanol: Acetic acid, 8:45:4

Table 4. Analysis of Free fatty Acids in plant tissue

	TISSUE	Free Fatty Acid fraction
Teak Clones		
MYSA 3	Leaf	Lauric, Oleic, Stearic, Linolenic, Myristic, Linoleic
APKKA 1	Leaf	Oleic, Myristic, Linoleic
APKKR 4	Leaf	Lauric, Oleic, Linoleic
Eucalyptus Clones		
IFGTB 19	Leaf	Lauric, Stearic, Linolenic, Myristic
IFGTB 7	Leaf	Stearic, Myristic, Linoleic
IFGTB 30	Leaf	Palmitic, Linolenic, Linoleic
IFGTB 31	Leaf	Oleic, Linoleic
ITC 3	Leaf	Myristic acid
ITC 10	Leaf	Palmitic, Myristic, Linoleic
ITC 7	Leaf	Palmitic, Oleic, Myristic, Linoleic

Eucalyptus species		
E citriodora	Leaf	Palmitic, Oleic, Linoleic
E. pellita	Leaf	Palmitic, Linoleic
Eucalyptus seed origin		
Sathyaveedu	Leaf	Myristic, Linoleic
Ongole red	Leaf	Palmitic, Myristic
Casuarina clones		
MALYSIA	Bark	Myristic
KENYA	Bark	Palmitic, Oleic, Myristic, Linoleic
VIETNAM	Bark	Myristic, Linoleic
CHANDIPUR	Bark	Linoleic
SARAWAK	Bark	Oleic, Myristic, Linoleic
NT- AUSTRALIA	Bark	Linoleic
C. junghuhniana	Bark	Palmitic, Lauric, Oleic, Stearic, Linolenic, Myristic, Linoleic
C. equisetifolia (Local)	Bark	Palmitic, Oleic, Linoleic

Comparatively lower levels of phenols and tannins in seedlings of Sathyavedu 1 seed source inhibit gall induction by *L. invasa*. In clonal plants, a strong inverse correlation has been indicated between the concentration of total phenols and suitability of a tree for gall development by aphids. Differences in condensed and hydrolyzable tannins, total phenolics, lignin, cellulose, hemicellulose and nitrogen influenced the occurrence of gall-inducing cynipids on species of *Quercus* (Abrahamson *et al.*, 2003). Phenolic compounds are considered to play an important role in plant defense influencing larval weight, survival of insects (Daniel *et al.*, 1990, Corcuera, 1993). Resistance of white Birch *Betula pubescens* to geometrid defoliator *Epirrita autumnatus* was found to be due to the phenolic content of leaves induced by insect damage (Haukioja, 1982). High phenolic content along with host of pheniloic acids like resorcinol, salicilic acid, phluroglucinol in resistant clones and *C. junghuhniana* tend to play an antibiotic role on *I. quadrinotata* by extending the larval duration and reducing the adult longevity. Ananthakrishnan *et al.* (1990) showed the chronic effects of such phenolic acids on growth, ingestion and utilization of food by *Heliothes armigera*. Salcyclic acid is another important signaling molecule involved in both locally and systemically induced disease resistance responses. A number of plants maintain high basic levels of salicylic acids when it plays a role in constitutive defense against insects. Levels of leaf oil content in *Eucalyptus* were shown to influence insect feeding and growth (Morrow and Fox, 1980). It has been shown that lipoxygenases catalyse the oxidation of fatty acids like linolenic or linoleic acids to hydroperoxidases which can be further converted into compounds like jasmonic acid which can elicit defence genes expression in plants (Ayers, 1992). Treatment of dietary proteins with linoleic acid and lipoxygenase significantly reduced the nutritive quality of soybean protein and soy foliar protein. Larval growth was reduced from 24 to 63% depending upon treatment. Feeding by *H. zea* on soybean plants caused damage-induced increases in foliar lipoxygenase and lipid peroxidation products. Larvae feeding on previously wounded plant tissue demonstrated decreased growth rates compared to larvae feeding on unwounded tissue (Felton *et al.*, 1994).

Increasing numbers of plant cultivars naturally call for more intensive approaches to a study of their chemodynamics in order to establish their susceptibility or resistance. Both as antibiotic agents against insects and as chemical signals, allelochemicals play an efficient role. A number of natural or synthetic compounds have been reported as inducers of systemic acquired resistance. Many of these chemicals inducers tend to function as systemic signals like salycilic acid. In conifers (Buchbauer *et al.*, 1994) and broadleaf tree species (Monsoon and Fall, 1989), an array of terpene hydrocarbons are released from plants. Elm bark beetles prefer some elm species over others for feeding, the possible existence of intraspecific preferences among individual trees or clones has remained mostly explored (Pajares, 2004). The ability of poplars to cope over many years of insect damage is reflected in a diverse set of constitutive and inducible defenses (Ryan and Jorg, 2007).

It is proposed that the combination of volatiles present in the clones/seed origin plants are strong enough to deter insects for feeding, oviposition (Pare and Tumlinu, 1996). Barata *et al.* (2000) identified range of host and non host semiochemicals of eucalyptus wood borer *Phoracantha semipunctata* and through behavioural assays volatile attractants from a range of host and non host plants. Zeng-hui *et al.* (2004) attributed differences in volatile profiles of poplar induced by various damages. The volatile signatures of clones and seed origin plants of teak, casurina and eucalyptus clearly differ qualitatively. This variation found between volatile profiles suggests that the host preference behaviour of *H. puera, L. invasa* and *I. quadrinotata* may be influenced by their differences and the complex variation in volatile blends is likely to interfere with drawing conclusion about the contribution of individual compounds. Volatiles such as (Z) 3 hexanal, (E) 2 haxonal, (Z) 3 hexanol and (Z) 3 hexenyl acetate are identified as volatiles of insect damaged cotton (Rose *et al.,* 1996). In insect induced resistance in Lodgepole pine, different (+)-3-carene synthase genes are expressed in the resistant tree genotype producing large amounts of (+)-3-carene, versus the susceptible tree genotype that produces trace amounts of (+)-3-carene. This information will support the identification and breeding of resistant Sitka spruce in order to re-introduce it as a viable, native commercial species (Trewhella *et al.,* 1997). Grant and Langevin (2002) found that aliphatic carboxylic acids applied to artificial substrate substantially increased oviposition preference of the spruce budworm for treated substrate. The presence of Cyclohexane in susceptible clones of teak, casuarina and eucalyptus could be the reason for oviposition preference by respective pests as observed in the present study. The structural similarity of the cyclohexane acids to simple aromatic acids such as benzoic and cinnamic acid derivatives, and to more complex phenolic acids suggesting that some aromatic acids might also act as oviposition stimuli for the spruce budworm. Eucalyptol (1,8-cineole) was moderately effective as a feeding repellent and highly effective as an ovipositional repellent against adult *A. aegypti* (Klocke *et al.*, 1987). The ovipositional repellency of 1,8-cineole may therefore account in large part for the observed suppression of oviposition and feeding by *H. puera* and *L. invasa* on resistant teak clones and seed origin plants of eucalyptus. Resistant provenances of *Pinus sylvestris* emitted high levels of limonene which influenced feeding by pine trunk borer *Diocryctria zimmermanni* (Sadof and Grant, 1997). Limonene was identified in resistant teak clone APKKA-1.

CONCLUSION

The natural variation in allelochemical profile between different clones/seed origin plants of eucalypts, teak and casuarinas provides some evidence of chemical basis of resistances to its key pests. Based on the extent of feeding by *H. puera* on teak, feeding preference by Bark feeder on casuarina clones and percentage of gall infestation by *L. invasa* as well as the tissue specific distribution of the gall within a plant indicates that teak clones APKKA1, MYSA3, Casuarina clones Kenya, NT Austarlia and *C. junghuhniana, E. camaldulensis* seedlings raised from Sathyavedu 1 seed source and eucalyptus clones ITC 7, IFGTB 7, 19, and 30 could be treated as germplasms that bear the least attack from key pests. Considerable evidence has been mounted to demonstrate the variation in the chemical characters in teak, casuarina and eucalyptus clones and several seed sources of *E. camadulensis* and *E. tereticornis* to varying levels of susceptibility to insect attack. Molecular verification of these traits should provide useful data in building resistance against insects, the populations of which are generally hard to manage in forest nurseries and plantations and because of its concealed and embedded nature within plant tissue by the gall inducers and bark feeders. Genetic engineering for the insertion of genes encoding enzymes involved in the synthesis of feeding inhibitors (allelochmicals) would render the trees less liable to become infected. Genetically modified plants expressing deterrents from other species or clones would likewise enhance the prospects of obtaining pest resistance candidates. Further investigations on whether the variation is also reflected in the transcript levels of genes that are associated with the formation of allelochemicals including volatiles need to be carried out. Genotypic variations have been demonstrated in many tree species. Use of newly available marker technologies will allow the charactersation and position of loci that control the types of traits. It would be interesting to screen recombinant imbred lines, populations of accessions that have very distinct volatile pattern for the individual volatiles compounds and subsequently to perform quantitative trait loci analysis. Superior individuals can be identified in the absence of insect attack. This will allow insect resistance to be incorporated into the suite of characters under consideration for genetics improvement of the species. Volatiles constitute an important array of substances governing the sensitivity of phytophagous insects in host plant selection. Sustainable management of insect pests in planted and natural forests will rely on the further development of knowledge of the natural defense and resistance mechanisms of forest trees and on the integration of such knowledge across multiple scales form the molecular level to the ecological levels and its application in tree breeding (Raffa *et al.*, 2005).

Number of candidate genes for the octadecanoid pathway in spruce is now available for further evaluation and possible manipulation of defense signaling in this gymnosperm system (Miller *et al.* 2005). Similarly defense signals likely involved in the orchestration of a complex allelochemical production and related volatile emissions of specific chemical(s) involved in species, clones, provenances subspecies, and geographic populations will provided convincing evidence for a role of specific compounds in resistance. Enhancing the inherent insect resistant qualities in a germplasm of forest trees through breeding will play a very important role in sustainable forestry and benefit if their use will become more prominent as the wood and pulp needs increase especially in developing countries. Refinement and increased use of tree breeding techniques including biotechnological approaches should be encouraged in order to accelerate the rate and accuracy of breeding plantation forest tree species for insect resistance through co-operative research efforts between entomologists, biochemists and plant breeders.

REFERENCES

Abrahamson W.G, Hunter M.D, Melika, G. and Price P.W. 2003. Cynipid gall-wasp communities correlate with oak chemistry. *Journal of Chemical Ecology* 29: 209–223.

Andrew, R.L., Wallis, I.R., Harwood, C.E, Henson, M. and Foley, W.J 2007. Heritable variation in the foliar secondary metabolite sideroxylonal in Eucalyptus confers cross resistance to herbivores *Oecologia* 153: 891-901.

Ananthakrishnan, T.N. 1990. Chemical ecology in Biological Control. Curr. Sci, 59: 1319-1322.

Ananthakrishnan, T.N. and Senrayan, R. 1991. Kairomones of *Heliothis armigera and Corycyra cephalonica* and their influence on the parasite potential of *Trichogramma chilonis*. (Trichogrammatidae: Hymentoptera). J. Biosci. 16:111-119.

Ananthakrishnan, T.N., Dhileepan, K. and Padmanabhan, B. 1985. Behavioural responses in terms of feeding and reproduction in some grasshoppers (Orthoptera: Insecta) Proceedings of Indian Academy of Sciences (Animal Science) 94: 443-461.

Ananthakrishnan, T.N., Senrayan, R., Annaduari, R.S. and Murugesan, S. 1990. Antibiotic effects of resourcinol, gallic acid and phloroglucinol on *Heliothis armigera* (Hunber) Proceedings of Indian Academy of Sciences (Animal Science) 99: 39-52.

Ahmad, M. 1987. Natural resistance in teak clones to leaf skeletonizer, *E. machaeralis* (Lepidoptera : Pyralidae) in South India. Indian J. Forestry, 14(3) : 228-231.

Ayers P.G. 1992. Pests and Pathogens (Ed.) Bios Scientific Publishers Oxford UK.

Barata E.N., Picket J.A. Wadhama L.J., Woodcock C.M. and Mustaparta H. 2000. Indentification of host and nonhost semiochemicals of eucalypts wood borer *Phoracantha semipunctata* by gas chromatorgraphy-electroantennography. J. Chem. Ecol 26: 1877-1895.

Croft, K.P., Juttner R. and Slusarenko A.J. 1993. Volatile products of the lipoxygenase pathway evolved from *Phaseolus vulgaris* (L) leaves inoculated with *Psedomonas syringae* sv *phaseoluicola*. *Plant Physiology* 101: 13-24.

Corcuera, L.J. 1993. Biochemical basis for the resistance of barley to aphids. *Phytochemistry* 33: 741-747.

Daniel, J., Gurrea J, Tom Cothren and Philiphs J.R. 1990. Influence of selected phenolic compounds on development of bollworm (Lepidoptera: Noctuidae) larvae. Journal of Economic Entomology 83: 2116- 2118.

Elzen, G.W., H.J. Williams, S.B. Vinson 1984. Isolation and identification of cotton synomones mediating searching behavior by parasitoid *Campoletis sonorensis*. J. Chem.Ecol. 10:1251-1254.

Feeny, P.P. 1968. Effect of oak leaf tannins on larval growth of the winter moth *Operophtera brumata*. *J. insect. Physiol*. 14: 805-817.

Felton G.W., C.B. Summers and A.J. Mueller 1994 Oxidative responses in soybean foliage to herbivory by bean leaf beetle and three-cornered alfalfa hopper *Journal of Chemical Ecology* 20 : 639-650.

Floate, K.D. and T.G. Whitham 1995. Insects as traits in plant systematics: their use in discriminating between hybrid cottonwoods. Canadian Journal of Botany 73:1-13.

Global. Eucalypts Map 2008 Eucalyptus Universalis- Culticvated eucalyptus forests global map 2008, GIT Forestry consulting www.git.forestry.com

Grant. G.G. and D. Langevin 2002 Structure-activity relationships of phenolic and nonphenolic aromatic acids as oviposition stimuli for the spruce budworm, *Choristoneura fumiferana* (Lepidoptera: Tortricidae) *Use of pheromones and other semiochemicals in integrated production IOBC wprs Bulletin* 25: 1-10

Haukioja. E. 1982 Inducible defence of white birch to a geometrid defoliator, *Epirrita autumnata Proceedings of 5th International Symposium on Insect Plant Relationship*. in J.H. Visser and A.K. Minks (editors) Wageningen, Netherlands

Jacob. J.P 2009 Invasive insect pest *(Leptocybe Invasa)* in *Eucalyptus* plantations in India ENVIS Forestry Bulletin : 9,(1) 65-70.

Jacob. J.P. 2010 Characterisation and evaluation of allelochemical profiles of some economically important, insect pest resistant tree species. Final Technical Report of the National Forestry Research Plan Project submitted to Indian Council of Forestry Research and Education. pp 32.

Jacob, J.P, A. Balu, S. Murugesan, B. Deeparaj and G. Srinivasan 2002. Variation in defoliator incidence on teak clones. In Proc. of the Natl.

Symp. Vistas of Entomological Research for the new Millennium (Eds) K. P. Sanjayan, V. Mahalingam and M.C. Muralirangan. 164-172

Jacob, J.P. and A. Balu 2007 Biochemical and cuticular variation in teak clone leaves and resistance to teak defoliator. *Indian Forester* 133: 197-205

Jacob, J.P. and A. Ramesh Kumar 2009. Incidence of Galls Induced by *Leptocybe invasa* on Seedlings of *Eucalyptus camaldulensis* and *E. tereticornis* from Different Seed Sources in Southern India. *International Journal of Ecology and Environmental Sciences* 35 (2-3): 187-198.

Jones T.H., Potts B.M., R.E.Vaillancourt and N.W. Davies 2002. Genetic resistance of *Eucalyptus globulus* to autumn gum moth defoliation and the role of cuticular waxes. *Can. J. For. Res.* **32**: 1961-1969.

Kaitaniemi, P., Kai Ruohomaki, Toomas Tammaru and Erkki Haukioja 1999. Induced resistance of host tree foliage and after a natural insect outbreak. *Journal of Animal Ecology* 68:382-386

Kedharnath, S. and Singh, P. 1975. Studies on natural variation in susceptibility of *Tectona grandis* to leaf skeletonizer *Pyrausta macharalis*. Paper presented at FAO/IUFRO Symposium on Forest Diseases and insects. New Delhi, India.

Kruel, W. and Teucher, G. 1958. Die tierischen feinde der Douglasie. In Gohre, K. Die Douglasie und ihr Holz. Akademie-Verlag, Berlin 403-436.

Klocke, J.A. Darlington, M.V. and Balandrin, M.F. 1987. 1, 8-Cineole (Eucalyptol) a mosquito feeding and ovipositional repellant from volatile oil of *Hemizonia fitchii* (Asteraceae) Journal of Chemical Ecology 13: 2131-2141.

Mann, J. 1987. *Secondary metabolites*. Clarendon Press. Oxford. England.

Markovic, I., Norris, D.M., Phillips, J.K. and Webster F.X. 1996. Volatiles involved in the nonhost rejection of *Fraxinus pennsylvanica* by *Lymantria dispar* larvae. *J Agric Food Chem* **44**: 929-935.

Mishra, S.C. 1992. Comparative natural resistance of different clones of Tectona grandis L.f. to teak skeletonizer, E. machaeralis Wlk. (Lepidoptera : Pyralidae). Indian Forester, 118(4) : 274-278.

Miller Barbara, Lufiani, L. Madilao, Steven, Ralph and Jörg, Bohlmann 2004. Insect-Induced Conifer Defense. White Pine Weevil and Methyl Jasmonate Induce Traumatic Resinosis, de Novo Formed Volatile Emissions, and Accumulation of Terpenoid Synthase and Putative Octadecanoid Pathway Transcripts in Sitka Spruce. *Plant Physiology* 137:369-382.

Mutikainen, P. Walls, M. Ovaska, J., Kenanea, M. Titto, R. J. and Vapavuori, E. 2002. Cost of herbivore resistance in clonal saplings of *Betula pendula. Oecologia* 133: 364-371.

Morewood W.D., Hoover, K., Neiner P. R., Mc Neil J. R. and Sellmer J. C. 2004. Host tree resistance against the polyphagous wood boring beetle *Anoplophora glubripennis. Entomologia Experimentalis et Apllicata* 110: 79-85.

Morrow, P.A., Whitham, T.G., Potts, B.M., Ladiges, P., Ashton, D.H. and Williams, J.B 1993. Gall forming insects concentrate on hybrid phenotypes of *Eucalyptus*. Pages 121-134. In: Price P.W, Mattson, W.J and Baranchikov T.N. (Eds.). The ecology and evolution of gall forming insects. General Technical Report no. NC-174, USDA, Forest Service, St Paul, Minnesota, USA.

Morrow P.A. and Fox L.R. 1980. Effects of variation in *Eucalyptus* essential oil yield on insect growth and grazing damage. *Oecologia* 45: 209–219

Monson, R.K. and Fall, R. 1989. Isoprene emission from aspen leaves. *Plant Physiol* **90**: 267-274.

Nair, K.S.S., Sudheendrakumar, V.V., Varma, R.V. and Chako, K.C. 1985. Studies on the seasonal incidence of defoliators and the effect of defoliation on volume, increment of Teak. KFRI Research Report No. 30. pp 78.

Nair, K.S.S., S. Kedharnath, D. Mathew and V.V. Sudheendrakumar, 1989. Search for natural resistance to insect pest, *Hybleae puera* in teak. KFRI Research Report, No.62 : 32 pp.

Nykanen, H. and Koricheva, J. 2004. Damage induced changes in woody plants and their effects on insect herbivore performance: a meta analysis. *Oikos* 104: 247-252.

Ollerstam, O., Rohfritsch, O., Hoglund, S. and Larsson, S. 2002. A rapid hypersensitive response associated with resistance in the willow *Salix viminalis* against the gall midge *Dasineura marginemtorquens*. Entomologia Experimentalis et Applicata 102: 153–162.

Pare, P.W. and Tumlinson, J.H. 1997. De novo biosynthesis of volatiles induced by insect herbivory in cotton plants. *Plant Physiol* **114**: 1161-1167.

Pare, P.W. and Tumlinson, J.H. 1996. Plant volatile signals in response to herbivore feeding Florida Entomologist 79: 93-103.

Painter, R.H. 1966 Lessons to be learned from past experiences in breeding plants fro insect resistance in *Breeding Pest Resistant Trees* Gerhold, H. D., Schreiner, E.J Mc Dermott, R.E., Winieski J. A., (Eds) Pergamon Press, London 505 pp.

Pajares, J.A. 2004. Elm breeding for resistance against bark beetles. *Invest Agrar: Recur. For.* **13** : 207-215.

Raffa, K.F., Aukema, B.H., Erbilgin, N., Klepzig, K.D. and Wallin, K.F. 2005. Interactions among conifer terpenoids and bark beetles across multiple levels of scale: an attempt to understand links between population patterns and physiological processes. *Recent Adv. Phytochem.* **39**: 79-118.

Raman, A. 1994. Genetic, developmental, and physiological variation in individual plants: strategies to counter insect herbivores. Phytophaga 6: 97-100.

Raman, A., Schaefer C.W. and Withers, T.M. 2005. Biology, ecology, and evolution of gall-inducing arthropods. 2 Volumes; Science Publishers, Inc. New Hampshire, USA. 817 pages.

Raupp, M.J. and Denno, R.F. 1983. Leaf age as a predictor of herbivore distribution and abundance. In: Denno, R.F. and McClure, M.S. (eds) Variable plants and herbivores in natural and managed systems. Academic, New York, pp 91–124

Roychoudhury, N., Avinash Jain, K.C. Joshi and R.B. Lal 1997. Natural resistance in teak clones to leaf skeletonizer *Eutectona machaeralis* Walker: An appraisal *Indian Forester*.123: 1027-1035.

Ryan N. Philippe and Jorg Bohlmann 2007. Poplar defense against insect herbivores. *Canadian Journal of Botany*. **85**: 1111-1126.

Sasidharan K.R., Balu, B. Deeparaj, A. Nicodemus and R.V. Varma 2005. Screening Casuarina equisetifolia provenances against the bark caterpillar Inderbala quadrinotata and the possible biochemical factors determining resistance. J. Tropical Forest Sci 17: 625-630.

Schuck, H.J. 1982. Monoterpenes and resistance of conifers to fungi. In: Heybroek, H.M., Stephan, B.R. and Weissenberg, K. von (eds.): Resistance to diseases and pests in forest trees. PUDOC. P. 169-175.

Slansky, F. Jr. 1982. Insect nutrition: An adaptationist's perspective. *Florida Entomologist*. 65, 45-71.

Sadof C.S. and Grant G.G. 1997. Monoterpene composition of *Pinus sylvestis* varieties resistant and susceptible to *Dioryctria zimmermani*. *Journal of Chemical Ecology*, **3**, 1917-1927,

Southwood S.R. 1986. Plant surfaces and insects – An Overview. Insects and Plant Surfaces (Juniper, B. and Southwood, R. Eds) Edward Arnold, London 1-22.

Thakur, M.L. and P.K. Sen-Serma, 1982. Current status of termites as pests of nurseries and plantations in India. J. Indian Acad. Wood Sci. 11(1) : 7-15.

Tikkanen, O.P and R.J. Tiitto 2003. Phenological variation as protection against defoliating insects: the case of *Quercus robur* and *Ophemeroptera brumata. Oecologia.* 136:244-251.

Tollsten, L. and Muller, P.M. 1996. Volatile organic compounds emitted from beech leaves. Phytochemistry 43: 759-762.

Trewhella, K.E., Leather, S.R. and Day, K.R. 1997. Insect induced resistance in Lodgepole pine: effects on two pine feeding insects. *Journal of Applied Entomology* 121:129 – 136.

Zeng-hui, Hu, Yang, Di and Shen, Ying-bai 2004. Difference in volatiles of poplar induced by various damages. *Journal of Forestry Research* **15**: 280-282.

Zavarin E. and Snajberk K. 1985 Monoterpenoid and morphological differentiation within *Pinus cembroides Biochemical Systematics and Ecology* 13: 89-104

Zucker W.V. 1982. How aphids choose leaves: The roles of phenolics in host selection by a galling aphid. Ecology 63: 972–981.

Dynamics of Insect Behaviour
T.N. Ananthakrishnan & K.G. Sivaramakrishnan
Scientific Publishers (India), Jodhpur, 2012, pp. 117-132

Chapter - 9

BIOLOGY, BEHAVIOURAL AND POPULATION DYNAMICS OF THE RICE LEAFFOLDER COMPLEX

S. Senthil-Nathan

1. INTRODUCTION

Insect pests attack the rice crop from the time the nursery bed is prepared until harvest, the actual species complex varying in abundance and distribution from locality to locality and from year to year (Kiritani, 1979). Among the rice insect pests Rice leaffolder complex is the major one attacking vegetative stage of the rice and they are polyphagous defoliators causing yield losses of 15% to 25%. Two or three crops a year, often overlapping, of heavily fertilized monocultures of "Green Revolution", high-yielding cultivars were considered a vulnerable pest breeding ground. Developing strongly pest-resistant rice cultivars was a high priority, but these were threatened by "resistance breakdowns," particularly to the *Cnaphalocrocis medinalis* (Guenée) (Lepidoptera: Pyralidae) and brown planthopper (BPH), *Nilaparvata lugens* (Stål) (Hemiptera: Delphacidae) (Kiritani, 1979)

Insecticides used according to economic thresholds were considered a valuable complement to varietal resistance and synchronized planting as the basis for integrated control. Unfortunately, pest problems increased with the intensification of irrigated rice production, which included increased investments such as insecticides. In particular insecticide use preceded outbreaks of secondary pests, notably the brown planthopper, *N.*

lugens, that was previously of minor significance. Although the problem of insecticide-induced secondary pests (sucking pest) was recognized and attributed to the devastation of natural enemies, insecticides used according to economic thresholds were considered an important complement to varietal resistance and synchronized planting as the basis for integrated control (Ananthakrishnan, 1999).

Crop intensification has been suggested as a reason for the changes in the insect pest fauna in rice (Loevinsohn, 1985). Loevinsohn (1985), defines intensification as "an increase in resources devoted to rice cultivation." Intensification involves (1) an increase in the number of crops grown per year, (2) an increase in the use of agricultural chemicals; fertilizer and pesticides, (3) planting of varieties responsive to fertilizers and pesticides, and (4) increased plant densities. Further Loevinsohn (1985), provides evidence that ecologically specialized species have been favored by crop intensification. Until the 1960s, the stem borers were considered the most serious pests of rice throughout the tropics. In recent years, however, damage from them has declined, rice leaffolder and plant hopper attained the status as major pest causing significant losses in crop productivity.

C. medinalis was believed to be the only pyralid rice leaffolder species. Since 1981, however, seven rice leaffolder species in the genus *Marasmia* were identified (Pathak and Khan, 1994). The shift from minor to major pest status has been attributed to the adoption of new rice growing practices that accompanied the introduction of high yielding varieties (Litsinger, 1989). *C. medinalis* were considered as minor or sporadic pests in the past in many Asian countries (Alam and Alam 1964; Dale, 1994) and became a major threat in the tropical and sub-tropical countries (Heinrichs *et al.,* 1985) with extension of rice cultivation with modern varieties in upland and lowlands area. Expanded rice areas with new irrigation systems, multiple rice cropping, reduced genetic variability of high-yielding semi-dwarf rice varieties, application of high levels of nitrogenous fertilizers, and insecticide-induced resurgence have further compounded the leaf folder problem (Khan *et al.,* 1989).

This review analyses literature relevant to the rice leaf folder complex, biology, population dynamics, behaviour of rice leaffolder, *C. medinalis* practical application of host-plant resistance and there analyses will be effective in implication of biocontrol agents in the integrated pest management.

2. RICE LEAFFOLDER COMPLEX-DIVERSITY AND DISTRIBUTION

The rice leaffolder complex is widely distributed pests of rice growing tracts of humid tropical and temperate countries in Asia, Oceania, Australia and Africa. (Bautista *et al.,* 1984; Khan *et al.,* 1989; Dale, 1994; Khan *et al.,* 1988; Senthil-Nathan, 2006; *et al.,* 2006).

The leaffolder complex of the rice comprises 8 species of pyralid moths. They are *Cnaphalocrocis medinalis* Guenée, *Marasmia patnalis* Bradley, *Marasmia (=Susumia) exigua* Butler, *Marasmia bilinealis* Hampson, *Marasmia ruralis* Walker, *Marasmia suspicalis* Walker, *Marasmia trapezalis* Guenée, *Marasmia venilialis* Guenée. Among them *C. medinalis*, *M. patnalis* and *M. exigua* attained the status of pest and former two are major widespread species (Khan *et al.,* 1988; Dale, 1994). C. *medinalis* traditionally has been accepted as the only leaf-folder pest in the lowland rice fields of Asia. But the discovery of a sympatric and morphologically related species, *M. patnalis* by Bradley, in 1981, has complicated the interpretation of past results.

3. MORPHOLOGY AND KEY IDENTIFICATION OF THE RICE LEAFFOLDER COMPLEX

3.1. *Cnaphalocrocis medinalis* Guenée

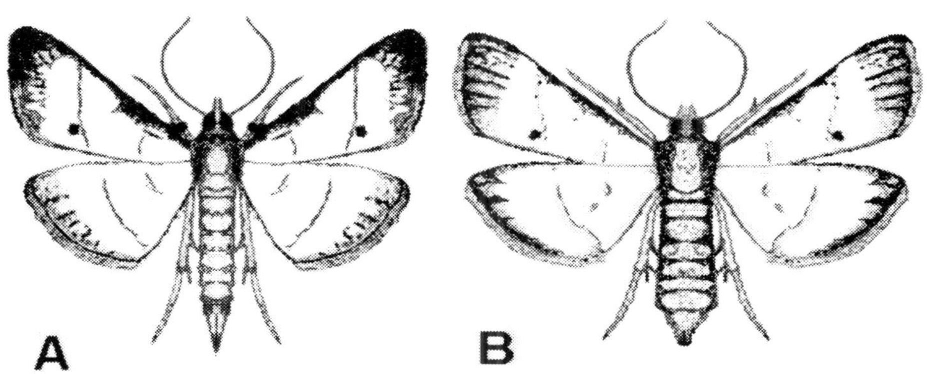

Fig. 1. *Cnaphalocrocis medinalis* Guenée adult moths (A-Male, B-Female)

Cnaphalocrocis medinalis (Guenée) (Fig. l) is monophagous, first described by Guenée (1854) from Indonesia and have been reported in many Asian countries. This moth resembles another leaffolder, *M. patnalis*. The male has a major patch of dark brown and shining androconial scales along the midcosta of forewing. Golden yellow with dark brown markings on both wings, black on the subterminal, antemedian and postmedian lines prominent, postmedian line extended straight to hind wing; median line very short, comma-like and curved outside; outer median line of hindwing much longer than the inner median line, ocelli present; antennae setaceous with 50 segments; compound eyes major portion of the head; Males similar to females except for the distinct dark brown band of androconial scales on the midcosta; thick black hair tuft on the foretibiae; and a thin but long longitudinal black band on the dorsal portion of the abdominal tip. Wing width is about 16 to 18 mm in the females and is about 14 to 16 mm in the males. Female genitalia is broad and not curved. Male genitalia is narrow, anteriorly projected and spiked cornuti on the vesica (After, Bradley, 1981; Barrion *et al.,* 1991; Khan *et al.,* 1988; Dale, 1994).

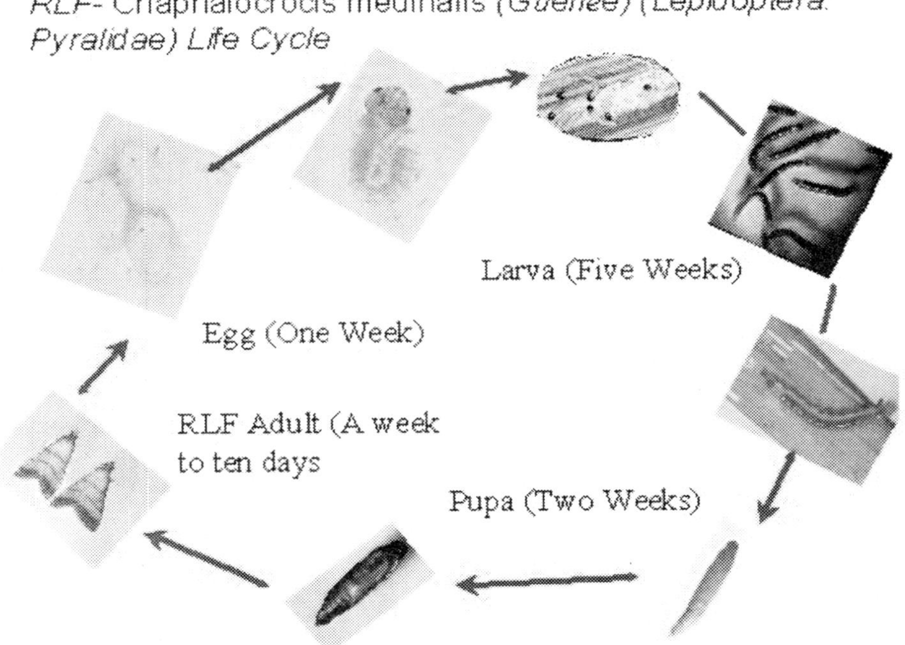

Fig. 2. Biology of the *C. medinalis*

The female attracts its mate with a pheromone call and generally mates at evenings. Adults hide on the rice plants during the day to escape from their natural enemies. Adult longevity is about a week or ten days. Egg-laying start after 24 hour after mating. Smooth, oval and whitish yellow eggs are laid singly or group (2 to 3 eggs), parallel on the both surfaces of young rice blade (Fig. 2). Each female moth lay about 300 to 350 eggs during its total life span. The incubation period vary from 5 days to one week. The newly hatched larva is whitish yellow (Fig. 2) in color, the body, turns transparent green once the larva starts feeding. The second instar migrates to healthy leaves and folds the leaf together.

4. GENERAL HABITS AND PROBLEMS OF THE RICE LEAFFOLDER COMPLEX

Several alternate hosts are reported for the rice leaffolders. Even though they have primary pest of rice, alternatively it attacks various grasses, sedges and cultivated crops like maize, sorghum, sugar-cane, wheat. *Brachiaria mutica* Forssk. Stapf, *Echinochloa colona* Linn. Link, *Eleusine coracana* Linn. Gaertn., *Isachne dispar* Trin., *Leersia hexandra* Sw., *Pennisetum pedicellatum* Trin., *Saccharum officinarum* Linn., *Sorghum bicolor* Linn.Moench., *Triticum aestivum* L., *Zea mays* L. (Dale, 1994) are reported to be alternative plant host of rice leaffolder.

The distribution of leaffolders can be seasonal. *C. medinalis* is a long-distant migrant into temperate China, Japan, and Taiwan (Wada, 1979; Khan *et al.,* 1988). Every year the population of this pest migrates to these temperate countries from tropical regions. The insect migrates north to south during the spring season and undergoes a reproductive diapause at the onset of seasonal emigratory periods. Under favorable conditions, leaffolders produce numerous generations. During high populations rice plants dry up and show parched. First and second instar larvae are usually feed within the slightly folded basal regions of the young leaves in a tiller. Starting from third instar, the larvae fold the rice blades and they become solitary (Fraenkel *et al.,* 1981). Generally, only one larva is present in the folded leaves and after feeding on one fold for about 2-3 days it moves to another leaf. Thus each larva demolishes a number of grown rice leaves during its growth period. The general problem is the photosynthetic ability of the rice plant becomes significantly reduced. The damaged leaves also serve as an entry points for microbial diseases (Pathak and Khan, 1994). The maximum yield loss caused by leaffolders is reported to be due to feeding on the flag leaf (Murugesan and Chelliah, 1983). As the leaffolders can attack the rice plant during the growth stage, fields should be monitored regularly.

5. FEEDING BEHAVIOUR

The feeding of rice leaffolders is out of sight and most of the feeding takes place inside the leaf roll. The young rice leaves are generally rich in nutrients and less tough. Adult rice leaffolders lay eggs on outer mature leaves while first-instar larvae feed at the base of the slightly folded young leaves (Fraenkel *et al.*, 1981). The feeding site of neonate larvae is away from the site where eggs are laid. The larvae start feeding before stitching or after spinning one or two stitches, where the leaf were not completely folded. The feeding occurs in longitudinal axis of the leaves between veins. The feeding damage caused by *C. medinalis* shows up characteristically in pale stripes in the leaf where the chlorophyll-bearing tissues have been eaten away, and only one epidermis and cuticle remains, Inside the folded leaves vascular bundles were left intact by the feeding larvae. All feeding damage is strictly linear, i.e., occurs in the longitudinal axis of the leaf between the veins. Feeding inside a roll occurs on the upper or underside of the leaf, dependent on whether the leaf is rolled up with the upper or underside inside. The feeding carried out by mandibles scraping sideways, which with the head already bent is along the longitudinal axis of the leaf, and on the upper side, where most feeding occurs, along the grooves between the ribs. Feeding in a fifth- or sixth-instar larva may proceed at the rate of 30-60 sec per 2 mm segment. Feeding is interrupted by stitching activities, especially when the leaf is not yet fully rolled up. This behaviour certainly suggests that feeding is induced by a chemical stimulus emanating from a feeding site. Feeding by *Cnaphalocrocis* is widespread in the *Gramineae* and even the related family of the *Cyperaceae*.

6. SPINNING AND FOLDING BEHAVIOUR

Spinning of silk threads occurs widely in many insect orders and especially in larvae of Lepidoptera, and particularly in the family *Pyralidae* to which the rice leaffolder belongs. This may take a very general aspect when random spinning often accompanies crawling and feeding, without regard to a particular webbing structure. The end product of the spinning activity is a leaf roll inside which the larva sits and feeds, and eventually pupates. For full development, a larva usually required more than one leaf (Fraenkel and Fallil, 1981; Ramachandran and Khan, 1991b; Islam and Karim, 1997). Neonate larvae of *C. medinalis* exhibit a behaviour of spinning down (or silking) from plants.

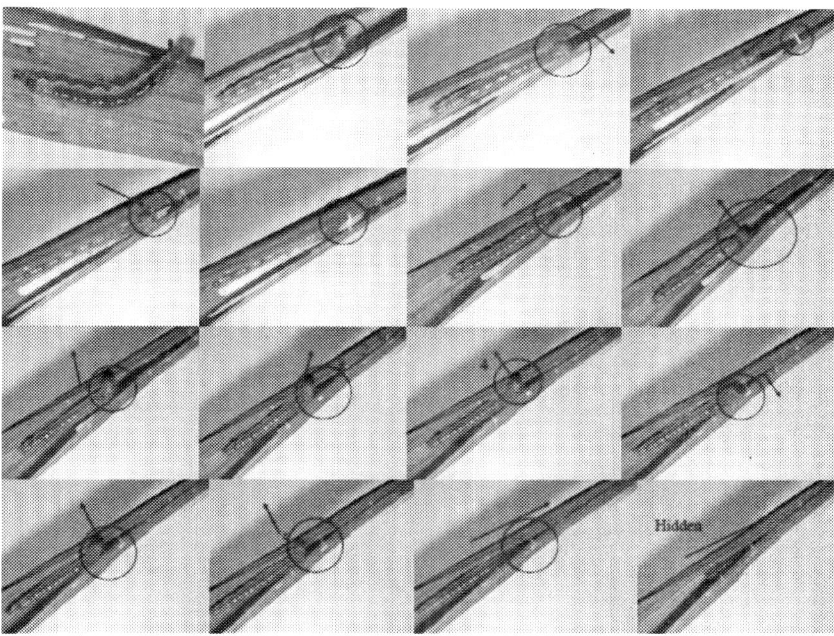

Fig. 3. Spinning and folding behaviour of the rice leaffolder, *C. medinalis*

Cnaphalocrocis and *Marasmia* exhibit the same leaf folding behaviour. After late second instar the folding behaviour is clearly visible. If the initial leaf is not suitable, the larvae moved to another leaf. In some cases, larvae examined maximum of five leaves before final selection. On average, larvae spend upto four minutes for leaf selection (Islam and Karim, 1997) and it usually prefers a wide rice blade rather than narrow one. Leaf folding usually began at the distal mid-quarter of the leaf blade, or sometimes at the distal or proximal mid-quarter.

Leaf folders are solitary in behaviour. In general, leaf folder larvae fold the upper leaves of rice plants. More than 70% of the folds were present on the top leaves. The length of leaf folds averaged 8 cm, only about 22% of the total length of the leaf blade. They preferred to make a fold along the distal and proximal mid-quarters rather than the distal and proximal quarters of the leaf blade. The majority of the leaves were folded longitudinally with the abaxial surface inside the fold. The first and second instar larvae seldom fold leaves, their smaller body size probably too small to handle the difficult task of folding rice leaf blades. Third to fifth instar larvae usually constructed folds. The selection of the leaves is based on the leaf character such as width, length, toughness and the condition of the edges either normal or twisted. Younger and more flexible are selected by the larvae, Narrower and tougher leaf blades were often rejected. Starting points of the folds mostly occurred in the distal (55%) and proximal mid-quarters (27%) of the leaf blades with a very few in the proximal and distal quarters. Young larvae usually rolled one edge of the leaf blade instead of binding both edges. The rice leaffolder make more than one fold during larval development. Immediately after release, larvae begin inspecting the leaf. If the first leaf was suitable, larvae did not inspect another leaf and began constructing a fold. If the first leaf was unsuitable, the larvae moved to another leaf and repeated the same selection procedure. In some cases, larvae examined up to four leaves before final selection: On an average, larvae spent about 4 min for leaf selection and usually selected a wide rather than narrow leaf. The larva attaches itself along the mid-rib by its prolegs and started to make the first bind. It initially stretched its head, contacted one edge of the leaf blade, and fixed a very fine, transparent and elastic silken thread spun by the leaffolder. Spinning process was repeated so that the thread became thicker, whitish and visible, and the leaf began to fold, forcing the edges closer (Fig. 3 and 4).

Securing the first few strands of thread was difficult; the strands often detached, probably due to leaf toughness. In such cases, the larva repeated the process, but if unsuccessful, the larva departed the leaf or tried to bend one edge by connecting it to the middle by the silken threads. By bending the tough edges, larvae then accomplished proper edge-to-edge bind to make the fold. Initially, binds were loose and the leaf edges were not in firm contact. As the number of binds increased the silken threads contracted, thickened and ultimately forced the edges together. The silken threads later lost elasticity and became hard. The folds were extended upwards, downwards or in both directions along the leaf which was usually folded abaxially with the upper leaf surface inside the fold. When the leaf edges were twisted or bent downwards, the leaf was folded abaxially with the lower surface inside the fold. After making two to three

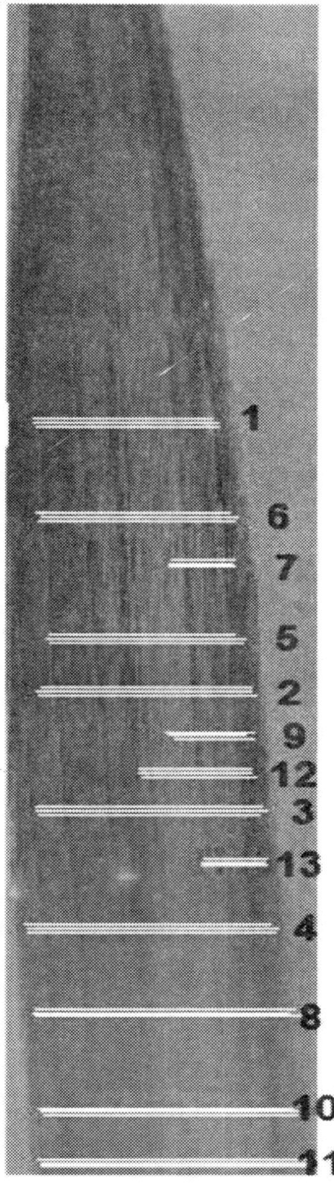

Fig. 4. Series of spinning (1 to 11) mechanism by late fourth instar *Cnaphalocrocis medinalis* larva and number of times its head swung for each spin (1-200 ± 15; 2-280 ± 10; 3-160±10; 4-175 ± 15;5-110 ± 15; 6-50 ± 10; 7-90 ± 15; 8-120 ± 15; 9-40; 10-105 ± 10; 11-125 ± 15; 12-55 ± 10; 13-15 ± 10 [Observation was made in green house larvae at SPKCES, Alwarkurichi and number of head swung for spins was average of three replicates] (After Fraenkel and Fallil, 1981; Ramachandran and Khan, 1991b; Islam and Karim, 1997)

binds, the larvae usually rested for 1-5 min inside the fold or sometimes fed on the green tissue. After each rest period, the larvae extended the folds by making additional binds and frequently inspected the existing folds. At times, the larvae reinforced the binds with additional strands of silken threads and/or added new binds among existing ones. In this way a larva constructed on average a 3.4 cm long fold, which is designated as a 'primary fold'. On average the construction of a primary fold required 20 min, after which the larva rested and/or fed within the primary fold. On average a larva swung its head 134 times to make one bind and 1226 times to make a primary fold with 9.15 binds (2.7 per cm) (Fig. 4). On average, a fold had 10 binds of silken threads with 1.4 binds per centimeter of fold. The abaxial sides of the leaves were folded longitudinally (96%); on a few occasions leaf tips were bent and folded Generally, only one leaf blade was involved in a fold (98%). Each fold was constructed of 9-10 binds and each bind and fold required an average of 134 and 1206-1340 head swings (Fig. 4). Generally, each larva made more than one fold since the food available in a single fold was probably insufficient for the completion of larval development. Larvae expended energy during fold construction by producing silken threads, selecting leaves, head swinging and inspecting binds and folds (Fraenkel and Fallil, 1981; Ramachandran and Khan, 1991b; Islam and Karim, 1997).

7. DAMAGE POTENTIAL

Infestation usually occurs during late vegetative growth stages of the rice crop. The larvae fold the leaves and scrape the green tissues of the leaves from within and cause scorching and leaf drying. 1.5 larvae per plant cause 4% of damaged leaves reduce the yield of rice variety IR36 by 200kg/ha and is considered to represent the economic injury level. In Asia pest populations causes >4% damaged leaves. The rice leaffolder make more than one fold during larval development. Under heavy infestation, each rice plant may have several rolled leaves, which severely restricts its photosynthetic activity. Folding and feeding behaviour of the rice leaffolder greatly reduce the general vigor, photosynthesis and transcription activity in plant. The total reduction in photosynthesis in folded leaves is the result of the combined effect of reduction of the maximum rate of net photosynthesis per unit green leaf area due to folding and reduction of the green leaf area due to feeding (de Jong, 1992; Yamamoto *et al.,* 1997). The damaged leaves also serve as entry points for fungal and bacterial infections. The yield was significantly decreased at 17.5% damaged leaves resulting in 16.5% yield loss, and a 21.3% yield loss occurred with 26.6% damaged leaves. Murugesan and Chelliah (1983) reported that a 10% increase in flag leaf damage by the leaffolder reduces grain yield by 0.13 g per tiller and the number of fully-filled grains by 4.5%. A close correlation between the intensity of leaffolder attack and loss in grain yield very often exists (Upadhyay *et al.*, 1975). Barrion *et al.* (1991) also reported a positive correlation between damage rating index and leaf width.

8. PLANT RESISTANCE

Resistant cultivars alter the physiology and behaviour of rice insects which in turn affects the insects susceptibility to chemical and biological controls. Resistant cultivars have an indirect adverse effect on natural enemies because of lower prey density. They have also been shown to adversely affect the development of parasites and predators by decreasing the suitability of the prey (insect pest) as a food source. However, resistant cultivars and biological controls are generally considered to be compatible. Resistant cultivars have a major impact on conserving natural enemies by decreasing the insecticide load. Wild relatives of cultivated crop plants often possess disease and insect resistance. Several wild relatives of *Oryza sativa* also were found to be resistant to rice leaffolders (Heinrichs *et al.*, 1985; Khan *et al.,* 1989), of which several accessions of *Oryza brachyantha* were identified as highly resistant (Heinrichs *et al.*, 1985; Ramachandran and Khan, 1991a). Egg deposition and establishment of neonate larvae is successful of establishment of insect in the plant and it is based on the crop resistance. Leaffolder resistance in the rice cultivars and the wild *Oryza* species associated with establishment of the neonate larvae on the plant and do not appear due to difference in ovipostion behaviour of adults but factors (Wada, 1979). Differences in plant factors associated with the successful location of neonate larval feeding sites has been suggested as the cause of resistance of plant. The mechanism of rice leaffolder resistance in TKM6 may be a combination of antixenosis and antibiosis. Rice varieties with narrow and long leaf blades likely offer resistance to rice leaffolder species. A plant resistance to insects could be due to the disruption of one or more of the physiological and behavioural responses of the insect to the plant characteristics. Behavioural and physiological responses are considered important during insect establishment on plants: Six main categories of insect (1) orientation and settling; (2) feeding; (3) metabolism of ingested food; (4) growth; (5) survival and fecundity; and (6) oviposition (Wada, 1979). Behavioural and physiology responses of insects alter to resistant plant.

The integration of host plant resistance with insecticides, cultural controls and biological control agents in a crop production system is more effective, environmentally safer, economically more profitable, and more stable than complete reliance on insecticides to control insects on susceptible cultivars. Research is needed to develop rice varieties with morphological and biochemical attributes conferring resistance to rice leaffolder species.

9. PHOTOTACTIC BEHAVIOUR

The lepidopteron insect pest responds to various environmental factors such as light and gravity. First-instar larvae of the rice leaffolder *Cnaphalocrocis medinalis* (Guenée) oriented downwards on a rice leaf when the light source was from above. The direction of orientation was reversed when the light source was from below. The opposite directions of orientation on these substrates persisted when the substrates were held horizontal with light source from the sides. The larvae posses positive phototactic and negative geotaxis. Orientation of larval stages of insects to light has been considered important in several insect species choice of feeding site (Ramachandran and Khan, 1991b). The responses of the first-instar larvae of to environmental factor (light or gravity) are altered by certain factors associated with the host plants. In leaffolders, the reduction in phototaxis occurs only in the presence of the volatiles. The rice plant volatiles along with some other unidentified rice plant factor cause a reversal of phototaxis, leading to downward orientation on the rice leaf. Host-plant induced changes in response to environmental factor may govern the insect's orientation and selection of ultimate feeding site on a plant. Understanding the factors that bring about such changes may be of value in the management and control of these insects.

10. OUTBREAKS

The rice leaffolder, *C. medinalis,* earlier considered as a minor and sporadic pest of rice in many Asian countries, appears to have become increasingly important with the spread of high-yielding rice varieties and accompanying changes in cultural practices. Misuse of insecticide and excessive use of nitrogenous fertilizers have been cited as the cause for high populations (Dhaliwal *et al.,* 1980). Outbreaks of *C. medinalis* frequently occur after a prolonged dry spell. Drought followed by rainfall often creates *C. medinalis* populations reaching outbreak proportions. There are varying reports on the effect of nitrogenous fertilizers on rice leaffolder incidence.

11. POPULATION DYNAMICS

Knowledge of rice leaffolder population dynamics, and insight into the factors determining leaffolder abundance, such as natural enemies, is essential as a basis for more rational pest management. Rice leaffolder population dynamics have been well-studied in China and Japan, where the pest is mainly migratory with distinct generations (Liang and Pang, 1987).

The mortality due to natural enemies and the changing attractiveness of the plant considerably affect the leaf-folder population dynamics in a rice crop. Population dynamics of rice leaf-folders was incorporated into the ecosystems model in order to study the plant-herbivore interactions in general and the effect of defoliation on rice yield formation in particular. Population changes due to birth, growth, mortality and migration. Simulation model, distributed delay models are model used to study the population dynamics. A simulation model for the population dynamics of rice leaf-folders (Lepidoptera: Pyralidae) interacting with rice (*Oryza sativa* L.) was designed to improve the understanding of the role of rice leaf-folders as elements of the rice ecosystem and to detect crucial knowledge gaps in view of a holistic assessment of their 1pest status. The model represents a synthesis of experimental results on leaf-folder biology and behaviour. A distributed delay model including attrition was applied to simulate dynamic population changes such as birth, ageing, mortality, migration and growth in terms of numbers and biomass. The host-plant on the one hand is assumed to affect leaf-folder survival, migration and growth rates while leaf-folders on the other hand influence plant growth and development by feeding and folding of leaves. The metabolic pool approach, leaf-folder feeding and hence leaf mass losses to the rice plant were described with a generalized functional response model which is 'source' and 'sink' driven. Rice leaffolder management can be improved when the effect of natural enemy densities on leaffolder infestations is known. Natural enemy abundance might be accounted for in the threshold for chemical control, or conservation measures could be taken to sustain natural control (Shepard and Ooi, 1992). Tropical rice fields harbour a large complex of natural enemies, that may keep rice leaffolder populations below damaging levels most of the time (Khan *et al.,* 1988; Barrion *et al.,* 1991). Natural enemies are the factors determining the population dynamics of the leaffolder complex and determine their temporal abundance and survival during a crop season. The dynamics of leaffolder natural enemies and leaffolder parasitism rates were determined in relation to leaffolder abundance and survival rates.

ACKNOWLEDGMENT

The author sincerely thanks the University Grants Commission, Government of India [Under Major Research Project No. 36-171/2008 (SR)] for financial support.

REFERENCES

Alam, M.Z. and Alam, S. 1964. Notes on biological studies of rice leaf-roller, *Cnaphalocrocis medinalis* Guenée, in East Pakistan. A Review of Research: Division of Entomology (1947-1964). Agricultural Information Service, Dhaka, East Pakistan, pp. 123-127.

Ananthakrishnan, T.N., 1999. Behavioural Dynamics in biological control of insects: Role of infochemicals. Current science 77, 33-37.

Barrion, A.T., Litsinger, J.A., Medina, E.B., Aguda, R.M., Bandong, J.P., Pantua, P.C., Viajante, Jr., V.D., Cruz, D.C.G. and Vega, C.R. 1991. The rice *Cnaphalocrocis* and *Marasmia* (Lepidoptera: Pyralidae) leaffolder complex in the Philippines: Taxonomy, bionomics and control. Philippines Entomology 8, 987–1074.

Bautista, R.C., Heinrichs, E.A. and Rejesus, R.S. 1984. Economic injury levels for the rice leaffolder *Cnaphalocrocis medinalis* (Lepidoptera: Pyralidae): insect infestation and artificial leaf removal. Environmental Entomology 13, 439–443.

Bradley, J.D. 1981. *Marasmia patnalis* sp. n. (Lepidoptera: Pyralidae) on rice in Southeast Asia. Bulletin of Entomological Research 323-327.

Dale, D. 1994. Insect pests of the rice plant – their biology and ecology. In: Heinrichs, E.A. (Eds). Biology and management of rice insects. Wiley Eastern Ltd, pp. 363–485.

de Jong, P.D. 1992. Effects of folding and feeding by *Cnaphalocrocis medinalis* on photosynthesis and transpiration of rice leaves. Entomologia Experimentalis et Applicata. 63, 101-102.

Dhaliwal, G.S., Shahi, H.N., Gill, P.S., Maskina and M.S. 1980. Effect of different levels of nitrogen and varieties on the incidence of hispa and leaf folder in rice. Pestology 4, 14-16.

Fraenkel, G., Fallil, F. and Kumarasinghe, K.S. 1981. The feeding behaviour of the rice leaffolder, *Cnaphlocrocis medinalis*. Entomologia Experimentalis et Applicata. 29, 147-161.

Fraenkel, G. and Fallil, F. 1981. The spinning (stitching) behaviour of the rice leaf folder, *Cnaphalocrocis medinalis*. Entomologia Experimentalis et Applicata. 29, 138-146.

Gueniee, A. 1854. Histoire naturelle des insectesspecies general des Lepidopteres. Beltoides Pyralides 8, 1-448.

Heinrichs, E.A., Camafiag, E. and Romena, A. 1985. Evaluation of rice cultivars for resistance to *Cnaphalocrocis medinalis* Guenée (Lepidoptera : Pyralidae). Journal of Economic Entomology 78, 274-278.

Islam, Z. and Karim, A.N.M.R. 1997. Leaf folding behaviour of *Cnaphalocrocis medinalis* (Guenée) and *Marasmia patnalis* Bradley, and the influence of rice leaf morphology on damage incidence. Crop Protection 16, 215-220.

Khan, Z.R., Barrion, A.T., Litsinger, J.A., Castilla, N.P. and Joshi, R.C. 1988. A bibliography of rice leaffolders (Lepidoptera: Pyralidae). Insect Science and its Application 9, 129-174.

Khan, Z.R., Rueda, B.P. and Caballero, P. 1989. Behavioral and physiological responses of rice leaffolder *Cnaphalocrocis medinalis* to selected wild rices. Entomologia Experimentalis et Applicata 52, 7-13.

Kiritani, K. 1979. Pest management in rice. Annual Review of Entomology. 24, 279–312

Kenmore, P.E., Carino, P.O., Perez, C.A., Dyck, V. A. and Gutierrez, A.P., 1984. Population regulation of the rice brown planthopper *(Nilaparvata lugens* Stal) within rice fields in the Philippines. Journal of Plant Protection in the Tropics 1, 19-37.

Liang, G.W. and Pang, X.F. 1987. Studies on the life system of rice leafroller. The developments in integrated control of rice diseases and insect pests in China, pp. 214–237

Litsinger, J.A. 1989. Second generation insect pest problems on high yielding rices. Tropical Pest Management 35, 235–242.

Loevinsohn, M.E. 1985. Agricultural intensification and rice pest ecology: Lessons and implications. A paper presented at the International Rice Research Conference 1-5 June, 1985. International Rice Research Institute. Los Baños, Philippines

Murugesan, S. and Chelliah, S. 1983. Rice yield loss caused by leaffolder damage to the flag leaf. International Rice Research Newsletter 8, 14.

Pathak, M.D. and Khan, Z.R. 1994. Insect Pests of Rice. International Rice Research Institute and International Centre of Insect Physiology and Ecology, pp. 33-36.

Ramachandran, R. and Khan, Z.R. 1991a. Mechanisms of resistance in wild rice *Oryza brachyantha* to rice leaffolder *Cnaphalocrocis medinalis* (Guenée) (Lepidoptera: Pyralidae) Journal of Chemical Ecology 17, 41-65

Ramachandran, R. and Khan Z.R. 1991b. Feeding site selection of first-instar larvae of *Cnaphalocrocis medinalis* on susceptible and resistant rice plants. Entomologia Experimentalis et Applicata 60, 43-49,

Senthil-Nathan, S. 2006. Effects of *Melia azedarach* Linn on nutrition physiology and enzyme activities of the rice leaffolder. Pesticide Biochemistry and Physiology 84 , 98-108

Senthil-Nathan, S., Kalaivani, K. and Murugan, K. 2006. Behavioural responses and changes in biology of riceleaffolder following treatment with a combination of bacterial toxins and botanical insecticides. Chemosphere 64, 1650–1658.

Shepard, B.M. and Ooi, P.A.C. 1992. Evaluating biocontrol in rice: Present and future considerations. In: Ooi, P.A.C., Lim, G.S., Teng, P.S., (Eds). Biological Control: Issues in the Tropics. Malaysian Plant Protection Society, Kualalumpur., pp. 93–99.

Upadhyay, V.R., Desai, N.D. and Shah, A.H. 1975. Extent of damage and varietal susceptibility by rice leaf folderrice leaf folder *Cnaphalocrocis medinalis* Guenée Gujarat Agricultural University Research Journal 6, 115-117

Wada, T. 1979. Influence of the temperature and the growing stages of the rice plant on the number of larval instars in the rice leaf roller, *Cnaphalocrocis medinalis* Guenée. Japanese Journal of Applied Entomology and Zoology 23, 178-182.

Yamamoto, H., Honda, Y., Hayakawa, S. and Ohgata, Y. 1997. Photosynthetic and respiration rates of leaves damaged by rice leaffolder *Cnaphalocrocis medinalis* Guenée. Japanese Journal of Applied Entomology and Zoology 41, 15-119.

Zhang, X.X. 1991. Migration of rice leaf roller, *Cnaphalocrocis medinalis* Guenée, in China. In Proceedings of International Seminar on Migration and Dispersal of Agricultural Insects, Tsukuba, Japan, September 1991, pp. 51–60.

Dynamics of Insect Behaviour
T.N. Ananthakrishnan & K.G. Sivaramakrishnan
Scientific Publishers (India), Jodhpur, 2012, pp. 133-157

Chapter - 10

BEHAVIOURAL DYNAMICS OF AQUATIC INSECTS: STRATEGIES AND MECHANISMS

K.G. Sivaramakrishnan

INTRODUCTION

Insects evolved about 350 million years ago at which time, several crustaceans were already successfully residing in the marine environment. Obviously, competitive exclusion prevented the Hexapoda to dominate the seas, although they were able to successfully colonize the diverse habitats and microhabitats of inland waters – mountain streams, cascades, rivers with either continuous or intermittent flow, mountain springs, wave-swept lakes, tranquil ponds and puddles and enter in the water trapped in tree holes and in pitfall traps of insectivorous pitcher plants. Insects of lotic waters evolved as early as the Permian and Triassic, whereas lentic insects appeared in the late Triassic and early Jurassic (Merritt and Wallace 2009).

Aquatic insect community offers a good model to study the multifaceted ethological intricacies of the animal kingdom. Aquatic insects are, by far the most abundant and speciose organisms inhabiting the freshwaters, which constitute 3% of the world's water occupying just about 1.9% of the land surface. They are vital components of the freshwater ecosystems in terms of being essential fish food or in their roles in structuring aquatic communities and in natural cycling. Though representing a small fraction of the total insect fauna, 13 orders of insects contain species with aquatic or semi aquatic stages, and in five of these (Ephemeroptera, Odonata, Plecoptera, Megaloptera and Trichoptera) almost all species are aquatic in their larval phase (Merritt and Wallace, 2009).

The diverse and often intriguing behavioural strategies of life stages of insects for survival in lentic and lotic waters and to a limited extent in the ocean are worth exploring. This diversity of aquatic insect behaviour is a reflection of the diversity of challenges posed by the aquatic environment to insect life. Furthermore, they face distinct classes of problems both in their aquatic and terrestrial adult phases of their life. They possess different suites of adaptations and also exhibit several categories of behaviour to face these challenges. These behavioural patterns expressed as diverse adaptations are products of natural selection operating over millions of years. Behaviours are often considered more evolutionarily labile than morphology or life history. Interestingly, a phylogenetic component can be discerned with regard to many insect behaviours as, for instance, the case construction in Trichoptera (Wiggins 2004). Factors influencing mutually dictated patterns of behaviours of larval and adult stages of aquatic insects are schematically presented in Fig. 1. They are highlighted in this chapter with concrete examples from various orders of aquatic insects.

I. LARVAL BEHAVIOUR

1. Habitat and microhabitat constraints

Southwood (1977) postulated that habitat provides the templet on which evolution shapes characteristic life history stages. The habitat is the templet for ecological responses, and habitat conditions select combinations of species traits, morphological, physiological and behavioural that enhances fitness (Statzner *et al.,* 2001).

Basically, freshwater insects exhibit specific behavioural adaptations to survive in flowing water (i.e., streams, rivers) and standing water (i.e., ponds, lakes, swamps, marshes) habitats. Within a given habitat, the modes by which individuals maintain their location or move about are provided in Table 1. Habitat heterogeneity and near-bed current velocities affect the behaviour, resource use and energetics of aquatic insects. Understanding not just where organisms move but how they move is important in integrating animal behaviour into landscape ecology. Gaining such knowledge provides critical insights into mechanisms underlying numerous key behavioural processes such as animal dispersal and foraging search strategies, which ultimately scale up to shape species distributions in respective microhabitats (Olden *et al.*, 2004).

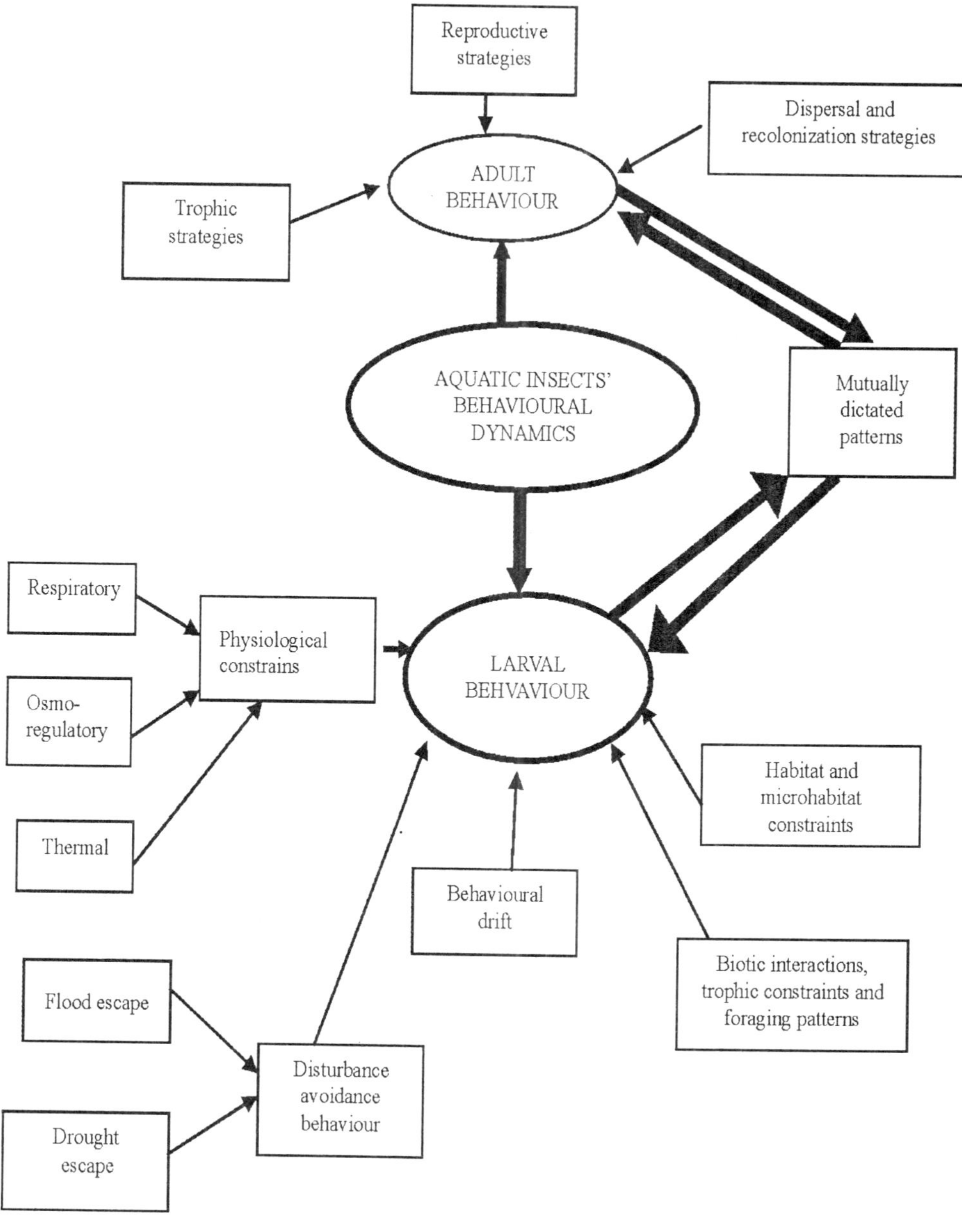

Fig. 1. Schematic presentation of factors influencing mutually dictated patterns of behaviour of larval and adult stages of aquatic insects.

Table 1. Habits of aquatic insects and their hahavioural adaptations to specific micro environment are briefly outlined in Table 1.

Habitat category	Behavioural adaptation
Skaters –Gerridae (water striders)	Skating on top of the elastic surface flow of water to feed on organisms trapped in the surface film
Floaters (Culucidae)	Float or swim in open water vertical diurinal migration common
Divers (Dystiscidae)	Swimming by "rowing" with hind legs in lentic habitats
Swimmers (leptophlebiid may flies, notonectids)	"fish like" swimming in lotic or lentic habits
Clingers (Psephenidae)	Fixed retreat making hydropsychids; clinging by psephenids
Sprawlers (inemid mayflies, Plecoptera)	Inhabiting and moving on the surface of floating leaves
Climbers (Zygoptera)	Aquatic macrophytes
Burrowers (ophemerid mayflies, Chironomids)	Constant discrete burrows of sand grains
Jumpers (Collembola)	Jumping

Two factors viz., habitat heterogeneity and more importantly the range of current velocities associated with various substrates increase habitat diversity. Insects adapted to erosional habitats include heptageniid and ephemerellid mayflies, perlid stoneflies, corydalids and leptocerid caddisflies. Rheophilic (current-loving) insects generally have flat body and a smooth streamlined dorsum. Mayflies with streamlined bodies such as *Baetis* sp., *Isonychia* sp., and *Ameletus* sp., are capable of short rapid bursts of swimming, but other lotic insects move by crawling or passive displacement. Many mayflies and stoneflies have legs that project laterally from the body, thereby reducing drag and simultaneously increasing friction with the substrate. Hydraulic suckers are found in blephariceríds which move in a 'zigzag' fashion, releasing the anterior three suckers, lifting the front portion of the body to a new position, and reattaching the anterior suckers before releasing and moving the posterior ones to a new position. These larvae are commonly found on smooth stones. The enlarged gills of *Epeorus* sp. and *Rhithrogena* sp. function as a friction pad. Some chironomids have 'pushing legs' in the form of a circlet of small spines that function as a false sucker when pressed to the substrate.

Although water flows creating current in streams and rivers, only a few insect larvae display rheotactic and thigmotactic (*Notophlebia jobi*) responses. Instead, they have developed behaviour patterns to avoid current. Larval blackflies (Simulidae) use a combination of hooks and silk for attachment. The last abdominal segment bears a circlet of hooks, which it uses to anchor itself to substrates. The larva moves forward, inch-wormlike, spins silk over the substrate, and attaches the proleg and then the posterior circlet of hooks to the silken web. Most blackfly larvae possess well developed 'cephalic fans', which are useful in filtering small food particles from suspension (Merritt and Wallace, 2009).

Olden *et al.* (2004) experimentally examined the larval movement of the case building caddisfly, *Agapetus boulderensis* in response to two current velocities in contrasting habitat types, one with smooth stones that facilitate movement and another with thick algal patches that constrain movement. Detailed behavioural observations showed that larvae employed two different velocities: faster crawling and slower pivoting. Individual decision-making between crawling and pivoting was related to the magnitude of current velocity across the streambed and the frequency at which larvae employed these behaviours translated into differential movement rates and directions.

Some aquatic insects (lepidostomatids, chironomids and tipulids) live in debris consisting of leaf packs and small woody materials. This debris offers both a food resource and a refuge for insects. Silk is used for attachment by a number of caddisflies (e.g., Hydropsychiidae, Philopotamidae and Psychomyiidae), which build fixed nets and retreats. Although the Philopotamidae are found in riffle habitats, their finemeshed, tubelike nets are usually found in crevices or underside of stones in low velocity microhabitats. The nets of the caddisfly, *Neureclipsis* sp., are limited to moderately slow velocities and the large (20 cm long), trumpet-shaped nets are used to capture small animals drifting downstream. Some case-making caddisflies (e.g., *Brachycentrus* sp.) also use silk for attaching their cases to the substrate in regions of fairly rapid flow. Free living caddisflies may use 'security threads' as they move over the substrate in fast currents (Merritt *et al.,* 2008).

Ephemerid mayflies are common inhabitants of sand and silt substrates. They construct shallow U-shaped burrows and use their dorsal gills to generate water currents through the burrow, while using their hairy mouthparts and legs to filter particles from the moving water. Some caddisflies such as *Molanna* sp., have elongate slender bodies and are adapted to sand and silt substrates by constructing a flanged, flat case. They are camouflaged by dull colour patterns and hairy integuments that accumulate a coating of silt (Merritt and Wallace, 2009).

The behavioural adaptations of the insects of the water surface, constituting the pleuston community are also equally significant. The gerrids striders skate over the surface of a pond on their middle and hind legs, the raptorial front pair reserved for taking prey. When skating on the water surface, their hind legs are held outstretched and their front legs are held close together on the water to provide support. *Gerris* spp. obtain information on their environment from ripples at the surface film. They detect a disturbance in their vicinity through sensors in their four longer legs. They also use their eyes as well as their vibration sensors to evaluate other situations. The shiny black whirligig beetles gyrate on the surface float on the water, supported by their buoyancy and also by the surface tension of beetles. Compound eyes are divided into two parts. The lower part of each eye detects events in the water and the upper part detects events on or about the water surface. Most of the information about their environment is gathered not with their eyes but with their two, short, 3-segmented antennae. Backswimmers can lie motionless at a shallow depth or move forward to a new position by swinging their outstretched hind legs. These oar-like hind legs can propel the backswimmer down through the water away from the surface film. They also get information from ripples in the surface film. The water-surface film is an environment of great antiquity. Wingless springtails (Collembola) pioneered to inhabit the film and moved across the water surface with the help of a springing organ (furcula). Very recent discovery by Rubinoff and Schmitz (2010) of multiple aquatic invasions of *Hypopmocoma* spp., an endemic terrestrial Hawaiian moth radiation is bound to reveal a series of innovative behavioural strategies while diversifying in many aquatic microhabitats. The abovementioned few examples highlight the fact that habitat is clearly a filter for biological traits and greater spatial habitat variability (i.e., greater filter diversity) is the reason for greater diversity of behavioural traits in aquatic insect communities (Statzner *et al.,* 2001).

2. PHYSIOLOGICAL CONSTRAINTS

Foremost among the problems of insects living in water are those concerned with respiration and osmoregulation. They also face the problem of thermoregulation. Exploration of how animals make decisions under conflicting demands and how patterns in species behaviour reflect the resolution of such trade-offs in evolutionary terms are fascinating aspects of modern behavioural ecology.

i. Respiratory constraints: Oxygen is obtained by aquatic insects either by taking that available in abundance above the water or by uptake from oxygen in solution in the surrounding water. Midge larvae generate respiratory pigment, dissolved in the plasma in abundance. Tracheated cuticular outgrowths in the form of tracheal gills and rectal gills are evolved in mayflies and odonates respectively. In contrast to gilled insects, aeropneustic insects (larval Culicidae) use respiratory siphons to gain direct connection to a source of air. This may be through the water surface as seen in many Culicidae or by tapping into specific aquatic plant aerenchyma tissue, as seen in larvae of *Mansonia* sp. Some aquatic adult beetles respire taking air bubbles which adhere to pleura due to hydrofuge hairs. This strategy is termed 'plastron respiration'. Water scorpions too have long respiratory siphons which are modified caudal filaments. Water surrounding the insect will become depleted especially in lentic waters and so many insects direct a flow of water over their bodies. *Elmis* sp. is a lentic insect which employs a passive current control by suitable orientation. Aquatic insects inhabiting soft sediments have different strategies to withstand hypoxial conditions. For instance, some chironomids may raise the openings of their tubes above the level of sediment in which they are living. Sediment depth has also a major influence on population density and tube shape in *Chironomus plumosus*.

Dryopid beetles have a layer of much shorter, more densely packed hairs which form a permanent plastron. The permanent plastrons seen in adult elmid beetles are believed to be supplemented, on occasion, by bubbles of oxygen produced by photosynthesizing benthic algae (Williams and Feltmate 1992). *Aphelocheirus* sp. has one of the most efficient permanent plastrons. It consists of very short hairs (5-6 um high) with a density of about 4.3 million hairs per mm^2, capable of withstanding a pressure of some four atmospheres before collapse. The spiracles connect with the plastron through small pores located along a series of canals in the cuticle and the canals are lined with hairs to prevent the entry of water. Under normal environmental conditions, the amount of oxygen supplied to *Aphelocheirus* sp. via this system is sufficient to allow the insect to remain permanently submerged (Chapman 1975). Naucorids sometimes ventilate the ventral portion of their plastron by rowing their hind legs under the abdomen (Mc Cafferty 1981). Van der Geest (2007) conducted a behavioural test, in which the tradeoff between different survival strategies after exposure to different oxygen concentrations is analyzed, using the caddisfly, *Hydropsyche angustipennis*, as a model organism. Exposure to hypoxia resulted in behavioural changes. Under low-oxygen conditions, larvae increased their ventilation rates, which would have increased oxygen uptake.

Overwintering populations of the mayfly, *Leptophlebia vespertina* respond in just this way to falling DO concentrations in ice covered ponds. Brittain and Nagell (1981) report that *L. vespertina* survive months of anoxic conditions in Norwegian humic ponds by migrating to the underside of the ice along the shore, the only location where oxygen concentrations are predictably high enough to permit survival. Extensive laboratory experiments demonstrated that under high DO condition nymphs were negatively phototactic. However, at low DO nymphs became very active, positively phototactic and negatively thermotactic. Shift in taxis orientation allowed the nymphs to move towards the pond's edge where temperature was lower, but oxygen concentrations were higher.

ii. Osmoregulatory constraints: Aquatic insects have evolved strategies to control the amounts of water and ions both entering and leaving the body for effective osmoregulation. The production of hypotonic urine is an important osmoregulatory mechanism in them, accomplished by specialized areas of the hindgut that reabsorb ions before wastes are eliminated.

Active absorption of ions from the environment occurs across specialized regions of the body, patches of chloride epithelia located on ventral or dorsal parts of the abdomen in caddisflies, adjacent to the anus in dipterans and within the rectal chamber of odonates. Specialized chloride cells scattered over the body of stoneflies, mayflies and some heteropterans, anal papillae of dipteran larvae and some protrusible anal 'gills' of helodids and non-tracheated processes on abdominal segment 9 in gyrinids

(Crowson, 1981) are also involved in the active absorption of ions from water. Species that can switch between freshwater and marine habitats have part of their rectum that specialising in water reabsorption and another part that is adapted for ion reabsorption (Mc Cafferty, 1981).

III. Thermal constraints: Behavioural thermoregulation in aquatic insects remains an important adaptation for life in water though thermal regimes in aquatic habitats are less variable than those in terrestrial habitats. Brittain and Nagell (1981) demonstrated that thermotaxis exists in mayfly nymphs, and that it can be used to locate microhabitats with favorable DO concentrations. The thermotactic response of *Leptophlebia vespertina* during winter would be, in part an adaptation to lower body temperature and consequently, metabolic demand for oxygen. Odonata possess both ecto- and endothermic thermoregulatory capabilities. The two basic behavioural styles, flyers and perchers, use different strategies to prolong activity under less than optimal ambient temperatures. Under cool conditions, fliers warm thorax by wing-whirring (endothermy), whereas perchers expose as much of the surface area as possible to solar radiation (Tennessen, 2009). Most aquatic insects survive unfavourable conditions by going into the inactive hibernation-like or aestivation-like state called diapause. But some insects, escape the cold of winter or the heat of summer by migrating. A few others, among them certain mayfly, dragonfly and stonefly nymphs, and dobsonfly, caddisfly and blackfly larvae do not undergo diapauses. They continue to feed and grow slowly and during temperature extremes they become quiescent and inactive although not in diapause. Diapausing insects protect themselves against freezing by producing an antifreeze. They permeate their bodies with an internally secreted antifr-eeze, glycerol, mannitol or sorbitol, which lowers the freezing point of water or proteins that prevent the formation of damaging ice crystals (Waldbauer, 2006).

3. DISTURBANCE AVOIDANCE

Flood and drought are very common natural disturbances in most streams and rivers, playing a central role in the regulation of populations, in the structuring of communities and in the functioning of ecosystems.

i. Flood escape: Aquatic insect strategies for surviving floods encompass several modes of adaptation: life-history, morphological and behavioural (Lytle and Poff, 2004). Life-history adaptations include emergence strategies that are synchronized to avoid seasons when floods are likely (Lytle, 2002). Morphological adaptations include body shapes that are streamlined to avoid shear stress during floods (Hart and Finelli 1999) and special organs that allow adhesion to substrates (Frutiger, 2002). Behaviours are context-dependent responses to different environmental cues and behavioural adaptations rely on proximate cues (rainfall, rising hydrograph etc.) to signal a flood, which is then avoided by moving to safe zones within the stream (Lancaster and Hildrew, 1993) or leaving the stream entirely (Lytle, 1999). Behavioural flood escape might be prevalent in taxa that are air-breathing (Hemiptera and adult Coleoptera) because these are able to persist out of water during floods.

Flood escape behaviours include the use of proximate environmental cues as a signal to abandon streams. This has been well documented in belostomatid bugs which use 'rainfall response behaviour (RRB)' to escape flash floods in desert streams. Lytle and White (2007) used controlled, replicated experiments to quantify the use of rainfall cues for flood escape in seven common desert stream insects. The hemipterans, *Curtcta pronotata* and *Aquarius remigis* responded consistently to rainfall cues by crawling vertically away from the water, in such a way that they may escape flash floods in nature. The coleopteran, *Gyrinus plicifer* showed no response to rainfall cues. The naucorid, *Ambrysus woodburyi* also showed no evidence of RRB but individuals did move underneath rock substrates during rainfall trials, possibly seeking shelter from floods within the stream channel. Naucorids are generally good swimmers capable of locating instream refugia or stream margins with slower flow that are less prone to flood scouring (Sites and Willig 1991; Herrmann *et al.,* 1993).

ii. Drought escape: For some taxa, behavioural escape from drought can be a viable strategy. Rivers and streams can experience sudden drops in flow caused by diel fluctuations in riparian plant evapotranspiration, lowering of the water table due to low rainfall inputs, and recession to baseflow flood events (Stanley *et al.,* 1997). Aquatic Coleoptera adults are particularly adept at drought escape. For example, the dytiscid beetles *Rhantus gutticollis*, *R. binotatis* and *Eretes stictus* have been observed taking flight en masse from drying desert ponds, and in some cases the emigration was preceded by loud, sustained vocalization (Zimmerman, 1959; Smith, 1973; Kingsley, 1985). Within-stream movement is also a mechanism for escaping the recession point of rapidly drying streams and rivers. Lytle *et al.,* (2008) document how two species of desert aquatic insects use positive rheotaxis to escape drought in desert rivers. They observed 3,600 adults of the long-toed water beetle, *Postelichus immsi* crawling upstream concurrent with upstream recession of surface water in the Santa Maria River, Arizona. At the same time, they observed larvae of the odonate, *Pogomphus borealis* burrowing and swimming upstream in large densities (690 larvae /m^2). Both taxa moved with sufficient speed to arrive at perennial reaches of the river before being overtaken by drought. Given the ability of many aquatic insects to optimally relocate within streams in response to local fine-scale differences in flow, for many taxa, behavioural drought escape may be a special case of a more generalized ability to tackle sudden changes in flow (Lytle, 2008).

4. BIOTIC INTERACTIONS, TROPHIC CONSTRAINTS AND FORAGING PATTERNS

Exploration of how aquatic insects make decisions under conflicting demands and how patterns in species behaviour reflect the resolution of such trade-offs in evolutionary terms is an interesting aspect of modern behavioural ecology.

i. Biotic interactions: Predator-prey interactions, parasitism, mutualism and processing chain commensalism displayed by aquatic insect communities are a few aspects of biotic interactions that involve strategic behavioural adaptations. Aquatic insects have numerous behavioural adaptations reflecting the co-evolution of predators and prey. Predators have a wide variety of search, capture and consumption mechanisms and prey demonstrate various physical, chemical and behavioural defenses.

Notable anti-predator defenses include using substrate as a refugium, substrate colour selection, drift, reductions in movement and feeding, case selection in some caddisflies, thanatosis, reflex bleeding deimatic behaviour and rarely cyclomorphosis. Brusven and Rose (1981) experimentally proved that predation on *Ephemerella subvaria* (mayfly), *Hesperoperla pacifica* (stonefly) and *Rhyacophila vaccua* (caddisfly) was generally low when cobbles were present, slightly higher with pebbles, and very high when only sand was present suggesting cobbles as preferred refugia. In combined field and laboratory studies, Feltmate and Williams (1989) found that substrate color selection by the dark stonefly, *Paragnetina media* functioned as a prey detection defense mechanism. Peckarsky (1980) studied the responses of mayfly to stonefly predators in two small streams in USA. Observations indicated that slow-moving mayflies such as *Ephemerella* avoided stonefly predators by using long-distance chemical cues, whereas faster-moving mayflies like *Baetis* avoided capture by drift-swimming, drifting, and/or crawling away from the predator. It has been shown that vertebrate predators prefer non-cased caddisfly larvae over case building species. However, there is also predator selection among cased larvae. For this reason, there is a high amount of intra-specific competition among trichopteran larvae for non-predator selected cases. Case building species have developed a mechanical defense with an advantage in occupying feeding patches and habitats which non case-building species would not inhabit for risk of predation (Weaver and Morse, 1986).

Agonistic behaviour to obtain resources is a well known phenomenon in a variety of animals. Nest sites and nests themselves are important resources for nest building caddisflies. Finakoshi (2005) examined in laboratory experiments, the effects of body size difference, species combination and intruder-resident relationship on the contest outcome in 2 closely related and co-occurring Japanese caddisflies, viz., *Stenopsyche marmorata* and *S. sauteri*. He concluded that the resident won when either *S. marmorata* or *S. sauteri* was an intruder. Some organisms use morphological structures obtained by behavioural processes to lower mortality by predation.

Some aquatic habitats of low stability like treeholes, pitcher plants, bromeliads and inflorescences broadly termed phytotelmata are microcosms harbouring aquatic insect assemblages which are ideal for studying behavioural patterns at community level. For instance, two types of common insect larvae in tree holes, scirtid beetles and ceratopogonid midges have a commensalistic relationship (Paradise and Dunson, 1997). Scirtids are leaf shredders, skeletonizing leaf litter. A portion of what they shred is not consumed by them and consequently made available to insects and microbes feeding on detritus. Midges benefit from this by growing to a large size and having a higher survival in the presence of scirtids. However, the presence of midges does not affect scirtids. This kind of 'processing chain commensalism' may be general feature of tree hole-like communities (phytotelmata). There are several instances of aquatic insects lodging in body parts of benthic macroinvertebrates like bivalves, with adaptations to cling on them (Muller-Liebenau and Heard, 1979; Gillies and Elouard, 1990; Subramanian and Sivaramakrishnan, 2009). This association may be commensalism beneficial to the insect and neutral to the mollusc (Muller-Liebenau and Heard, 1979), although in view of the internal lodgement of the nymph among the fimbriated lamellae of the gills, the mayfly might well be regarded as an inquiline (Gillies and Elouard, 1990).

ii. Trophic constraints: Several modes of handling food material are distinguished in aquatic insects. There are shredders, collectors, scrapers, piercers and engulfers (Cummins and Klug, 1979). Feeding behaviour is influenced by trophic constraints and several unique strategies are evolved in different taxa. For example, the prehensile, protractile labium is unique to the odonate nymphs and it strikes within milliseconds to capture prey. The labium grasps the prey and brings it to the mandibles where it is chewed or engulfed whole. Tactile detection is more important in earlier instars, vision becoming more keen in later instars (Tennessen, 2009). Larvae of mosquitoes and blackflies are active filter feeders, whose mouthparts or other appendages are in some way modified for sieving plankton and detritus particles as they fan through the water. Many caddisflies, the 'underwater architects' are passive filter feeders, which live in flowing water and weave stationary nets of silken filaments that are fixed in place and trap food from the current that flows through them. They engage in 'ecosystem engineering' activities like bioturbation, bioconsolidation and particulate matter processing.

iii. Foraging strategies: Food resources of aquatic insects are patchily distributed and their foraging behaviour is often determined by the conflicting benefits of energy gain and the risk of mortality from predation or other causes. Animals should have lower activity levels when either at the risk of predation or the availability of resources in the environment is high. Schaffner and Anholt (1998) investigated the adjustment of the behaviour of *Ischnura elegans* larvae to predator presence (*Anax imperator*) and prey density (*Daphnia* sp.). The foraging activity of the *I. elegans* larvae was significantly reduced in the presence of a free-living predator but not a caged predator. Abdominal movements were significantly reduced by the presence of a free-swimming predator and low prey densities. It is quite obvious that these damselfly larvae adjust their behaviour to the presence of predators to increase their survival at the expense of reduced growth and development.

5. BEHAVIOURAL DRIFT

Drift or the downstream transport of animals in the current also involves flows of numbers and biomass through time and space. It is a very important aspect of life in running water, particularly in terms of the relocation of benthic insects in response to competition for food and space. Behavioural drift includes insects that either actively enter the water column, for example to escape from a predator, or are dislodged as a consequence of an increased period of general activity on rock surfaces, as in foraging (Williams and Feltmate, 1992). Species of *Baetis* are well-known drifters in streams all over the world. In laboratory experiments, Kohler (1985) showed that *Baetis tricaudatus* actively enters the water column at night, after abandoning patches of stream bed when habitat quality, primarily assessed as food availability, falls to some threashold level. Drift by insects is typically nocturnal and while some species show a 'bigeminus' pattern of one large peak after sunset and another, smaller, one just before sunrise, other species show the converse 'alterans' pattern (Muller, 1966). There is also a seasonal component to drift. Typically, in temperate regions, the rate of drift is low during the winter and high in summer whereas, in the tropics, differences are less marked (Williams, 1980). Termination of drift by individuals is thought to be largely a function of behaviour rather than of morphology (Walton, 1978). Drift response has been shown to vary according to a number of features in the stream environment including photoperiod, water temperature, discharge and the presence of large predators such as trout (Williams, 1990).

II. ADULT BEHAVIOUR

It cannot be gainsaid that the surrounding environment has moulded through natural selection, life history patterns exhibited by aquatic insects. The influence of the terrestrial environment on emerging aquatic insects can be deduced by comparing life history traits of adult insects belonging to different orders from different environments. The abundance of aquatic insects in the forests bordering streams suggests that adult aquatic insects provide an important conduit for energy and nutrients from the aquatic system to the terrestrial food web. Adult aquatic insects engage in several activities, the most common of which are reproduction, dispersal and feeding. Relating how the ecology of larval aquatic insect dictates, or is dictated by adult ecology is vital to reveal much about the evolutionary drivers of adult behavioural ecology, which aspect has been well explored by Harrison and Dobson (2008). Interestingly, Harrison and Dobson (2008) have distinguished three broad categories of aquatic insects, each with distinct suites of species traits in larval stage with corresponding species traits in the adult stage: 1. Adults belonging to Ephemeroptera, Plecoptera, Trichoptera and Diptera (hereafter referred to as EPTD) are usually inconspicuous, short-lived, cryptically coloured, non-feeding or carbohydrate feeding and with poor powers of flight. The suite of adaptations shown by the adults of the EPTD may be a function of the lack of directional dispersal between habitats and may reflect a reduction in the metabolic cost of dispersing and avoiding predators, in favour of allocating resources to fecundity. 2. Adults of the Odonata and larger predatory aquatic Coleoptera and Hemiptera (hereafter referred to as OCH) are typically large, brightly coloured, with good flight and dispersal abilities, and relatively long adult lifespan. The very different adult feeding and flight adaptations of OCH may be a response to density-dependent food limitation and mortality of immature aquatic stages within spatially constrained habitats. Protein feeding as adults allows these to subsidize the resource-limited habitat of predatory larvae, which are often top predators. Such predatory feeding necessitates relatively long-lived adults with good flight and dispersal ability. Cannibalism among larvae will also promote adult dispersal to locate oviposition sites with low larval density. 3. Blood-feeding Diptera (hereafter referred to as BFD) provide a further contrast in that, although actively feeding as adults, like the OCH group, their larval stages are more like those of the EPTD, in that they are usually non-predatory and often consume poor quality detrital resources. For BFD, blood of vertebrates coming to drink water at nearby lentic or lotic bodies represents a predictable and nutritious food source for adults. This adult food may allow species to adopt larval feeding habits where food supply is restricted in quantity or quality, but also where competition for food is likely to be weak. The superior quality of adult food also may allow adult Diptera to adopt a metabolically costly, highly discriminatory oviposition behaviour, in contrast to other aquatic insects (Harrison and Dobson 2008). Exceptions do occur which come outside these categories. Notably, smaller aquatic Coleoptera and Hemiptera may be intermediate between EPTD and OCH. Many of these species feed on plant material and animal prey as both immature and adult stages have weaker dispersal and flight ability, and less well-developed antipredator adaptations compared to the larger OCH (Harrison and Dobson, 2008).

i. Trophic strategies: Feeding as an adult prolongs life, provides energy for flight and dispersal, and provides protein for oogenesis. Ephemeroptera with short adult life span, with conservative dispersal and with eggs fully formed at emergence obviously do not feed. The mouthparts are atrophied. Among the Plecoptera, many species within the superfamily Nemouroidea gain considerable weight as adults by feeding on terrestrial pollen, fungi and lichens (Zwick, 1990). Odonata feed on living prey throughout their life. When foraging, dragonflies are categorized as 'perchers' or 'fliers'. Perchers spend much of their time stationary, making short flights to capture prey and perching to consume it. In contrast, fliers are on the wing for a large part of their feeding activity, capturing their prey in the air and swallowing small prey while in flight; they do perch to consume larger prey (Tennessen 2009). Those caddisflies endowed with longer adult life span have an especially well-developed haustellum for imbibing liquids; apparently they ingest water or nectar for sustenance (Morse, 2009). Crichton (1957) found that adult fluid feeding was recorded most frequently in the widely-dispersing Limnephilidae and Phryganeidae. These two families also have the most highly developed mouthparts and some species are reported to feed on flowers at night. All gerromorphs are predaceous, those living on the water surface feeding on land arthropods that fall onto and are caught by the water's surface film. Prey is located by many bugs by ripples caused by their struggling. The bugs probably secrete something unpleasant because they themselves are occasionally preyed upon by fish (Schaefer, 2009). Nepomorphs are eager predators and they prey near the water surface, whence they may drive their prey, a procedure aided by the bug's buoyancy. The stimulus causing a *Ranatra* to grab at passing prey is partly visual and partly due to information about water vibrations received from sensillae on the spine on the anterior femur (Clorec, 1976). Ordinarily the insect remains at rest on dead vegetation until a suitable prey happens to swim into its sensory field. If the prey stimulates the sensory apparatus of the *Ranatra* , the latter may rapidly stretch out the prothoracic legs in an attempt to take the prey. This attempt is a 'strike', and if the strike is successful, it becomes a 'hit'. *Ranatra fusca* stridulates by moving a roughened spot on the anterior coxa against a stridular area on the outer lobe of the prothorax. The sound is produced when the animal is removed from the water and is presumably an alarm call. *Lethocerus* is a fairly generalized predator feeding on moving animals. Like Nepinae, it is an ambush predator, lying in wait for its prey and seizing it as it passes. Its behaviour is based on visual stimuli and apparently stereoscopic vision is involved (Hutchinson, 1993). Adults of female mosquitoes feed on vertebrate blood for the development of eggs. Blood feeding in insects is believed to have evolved several times independently from ancestral forms adapted for sucking plant juices or for preying on other insects. Female mosquitoes locate suitable hosts for blood feeding attracted by warmth, moisture and carbon dioxide from hosts and also by lactic acid, a component of human sweat (Eldridge, 2009). Finally, resource theft appears to be yet another important feeding strategy. For instance, the heteropteran, *Hydrometra championiana* has been reported to creep up behind feeding water striders (*Gerris* and *Velia*) and loot their prey. The bug plunges its long beak into their captured prey, stealing food which its competitors have captured for themselves (Schaeffer, 1978).

ii. Reproductive strategies: Most adult aquatic insects emerge from lentic and lotic water bodies to mate. Some insects like mayflies swarm before mating. Females then oviposit in the aquatic habitat. Successful reproduction depends, in part, on the adaptive physiological and behavioural traits that adults possess for life in the terrestrial environment. Three important traits of the adult stage are: 1. time of emergence (season or time of day); 2. location, time and behaviour of swarming and mating; and 3. longevity of winged forms (subimagos and imagos). These features influence the individual reproductive success by determining physical conditions, abundance of potential mates and abundance and efficiency of predators encountered by emergent adults. Adult life span also affects the adult's potential contribution to recolonization after floods.

Behavioural strategies of emergence, mating, swarming and oviposition in aquatic insects are geared to guarantee maximum survival, protection from prospective predation and propagation of the species (Sivaramakrishnan and Venkataraman, 1985). Edmunds and Edmunds (1979), working in Malaysia, found that practically all mayfly subimagos, the most vulnerable stage, emerged during the first two hours of darkness and moulted to imagos that could reproduce in the early morning. Afternoon emergence was practically rare unlike in temperate regions. The difference is explained as due to selection pressure by predators being greater in the tropics than in temperate regions, while the greater reduction of temperature at night in higher latitudes acts unfavorably, tending to inhibit the transformation of the subimago. High altitudes in low latitudes may exert the same general effect as high latitudes. Mayflies have distinct and finite emergence periods, especially in temperate and arctic areas. In the tropics, emergence is often nonseasonal, although some species have clear emergence patterns. The lunar emergence of the African species, *Povilla adusta*, from a number of lakes is well known (Hartland-Rowe, 1958). The burrowing mayflies of the Ephemeridae, Polymitarcyidae and Oligoneuriidae are noted for their sporadic mass emergence. Swarming in adults is a male activity, apart from the Caenidae and Tricorythidae, where both males and females may participate. The females fly into these swarms, and mating occurs almost immediately and usually in flight. Swarming may take place over the water itself, over the shore area, or even away from the water. The time of swarming varies considerably, although dusk is the most common time of day in temperate regions. The majority of mayflies, including most Ephemeridae, Heptageniidae, and Leptophlebiidae, oviposit by descending to the water and releasing a few eggs at a time by dipping their abdomen into the water. In several species of *Baetis* , female actually goes under water and lays her eggs on suitable substrate, often under stones (Brittain and Sartori, 2009). Northern Hemisphere stoneflies (Arctoperlaria) have evolved a diverse and complex system of vibrational signaling through substrates for locating mates. The drum calls of male stoneflies range from a simple series of taps, probably the ancestral call, to complex patterns of bursts and rubbing. The vibrations are produced by tapping or rubbing the abdomen on the substrate or by body tremulations transferred to the substrate. But the responsive drumming of females is far simpler usually nothing more than a few taps. Because females will respond only to the drumming calls of males of their own species, a male knows that a responding female must be a member of its own species. Consequently, there is no need for the reply call to be complex and elaborate (Waldbauer, 2006). Southern Hemisphere stoneflies (Antarctoperlaria) have never been documented to drum. They may have evolved a highly specific encounter site aggregation behaviour that enables sufficient mate locating ability without vibrational or other forms of intersexual communication (Stewart, 2009). Mating behaviour for some species of Trichoptera is mediated by sexual pheromones, with females emitting odors that attract males (Morse, 2009).

All species of the heteropterans, *Anisops* and *Buenova* can stridulate to produce mating calls. Stridulation may be tibial by tibial combs or femoral by *pars stridens* of sclerotized ridges in femora. Stridulatory activity is temperature dependent and species specific to signal mating of males and females of the same species. Exclusive male parental care in belostomatids, *Diplonychus* and *Lethocerus* is an extremely rare trait among animals (Dudgeon, 1999). After mating, the female of *Diplonychus* lays her eggs on the back of the male whence they are carried around until hatching. Male *Lethocerus* attend eggs laid above the water level on emergent vegetation. During the day, they remain in the water at the base of plant stems but, at night, climb up and cover the eggs with their wet bodies thereby moistening them (Ichikawa, 1988). The eggs fail to hatch if they are not supplied with water. When mature females encounter a brooding male in the water, they destroy the male's egg mass and take over the mate thereby gaining a nurse for their own egg mass (Ichikawa, 1995). Females are unable to detect brooding males when they are out of the water on the vegetation. A male water strider guards his mate staying coupled with her for as long as 24 hours, thereby preventing her from being inseminated by other males, making certain that it will be his sperm that fertilize his mate's eggs. A coupled female, not greatly inconvenienced by the small male perched on her back, moves about freely and continues to hunt for food without the harassment of sex-starved males. The mounted male also succeeds in protecting his parentage. Reproductive competition among males may select for a variety of traits or strategies that cause direct harm, or are otherwise costly to their mates. In Zeus bugs, *Phoreticovelia* spp., a genus of small and elusive semiaquatic veliids, sexual size dimorphism is very pronounced.

The much smaller males ride on the back of their mates for several weeks. Whereas it is common that males of many animals provide females with 'nuptial gifts' such as prey items or glandular secretions during courtship and mating, zeus bugs exhibit a reversal of gender roles in that females provide food for males during mating. Remarkably enough, females but not males are equipped with a pair of dorsal glands that produce a wax-like secretion that males feed from when riding on the backs of their mates (Arnqvist *et al.,* 2003).

iii. Dispersal and recolonization patterns: Adult dispersal, recolonization and successful completion of life cycle, be it univoltine, bivoltine or multivoltine of an aquatic insect species are essential for the persistence of populations, colonization of new habitats and microhabitats and maintenance of genetic diversity. Advantages of dispersal include avoiding inbreeding, locating to a site with more resources, fewer intra- and interspecific competitors and predators, and escaping unsuitable habitat conditions. Disadvantages include the frequent failure by individuals to locate a suitable new site, predation *en route,* failure to locate a mate and outbreeding depression. Critically, dispersal and flight necessitate a suite of metabolically costly adaptations in an adult insect, including adaptations to counter predation, desiccation and inclement weather. The evolutionary stability and predictability of aquatic habitats, compared with that of terrestrial habitats, may thus be expected to forge ecological traits that favour fecundity over dispersal. In insects with non-protein feeding adults, larval resources determine both adult size and fecundity, as the maximum number of eggs is present at adult emergence. In insects with feeding adults, however, egg mass and total number of egg are largely determined by resources available to adults. Thus, the trade-off between fecundity and dispersal ability should be biased towards fecundity in insects with non-feeding adults and towards dispersal in insects with feeding adults (Harrison and Dobson, 2008).

Recolonization can be viewed as the sequence of events that leads to the buildup of individuals, populations, species or higher taxa in places from where they have drifted downstream and to return through flight to upstream areas by adults or through upstream transport of larvae. Most aquatic insect orders have species that are known to disperse as adults. Resh *et al.* (1981) showed that adult oviposition was responsible for recolonization of a caddisfly population in a California spring following complete elimination and subsequent recovery of the habitat. In a desert stream, most of the colonizing insects arrived as adults, and the life histories of the majority of species in this habitat featured long-lived adults or continuous emergence (Gray, 1981). While taxa differ in their dispersal abilities, it is obvious that adult insects do move about though restrictions imposed by adult behaviour (e.g. in the selection of oviposition sites) and larval requirements drastically reduce the numbers of successful colonists. Dispersal flights are often synchronous within species and proximally controlled by temperature, light intensity and wind velocity (Sheldon, 1984).

FUTURE DIMENSIONS

As pointed out by Sheldon (1984), comprehensive study of aquatic insect behaviour transcends any particular topical area and a strong foundation of detailed observation coupled with tests of behavioural and evolutionary hypotheses through carefully crafted experimental designs will go a long way to develop a coherent theory ultimately to provide an integrated picture of behavioural adaptations. In this context, Harrison and Dobson (2008) very pertinently have prioritized the future research in this field to explore links between larval and adult behavioural ecology of aquatic insects such as Odonata, where both larvae and adults are well studied and Trichoptera, with a wide range of larval feeding modes. This will further our understanding of the evolutionary drivers of adult ecology and of the mutually dictated patterns of larval and imaginal behavioural strategies in respective species and species groups.

REFERENCES

Arnqvist, G., Jones, T.M. and Elgar, M.A. 2003. Insect Behavior: Reversal of sex roles in nuptial feeding. Nature 424, 387.

Brittain, J.E. and Nagell, B. 1981. Overwintering at low oxygen concentrations in the mayfly, *Leptophlebia vespertina*. Oikos 36: 45-50.

Brittain, J.E. and Sartori, M. 2009. Ephemeroptera. In: *Encyclopedia of Insects* edited by V.H. Resh and R.T. Carde (II Ed.), NY, pages 328-333.

Brusven, M.A. and Rose, S.T. 1981. Influence of substrate composition and suspended sediment on insect predation by the torrent sculpin, *Cottus rhotheus.* Canadian Journal of Fisheries Aquatic Science 38: 1444-1448.

Chapman, R.F. 1975. The insects: Structure and Function. English Univ. press, London, pages 819.

Crichton, M.I. 1957. The structure and function of the mouth parts of adults caddis flies (Drichoptera). Philosophical transactions of the Royal Society B241: 45-91

Cloarec, A. 1976. Interaction between different receptors involved in prey capture in *Ranatra linearis* (Insecta: Heteroptera). Biological Behaviour 1: 251-266.

Crowson, R.A. 1981. The Biology of the Coleoptera. Academic Press, London, pages 802.

Dudgeon, D. 1999. Tropical Asian streams: Zoobenthos, Ecology and Conservation. Hong Hong University Press, pages 830.

Edmunds, G.F. Jr. and Edmunds, C.H. 1980. Predation, climate and emergence and mating of mayflies. In *Advances in Ephemeroptera Biology* edited by JF Flannagon and KE Marshall, Plenum Press, NY, pages 277-285.

Eldridge, B.F. 2009. Mosquitoes. In *Encyclopedia of Insects* (II Ed.) edited by V.H. Resh and R.T. Carde, Academic Press, New York, pages 658-663.

Feltmate, B.W. and Williams, D.D. 1989. Influence of rainbow trout on density and feeding behaviour of a perlid stonefly. Canadian Journal of fisheries and aquatic science 46: 1575-1580.

Finakoshi, S. 2005. Intraspecific and interspecific competition for larval nests of the caddisflies, *Stenopsyche marmorata* and *Stenopsyche sauteri*. Entomological science (8): 339-345.

Frutiger, A. 2002. The function of the suckers of larval net-winged midges (Diptera: blephariceridae). Freshwater Biology 47: 293-302.

Gillies, M.T. and Elouard, J.M. 1990. The mayfly – mussel association, a new example from the river Niger basin. In *Mayflies and Stoneflies* edited by IC Cambell, Kluwer Academic Publishers, USA, pages 289-287.

Gray, L.J. 1981. Species composition and life histories of aquatic insects in a lowland Sonavan desert stream. American Midland Naturalist 106: 229-242.

Harrison, S.S.C. and Dobson, M. 2008. Evolutionary drivers nd to cological traits of adult aquatic insects. In Aquatic insects: Challenges to populations edited by J. Lancaster and R.A. Briers, CAB International, UK. pages 250-267.

Hart, D.D. and Finelli, C.M. 1999. Physical-biological coupling in streams: the pervasive effects of flow on benthic organisms. Annual Review of Ecology and Systematics 30: 363-395.

Hartland-Rowe, R. 1958. The biology of a tropical mayfly, *Povilla adusta* Nawas (Ephemeroptera: Polymitarcidae) with special reference to the lunar rhythm of emergence. Review Zoological Botany of Africa 58: 185-202.

Herrmann, D.P., Sites, R.W. and Willig, M.R. 1993. Influence of current velocity on substratum selection by Naucoridae (Hemiptera): an experimental approach via stream simulation. Environmental Entomology 22: 571-576.

Hutchinson, G.E. 1993. A treatise a limnology. In *The Zoobenthos* edited by Y.H. Edmonton, John Wiley and Sons Inc, NY, Vol. IV, pages 944.

Ichikawa, N. 1988. Male brooding behaviour of the giant water bug *Lethocerus deyrollei* Vuillefroy (Hemiptera: Belostomatidae). Journal of Ethology 6: 121-127.

Ichikawa, N. 1995. Male counter strategy against infanticide of the female giant water bug *Lethocerus deyrollei* (Hemiptera: Belostomatidae). Journal of Insect Behaviour 8: 181-188.

Kholer, S.L. 1985. Identification of stream drift mechanisms an experimental and observational approach. Ecology 66: 1749-1761.

Kingsley, K.J. 1985. *Ertes sticticus* (L.) (Coleoptera: Dytiscidae): Life history observations and an account of a remarkable event of synchronous emigration from a temporary desert pond. Coleopterists Bulletin 39:7-10.

Lancaster, J. and Hildrew, A.G. 1993. Flow refugia and the microdistribution of lotic invertebrates. Journal of the North American Benthological Society 12: 385-393.

Lytle, D.A. and Poff, N.L. 2004. Adaptation to natural flow regimes. Trends in Ecology and evolution 19: 94-100.

Lytle, D.A. 1999. Use of rainfall cues by *Abedus herberti* (Hemiptera: Belostomatidae): a mechanism for avoiding flash floods. Journal Insect Behaviour 12: 1-12.

Lytle, D.A. 2002. Flash floods and aquatic insect life history evolution: evaluation of multiple models. Ecology 83: 370-385

Lytle, D.A., Bogan, M.T. and Finn, D.S. 2008. Evolution of aquatic insect behaviours across a gradient of disturbance predictability. Proceedings of the Royal Society of Series B 275: 453-462.

Mc Cafferty, W.P. 1981. Aquatic Entomology. Science Books International, Boston, Massachusetts, pages 448.

Merrit, R.W., Cummins, K.W. and Berg, M.B. (eds.) 2008. An introduction to the aquatic insects of North Americh (4th Ed.), Kandall/Hunt, Dubuque, IA, USA.

Merritt, R.W. and Wallace, J.B. 2009. Aquatic habitats. In *Encyclopedia of Insects* edited by V.H. Resh and R.T. Cardi, Academic Press, New York, pages 38-48.

Morse, J.C. 2009. Trichoptera (Caddislies). In *Encyclopedia of Insects* (II Edition) edited by V.H. Resh and R.T. Carde, Academic Press, New York, pages 1015-1019.

Muller, K. 1966. Die Tagesperiodik von Fliesswasser- organismen. Z Morph. Okol. Ticine 56: 93142.

Muller-Liebenan, I. and Heard, W.H. 1979. *Symbioclocon:* A new genus of Bactidae from Thailand (Insecta: Ephemeroptera). In *Proceedings of the second Internatioanl Conference on Ephemeroptera* edited by K. Pasternak and R. Sowa, Panstwowe Wydawnictwo Naukowe, Warsigawa-Krakow, pages 57-65.

Olden, J.D., Hoffman, A.L., Monroe, J.B. and Le Roy, Puff, N. 2004. Movement behavior and dynamics of an aquatic insect in a stream benthic landscape. Canadian Journal of Zoology 82: 1135-1146.

Paradise, C.J. and Dunson, W.A. 1997. Insect species interactions and resource effects in tree holes: Are helodid beetles bottom-up facilitators of midge populations? Oecologia 109: 306-312.

Peckarrsky, B.L. 1980. Predator-prey interactions between stoneflies and mayflies: behavioural observations. Ecology 61: 932-943.

Rubinoff, D. and Schmitz, P. 2010 Multiple aquatic invasions by an endemic terrestrial Hawaiian moth radiation. Proceedings of the National Academy of Sciences, USA 107: 5903-5906.

Schaefer, C.W. 2009. Prosorrhyncha (Heteroptea and Coleoptra). In *Encyclopedia of Insects* (II Ed.) edited by V.H. Resh and R.T. Carde, Academic Press, New York, pages 839-855.

Schaeffer, C.W. 1978. Hydrometrid-gerrid interactions: a further note (Hemiptera). Journal of the Kansas Entomological Society 51: 286.

Schaffner, A.K. and Anholt, B.R. 1998. Influence of predator presence and prey density on behaviour and growth of damselfly larvae *(Ischnura elegans)* (Odonata: Zygoptera). Journal of Insect Behaviour II: 739-809.

Sheldon, A.L. 1984. Clonizaiton dynamics of aquatic insects. In *The Ecology of Aquatic Insects* edited by V.H. Resh and D.M. Rosenberg, Praeger Scientific, NY, pages 401-429.

Sites, R.W. and Willig, M.R. 1991. Microhabitat associations of three sympatric species of Naucoridae (Insect: Hemiptera). Environmental Entomology 20: 127-134.

Sivaramakrishnan, K.G. and Venkataraman, K. 1985. Behavioral strategies of emergence, swarming, mating and oviposition in mayflies. Proceedings of Indian Academy of Science (Animal Science) 94: 351-357.

Smith, R.L. 1973. Aspects of the biology of three species of the genus *Rhantus* (Coleoptera: Dytiscidae) with special reference to the acoustical behavior of two. Canadian Entomologist 105: 909-919.

Southwood, T.R.E. 1977. Habitat, the templet for ecological strategies. Journal animal Ecology 46: 337-365.

Stanley, E.G., Fisher, S.G. and Grimm, N.B. 1997. Ecosystem expansion and contraction in streams. Bioscience 47: 427-435.

Statzner, B., Hilder, A.G. and Resh, V.H. 2001. Species traits and Environmental Constraints: Entomological Research and the History of Ecological Theory. Annual review of Entomology 46: 291-316.

Stewart, K.W. 2009. Plecoptera (Stoneflies). In *Encyclopedia of Insects* (II Ed.) edited by V.H. Resh and R.T. Carde, Academic Press, New York, pages 810-813.

Subramanian, K.A. and Sivaramakrishnan, K.G. 2009. A new species of *Symbioclocon* (Ephemeroptera: Bactidae) associated with a freshwater mussel from India. Oriental Insects 43: 71-76.

Tennessen, K.J. 2009. Odonata (Dragonflies, Damselflies). In *Encyclopedia of Insects* edited by V.H. Resh and R.T. Carde, Academic Press, New York, pages 1132.

Van der Geest, H.G. 2007. Behaviioural responses of caddisfly larvae (*Hydropsyche angustipennis*) to hypoxia. Contributions to Zoology 76(4): 255-260.

Waldbauer, G. 2006. A walk around the Pond: Insects in and over the water. Harvard University Press, Massachusetts, USA, pages 286.

Weaver, J.S. and Morse, H.C. 1986. Evolution of feeding and case-making behaviour in Trichoptera. Journal of North America Benthological Society 5: 150-158.

Wiggins, G.B. 2004. Caddisflies: The underwater architects. University of Toronto Press, Toronto, Canada.

Williams, D.D. and Feltmate, B.W. 1992. Aquatic insects. CAB International, Wallingford Oxon, UK, pages 358.

Zimmerman, J.R. 1959. A note on flights emigrations of water beetles from a temporary pond. Coleopterists Bulletin 13: 102.

Zwick, P. 1990. Emergence, maturation and upstream oviposition flights of Plecoptera from the Breitenbach, with notes on the adult phase as a possible control of stream insect populations. Hydrobiologia 194; 207-223.

Dynamics of Insect Behaviour
T.N. Ananthakrishnan & K.G. Sivaramakrishnan
Scientific Publishers (India), Jodhpur, 2012, pp. 158-171

Chapter - 11

FORAGING AND BREEDING BEHAVIOUR OF PENINSULAR INDIAN ODONATA

K.A. Subramanian

INTRODUCTION

The order Odonata comprising Anisoptera (dragonflies), Anisozygoptera and Zygoptera (Damselflies) together called as odonates are diverse and prominent insects of the wetlands. These colourful primitive predators are phylogenetically very ancient and precursor of Odonata-*Namurotypus sippeli* (Protoodonata) appeared during Upper Carboniferous *ca.* 325 million years ago (Corbet, 1999). The earliest representatives of modern Odonata appeared during Lower Permian (*ca.* 250 million years ago). Morphologically they have largely remained unchanged in their evolutionary history. However, the ancestors of extant odonates were much larger in size, for example the largest fossil Odonata *Meganeura moneyi* and *Meganeuropsis permiana* with a wing span of over 75 cm are the largest species of insects ever known. The Anisozygoptera with two relict species was earlier recognized as a third suborder of Odonata. However, recent studies group Anisozygoptera with Anisoptera or some authors brings them together under a new name Epiprocta (Anisoptera + Anisozygoptera) (Trueman, 1996; Lohmann, 1996; Rehn, 2003). Odonates are predators, spending considerable part of their life in close association with wetlands as larval habitats or as adult foraging and breeding sites. Species are highly specific to particular aquatic habitat and associated with this, species exhibit a diverse array or foraging and breeding repertoire. The present paper aims to summarize the foraging and breeding behaviour of odonates of peninsular India.

DIVERSITY

Odonates of the Indian region are well documented and comprehensive systematic accounts have been published in the form of fauna volumes (Fraser, 1933-36). In these volumes, Fraser documented 536 species and subspecies from the Indian region which includes present day India, Sri Lanka, Pakistan, Nepal, Bhutan, Bangladesh and Myanmar. Since the publication of Fraser's volume, many species were described and more clarity was achieved in understanding the phylogenetic relationships between higher taxa. Globally 5,740 species of odonates are known of this 463 species in 140 genera and 19 families exist within Indian political limits (Table 1) (Subramanian, 2009). Within India, the regions such as Western Ghats, Eastern and Western Himalaya harbour high diversity and endemism.

Table 1. Global, Oriental and Indian diversity of Odonata.

S.N.	Suborder	Family	No. Genera			No. Speices		
			World	Oriental	India	World	Oriental	India
	Anisoptera							
1		Aeshnidae	52	18	13	436	138	45
2		Austropetaliidae#	4	0	0	11	0	0
3		Chlorogomphidae	4	1	1	45	40	10
4		Cordulegastridae	4	5	3	51	27	9
5		Corduliidae	39	7	4	249	57	16
6		Gomphidae	95	43	25	958	358	85
7		Libellulidae	142	56	38	972	190	85
8		Macromiidae	4	2	2	123	50	17
9		Neopetaliidae#	1	0	0	1	0	0
10		Petaluridae#	5	0	0	11	0	0
11		Synthemistidae#	7	0	0	43	0	0
	Anisozyg optera							
12		Epiophlebiidae	1	1	1	2	1	1
	Zygoptera							
13		Amphipterygidae#	4	1	0	10	5	0
14		Calopterygidae	21	10	6	176	60	7

15	Chlorocyphidae	16	14	4	149	80	21
16	Coenagrionidae	94	23	14	1092	185	58
17	Dicteriadidae#	2	0	0	2	0	0
18	Euphaeidae	9	12	6	69	65	18
19	Hemiphlebiidae#	1	0	0	1	0	0
20	Isostictidae#	12	0	0	45	0	0
21	Lestidae	9	5	4	152	39	23
22	Lestoideidae#	1	0	0	4	0	0
23	Megapodagrioniidae	43	10	1	303	28	1
24	Perilestidae#	3	0	0	19	0	0
25	Philogangidae*	2	1	1	9	4	1
26	Platycnemididae	26	8	5	224	130	27
27	Platystictidae	6	5	3	210	119	13
28	Polythoridae#	7	0	0	59	0	0
29	Protoneuridae	25	8	7	258	57	22
30	Pseudolestidae#	1	3	0	1	15	0
31	Pseudostigmatidae#	6	0	0	19	0	0
32	Synlestidae	7	2	1	35	17	4
	Total	654	236	140	5740	1669	463

Note: Families in '#' are not represented in India.
World and Oriental Odonata data after Schorr *et al.*, 2009 and Kalkaman *et al.*, 2008.

HABITATS

Odonates use diverse natural and man made wetland habitats such as streams, rivers, temporary rain water pools, ponds, storage wells, overhead tanks, irrigation canals, paddy fields and reservoirs. Community composition of odonates in a landscape is determined by available wetlands type and riparian or catchment landuse. In general, erosive habitats such as streams and rivers have higher diversity and endemism than depositional habitats such as flood plains, ponds, lakes etc. Each odonate species has specific habitat requirement for larval development. Odonates lay eggs in wetland habitats and larvae develop there. These oviposition sites with specific physical habitat structure and water quality are fiercely guarded and sought after by breeding adults.

Foraging Behaviour

Odonates are predators in their larval and adult stages of life. Their morphology and behaviour is fine tuned to this predatory mode of life. The Odonata larvae are specialized to an extent that different species occupy a wide range of habitats and microhabitats within a wetland. The larvae have unique extensible labium for capturing and killing the prey. As a feeding guild they have been classified as *swallowers* (Cummins, 1973). Larvae detect its prey by its movement and catch them alive and rarely do they feed on dead or immobile prey. The larvae generally use *the ambush* or *sit-and-wait* strategy to capture the prey. They remain immobile until the prey comes within range of the extended labium. The larvae also actively search or stalk the prey. The body colouration, markings and appendages perfectly blend with the microhabitat, enhancing the efficiency of predation. Based on the mode of prey detection, the larvae are classified as *climbers* or *sprawlers* (Pritchard, G., 1965). The climbers move among aquatic vegetation near surface and use their large compound eyes to detect prey at a distance, and sprawlers living near bottom largely use tactile simulation for prey capture.

The Larvae of stream breeding species such as *Neurobasis chinensis*, *Rhinocypha bisignata* or *Euphaea frasaeri* feed on aquatic insects among cobbles and pebbles in cascades and riffles. Species of *Macrogomphus*, *Zygonyx* and *Epopthalmia* usually burry themselves in fine gravel or sand in deep stream pools and feed on larger aquatic invertebrates and small tadpoles and fishes. In ponds and lakes, species of *Coenagrion, Pseudagrion, Ischneura*, *Trithemis*, *Crocothemis*, *Ictinogomphus*, *Epopthalmia* etc. are commonly found among submerged aquatic vegetation and they feed on other invertebrates in that habitat. Studies show that larvae are euryphagous, opportunistic foragers. Larvae of Diptera and planktonic Crustacea such as Cladocera and Ostracoda are primary prey during early larval stages. Larger larvae feed on large aquatic insects such as *Gerris* sp. Larvae of *Aeshna*, *Anax*, *Tramea* and *Pantala* feed on tadpoles and juvenile fishes (Corbet, 1999).

Adult odonates are sophisticated aerial predators. Their visual system and body morphology is especially well developed for this mode of life. Different species specialize and occupy perticular vertical strata in air to minimize interspecies competition. Species of Coenagrionidae, Calopterygidae, Cholorocyphidae and Euphaeidae occupy and forage in an aerial strata about 5m above the ground. Anisoptera occupy wider foraging strata than Zygoptera. Species of Libellulidae like *Diplacodes trivialis*, *Brachythemis contaminata* and *Bradinopyga geminata*, forage very close to the ground. Medium sized species like *Pantala flavescens*, *Zygonyx iris* and *Orthetrum* sp. usually occupy below 10m. Large gomphids and chlorogomphids usually forage above canopy, sometimes over 40m. Species of *Idionyx* (Corduliidae) are specialized to forage below one meter over forest floor with good canopy cover and low light conditions.

The body of Odonata, especially Anisoptera is designed for agile flight. Odonata have large thoracic weight, sometimes reaching up to 63% of body weight and low wing loading (0.02-0.04 g/cm^2). Females generally have lower thoracic weight than males. The nodus of the wing permits elastic tension of the wing's leading edge, allowing strong twisting. The pterostigma, serves as inertial regulator and can raise the critical gliding speed. Pterostigma though contributing only 0.1% of the total body weight, it can raise the critical gliding speed by 25% (Corbet, 1999). The flight performance and flight speed is regulated and monitored by antennae and trichoid sensilla on the head. In flight dragonflies can hover and soar using thoracic ventilation and by adjusting hemolymph circulation between the thorax and abdomen enhancing the metabolic rate.

Odonata are characterized by large compound eyes, in fact whole of head is covered by two large compound eyes. The visual field of the adult is almost 360°. The compound eye can detect colour, ultraviolet light, the plane of polarizaion of light and movement. The sensitivity to ultraviolet and plane of polarized light are restricted to certain regions. Similarly, movement detection occurs in the acute zones. The fovea forms the centre of pseudopupil or the black spot, size of which is a good indicator of how well a dragonfly can see in a particular direction. The distribution of acute zones in eyes are related to the habitat in which adults fly and forage and differ in perchers and fliers (Corbet, 1999).

Odonata obtain food by catching living prey –either in mid air or by gleaning and these modes of foraging is believed to have persisted since Carboniferous times. Based on foraging styles the Odonata is generally classified as perchers or fliers. Species like *Orthetrum sabina*, *O. pruinosum*, *O.chrysis* and many other libellulids occupying open wetlands like lakes and ponds are perchers. They return to their perch with their prey after a short sallying flight. Other libellulids and aeshinids such as *Pantala flavescens*, *Tramea limbata*, *T.basilaris*, *Anax immaculifrons*, and *Gynacantha* sp. continuesly fly and capture their prey in flight. Both fliers and perchers obtain their either by mid air foraging or gleaning. Gleaning is more common in Zygoptera, they routinely glean on small homoptera attached to plants.

Field observations on Odonata of peninsular India show that they routinely feed on wide variety of insects. Large Anisoptera like *Anax immaculifrons, Anax guttatus* and *Ictinogomphus rapax* feed on butterflies, especially species of *Euremea* sp. and *Catopsilla* sp.; many species of Diptera and other smaller Anisoptera like *Pantala flavescens* and *Trithemis aurora.* However, Diptera constitute the bulk of food. Species like *Gynacantha dravida*, *Zyxomma petiolatum* and *Tholymis tillarga* which are crepuscular foragers, specilizes in foraging swarming Culicidae and Chironomidae.

Foraging habitats of the species is highly specific. Fliers such as *Pantala flavescens*, *Tramea limbata*, *Tramea basilaris* and *Anax guttatus* forage over open wetlands, gardens and agricultural fields. Species such as *Anax immaculifrons*, *Macromia* sp., and *Epopthalmia* sp. forage over partially or fully shaded streams. Many species of Anisoptera and all most all of the Zygoptera are perchers. Species of Coenagrionidae forage among emergent and riparian vegetation in open wetlands. Species such as *Neurobasis chinensis*, *Rhinocypha bisignata, Euphaea sp., Caconeura sp.,* and *Protosticta sp.,* are exclusive to forested streams and forage there. Perching Anisoptera and Zygoptera of forests such as *Neurothemis fulvia*, *Hylaeothemis indica*, *Epithemis mariae*, *Vestalis gracilis* and *Vestalis appicalis* prefer to occupy canopy and tree fall gaps, trek paths and forest edges.

Breeding Behaviour

Odonata are sexually dimorphic. Generally Males are more brightly coloured than females. Behaviourly, males are more aggressive and territorial. Males of perching species occupy very prominant vantage points such as bare tips of branches, patches of boulders etc. Fliers perambulate their territory repeateadly. Both fliers and perchers chase away conspecific males from their territory. Species like *Rhynocypha bisignata*, *Neurobasis chinensis* and *Libellago lineata* hover infront of intruding males and continue to do so, till the invader leaves the territory. Territory size of the males vary widely. In stream damselflies like *Rhynocypha bisignata* and *Neurobasis chinensis* it is few square meters. In large stream dwelling dragonflies like *Macromia ellisoni* or *Gomphidia kodaguensis* single male occupies 50-100m of stream stretch.

Females usually occupy and forage in riparian landscapes. Females of species such as *Trithemis aurora*, *Neurothemis fulvia*, *Vestalis gracilis* and *Vestalis appicalis* sometimes congregate in large numbers in riparian vegetation. On the other hand, females of species such as *Anax immaculifrons*, *Ictinogomphus rapax*, *Macromia* sp. and *Gomphidia* sp. are solitary hunters and visit territories of more than one male.

At the breeding habitats mature male spend considerable amount of time searching for females. Males search females by surveying from a perch and patrolling. Most of the species recognize sex and species by visual cues such as size, flight style, colour, UV reflection, and pattern. In general Anisoptera can visually discriminate conspecifics at a longer distance than Zygoptera. Movement of female is one of the important cues for initiating sexual response in males. In Zygoptera, colour and pattern are primary cues used by males to recognize females, whereas flight style and body shape are important for Anisoptera (Corbet, 1999).

Immediately after the male-female encounter, the sexes communicate visually to each other. This interaction normally precedes copulation with display and posturing by the males. Species such as *Rhinocypha bisignata*, *Euphaea fraseri*, *E.dispar* and *E.cardinalis* hover infront females displaying irridiscent wing markings. Courtship displays are common among species of Calopterygidae, Chlorocyphidae and Euphaeidae (Photo 1).

Courtship in Anisoptera is so far been recorded only for few genera such as *Palpopleura* sp. and *Brachythemis* sp. (Libellulidae). Clear courtship behavriour is known only from territorial Odonata. It is hypothesized that evolution of courtship and territoriality is causally linked. Possible functions of courtship include prevention of heterospecific tandem formation, receptivity assessment of female by male, female assessment of male quality and oviposition site (Corbet, 1999).

Photo 1. Courtship Diplay of *Euphaea dispar* (Photo: K.A. Subramanian)

Photo 2. Intramale sperm transfer in *Euphaea dispar* (Photo: K.A. Subramanian)

The courtship preceding tandem link assures that tandem with unreceptive female, confspecific male or hetero specific females are reduced. Precopulatory tandem is the first in the sequence of coupulation (Photo 2). This involves the attachment of the male's anal appendages to the female's head or thorax. Tandem linkage is achieved when the anal appendages of the male interlock reciprocally the female's pro and mesothorax in Zygoptera and with head and prothorax in Anisoptera. The morphology of male anal appendage and morphology of female pro thorax, especially mesostigmal laminae matches perfectly. These complementary interlocking structures are strong enough to prevent heterospecific mating.

In male Odonata primary genital orifice and the intromittent organ are located at opposite ends of the abdomen. This feature is unique among exopterygote insects. The primary gonopore is situated in 9th abdominal segment and secondary genitalia are on the second and third segements. Before copulation male charges his secondary genitalia with fresh sperm, a process termed intramale sperm translocation. Within a species sperm translocation occurs both in solitary males and during precopulatory tandem. In Zygoptera sperm translocation almost always occurs during precopulatory tandem except in certain Calopterygidae and Euphaeidae such as *Neurobasis chinensis* and *Euphaea cardinalis* (pers.observations) (Photo 3).

Photo 3. Tandem formation in *Disparoneura quadrimaculata*
(Photo: Shyam Prasad)

Wheel formation, a position in which females genital aperture is secured to the male's secondary genitalia always follows precopulatory tandem. Sperm competition is widely reported among insects and was first reported in Odonata. Odonate penis both inseminates the female and displaces other male's sperm. Morphological specilization of penis and complementary modifications in the female genital tract permits sperm displacement. Sperm competation by sperm displacement has been inferred in on the basis of four citeria: (1) multiple mating by the female (2) sperm storage organ in female, (3) modification of male and female genitalia (4) guarding of female by male after copulation (Waage, 1986). Sperm displacement occurs widely among Zygoptera, Libellulidae and few Aeshnidae.

Photo 4. Endophytic oviposition in *Pseudagrion microcephalum* (Photo: Shibu Bhaskar)

Post copulatory guarding, is a behaviour exhibited by male odonata that accompanies an ovipositing female. If the male is guarding, it is either by tandem or by hovering or settling near the female. However guarding is not exhibited by all families of Odonata and male may or may not guard the female after copulation. Contact guarding occurs in Coenagrionidae, Lestidae, Platycnemididae, Protoneuridae, Libellulidae and certain genera of Aeshnidae. Non contact guarding occurs in many families of Zygoptera, few Libellulidae and other Anisoptera.

Photo 5. Epiphytic oviposition in *Anax immaculifrons* (Photo: Giby, K)

Photo 6. Exophytic oviposition in *Tetrathemis platyptera* (Photo: Shibu Bhaskar)

Depending on species oviposition could be endophytic, epiphytic and exophytic. Female commences oviposition immediately after copulation or in species of *Orthetrum* male and female perch closely and rest briefly before oviposition. Many species of Zygoptera, especially Coenagrionidae oviposit endophytically (Photo 4). Mated pair search for suitable oviposition site in tandem. Male perches just above water surface and lowers female. Sometime female submerge under-water to oviposit. Species such as *Tholymis tillarga* and *Brachythemis contaminiata* oviposit ephiphytically. In the former species oviposition occurs during early morning or late evening. Both the species oviposit over submerged leaves of aquatic plants which are just below the water surface (Photo 5). Species such as *Tetrathemis platyptera* oviposit exophytically over dry twigs hanging few centimeters over water bodies (Photo 6). In the first rains, the larvae hatch and enter water.

Oviposition of conspefic groups have been observed in many species of Zygoptera. This mode of oviposition is common in *Pseudagrion rubriceps* and *Pseudagrion microcephalaum*. Among Anisoptera, females of *Idionyx galeata* (Corduliidae) oviposit in groups of 10-12 individuals (Subramanian *pers. obser.*). Diverse modes of oviposition found in peninsular Indian Odonata is given in Table 2.

Table 2. Modes of Oviposition in Peninsular Indian Odonata (modified after Eda, S (1960, 1975) and Corbet (1999)).

SN	Position of Female	Ovipostion Mode	Family/Genera/Species
1	Female settled with contact between ovipositor and substrate.	Attached to Plants	*Brachythemis contaminata, Anax immaculifrons*
		Within Plants	Lestidae, Coenagrionidae
		Into or onto mud, sand, detritus or soil	*Anaciaeschna, Anax, Hemianax*
		Onto rock	*Zygonyx iris,*
		Dead Wood	*Libellago lineata, Euphaea, Rhinocypha bisignata*
		Onto Water	*Ictinogomphus rapax*
2	Female flying with contact between ovipositor and substrate.	Attached to Plants	*Brachythemis contaminata, Tholymis tillarga, Tetrathemis platyptera, Urothemis signata, Ictinogomphus rapax, Zyxomma petiolatum,*

		Into or onto mud, sand or soil	*Epopthalmia, Idionyx,*
		Onto water	*Pantala flavescens, Trithemis, Orthetrum, Tramea,*
3	Female flying without contact between ovipositor and substrate.	Onto water	*Onychogomphus niligiriensis*

SUMMARY

The order Odonata, exhibiting high diversity in foraging and breeding behaviour are excellent model system to address fundamental in questions ecology and behaviour. Though peninsular Indian Odonata are taxonomically well studied, very few studies explore the larval ecology and behaviour of adult Odonata in field. Breeding biology of most of the species, especially the endemic forest species is little known. Another dimension of Odonata behaviour which is least studied is migration. Field observations suggest that many species migrate with monsoon winds across Indian peninsula. Future studies should focus on larval ecology, adult breeding and migratory bahaviour of Odonata. Odonates being an established indicator of landscape health, the results from these studies are fundamental for effective conservation planning and management.

ACKNOWLEDGEMENTS

The author acknowledges Dr. K. Venkataraman, Director, Zoological Survey of India, Kolkata for extending facilities for writing this manuscript.

REFERENCES

Corbet, P.S. 1999. Dragonflies. Behaviour and Ecology of Odonata. Cornell University Press, Ithaca, New York. 829 pp.

Cummins, K.W. 1973. Trophic relations of aquatic insects. *Annual Review of Entomology*. 18: 183-206.

Eda, S. 1960. Oviposition of *Gomphus pryeri* Selys, with a note on the ovipositon of dragonflies. Insect Ecology. 8:82-88.

Eda, S. 1975. On the oviposition behaviour of the dragonflies of the genus *Sympetrum*. Tombo 18:2-9

Fraser, 1933-36. The fauna of British India, including Ceylon and Burma. Odonata. Vols.1-3. Taylor and Francis, London.

Kalkman, V.J., Clausnitzer, V., Dijkstra, K.D.B., Orr, G.A., Paulson, R.D., Kumar, A. and Prasad, M. 1977. Reproductive behaviour in *Neurobasis chinensis chinensis* (Linnaeus) (Zygoptera: Calopterygidae). *Odonatologica*. 6: 163-171.

Lohmann, H. 1996. Das phylogenetische System der Anisoptera (Odonata). *Entomologische Zeitschrift* 106, 209-252.

Pritchard, G. 1965. Prey capture by dragonfly larvae (Odonata; Anisoptera). *Candian Journal of Zoology*. 43: 333-336.

Rehn, A.C. 2003. Phylogenetic analysis of higher-level relationships of Odonata. *Systematic Entomology. 28*,181-239.

Schorr, M., Lindeboom, M. and Paulson, D. 2009. World Odonata List. http://www.pugetsound.edu/x6140.xml.

Subramanian, K.A. 2009. A checklist of Odonata of India. http://zsi.gov.in/checklist/Odonata_Indica_151209.pdf

Trueman, J.W.H. 1996. A preliminary cladistic analysis of odonate wing venation. *Odonatologica*. 25, 59-72.

Waage, 1986. Evidence for widespread sperm displacement ability among Zygoptera (Odonata) and the means for predicting its presence. Biological Journal of Linnean Society of London. 28:285-300.

Dynamics of Insect Behaviour
T.N. Ananthakrishnan & K.G. Sivaramakrishnan
Scientific Publishers (India), Jodhpur, 2012, pp. 172-188

Chapter - 12

CLIMATE CHANGE AND INSECT BIODIVERSITY - BEHAVIOURAL PATTERNS AND IMPLICATIONS IN CONSERVATION/CONTROL

K.G. Sivaramakrishnan

INTRODUCTION

Insects constitute the most diverse form of animal life in terrestrial and freshwater ecosystems. Most species are the indispensable components of natural ecosystems. Because they are cold-blooded, the rates of key physiological and behavioural processes in their life cycles are determined by environmental conditions, especially temperature and precipitation. In general, they have short generation times, high fecundity and high mobility. Many insects are sensitive to extreme weather events (droughts, heat waves, cold spells). As a result of climate change and deforestation, tropical environments that harbor the bulk of Earth's biodiversity could very well become too hot, dry or fragmented for many insect species to persist (Williams *et al.,* 2003). Species that exhibit highly evolved host plant interactions or inhabit specialized microhabitats are at high risk of extinction, especially in tropical areas (Lewis, 2006). On the other hand, climates of temperate and subarctic regions are becoming increasingly hospitable to plant and insect life, raising concerns about the behaviour of indigenous species and about the risk of invasion by exotic species, which could result in disruption of normal ecosystem functions (Regniere, 2009). Aspects of climate change-induced behavioural responses of insects and their implications in conservation of useful insects as well as control of pests and vectors of diseases are briefly reviewed.

CLIMATE CHANGE SCENARIOS

i. Global Scenario

Climate has changed many times over the course of the Earth's history. These Paleoclimatic changes can be constructed using direct measurements from land and ocean weather stations. Over the past two million years, there have been numerous Ice Ages, but at present we are in a warm interglacial period. The geological record reveals that past climate change has not always been gradual. There have been many changes that were rapid, over periods as short as centuries or decades (Bonam, 2002). Beginning around 1550 A.D. and lasting until 1850 A.D. was a period of cold temperatures, known as the **'little Ice Age'** (Lamb, 1995). Temperatures warmed by over 0.5°C between the mid 1800s and 1900s, but cooled again over the next 25 years. On an average, global surface temperatures increased by 0.6°C during the 20th century. We are presently in an **'Era of anthropogenic climate change'** with unprecedented warming mainly due to land use changes, nitrogen fertilization, and direct effects of increased atmospheric CO_2 on plant and insect ecophysiology (IPCC, 2001). The 2007 Fourth Assessment Report of the Intergovernmental Panel on Climate Change (IPCC) details the unequivocal warming of the Earth's climate over the last 50 years, most of which is very likely a result of increases in **anthropogenic greenhouse gas (GHG)** emissions. There has been a rapid 35% rise in atmospheric GHG concentrations since preindustrial times, and in 2005, atmospheric GHC concentrations were higher than any levels recorded or estimated for the previous 6,50,000 years. Although the magnitude of warming has varied across the Earth's surface, globally averaged temperature has risen 0.6°C over the last 50 years. Precipitation trends have been more variable, with some regions of the world showing significant increases and others significant decreases. Using a range of projected GHG emissions scenarios, the IPCC's best estimates of global anthropogenic warming will range from 1.8 to 4.0°C over the present century, with an even greater magnitude of warming expected at high northern latitudes. Annual precipitation is very likely to increase in high latitudes, and parts of subtropics and tropics are likely to experience decreased precipitation over the next century (Debinski and Cross, 2008). In addition to changes in temperature and precipitation, greenhouse warming will also lead to increased climate variability and the occurrence of climatic extremes. For example, many areas across the globe are predicted to experience more severe droughts, floods, large-scale erosion, landslides and extreme heat events. Obviously, the insect conservation community has become focused on developing strategies that allow species to adapt and ecosystem processes to function in the light of inevitable climate change.

ii. Indian Scenario

An approximately 1.5°C increase in monsoon seasonal mean surface temperature, and a 2 mm increase in rainfall per day, between 2040-2049 with respect to the 1980s is predicted using the IPCC model. The plant usable concentration of carbon dioxide is predicted to be about 60 ppm by the middle of this century (Saseendran *et al.,* 2000).

BEHAVIOURAL RESPONSES OF INSECTS TO CLIMATE CHANGE

1. OVERALL PATTERNS

Climate is one of the most important factors that influence the behaviour, abundance and distribution of insect species, as well as having a strong influence on the ecology of their habitats and ecosystems. Changes in the behaviour, abundance and distribution of insect species are already being observed and linked to climate. Over time, these and other changes are likely to become increasingly profound. New (2009) recognizes three broad categories of response in insects for climate change: 1. **Extinction**, if climate changes to eliminate the specific regimes needed by a species throughout its current or potential change. 2. **Behavioural adaptations** within the current range, if the species is sufficiently 'flexible' to thrive under the changed conditions and 3. **Migration,** so that the species' range changes to encompass areas now expressing the tolerable climate regime and manifest in range expansion and sometimes loss from part or all of a former range. In a nutshell, a globally coherent **fingerprint of climate change** in the overall patterns across natural systems inclusive of insect communities (Parmesan and Yohe, 2003) is decipherable in terms of insects' phenology, physiology and behaviour, the composition of and interactions within their communities and the structure and dynamics of the ecosystems of which they form a part (Walther *et al.,* 2002).

i. Climate as a Determinant of Geographical Range in Insects

Because the metabolism of insects is directly affected by ambient temperature, climate change can directly affect the population densities and geographic distributions of insects by changing the ambient temperatures within their ranges. An increase in local temperature, for example, can speed up insect metabolism, with resultant effects on local population density. Predicting effects on insects is difficult because insects can buffer the effects of temperature behaviourally by seeking shelter. Herbivorous insects are also limited by the presence of host plants they need.

According to Rubenstein (1992), insects' metabolic activity and resultant behaviour patterns fall into distinct types, depending on how their performance, as measured by traits such as basic metabolism and fecundity, responds to climate variables, notably temperature and humidity. The first type of response is linear, where continued increase in a climate variable such as temperature, results in continued increase in the biological response.

ii. Community and Ecosystem Responses

Ecobehavioural responses by individual species to climate change are not isolated. They are connected through interactions with others at the same or adjacent trophic levels (Tylianakis *et al.*, 2008; Van der Putten *et al.*, 2010). Species do not respond synchronously in time. Even in cases where climate is considered the dominant trigger that drives the phenology of species, different species may respond to different climatic parameters (Visser and Holleman, 2001). The consequences of such differential responsiveness are revealed in differences in the timing of species interactions within ecological networks, such as food webs, host-parasitoid webs and/ or mutualistic webs. In host-parasitoid webs, species of the higher trophic level (parasitoids) may be affected by both the climatic variability itself and indirectly by the amplified variance in host dynamics in response to climate (Stireman *et al.*, 2005). Climatic fluctuations may be responsible for increasing the risk of host outbreaks through disruption of synchronization in host-parasitoid systems (Hance *et al.*, 2007). In mutualistic webs, as exemplified by plant-pollinator interactions, Hegland *et al.,* (2009) observed a linear relationship between climate warming and appearance dates in pollinators and plants. However, they emphasized that such linear responses cannot continue perpetually. A few studies dealing with phenological decoupling revealed varying responses among species and regions (Walther, 2011). In a nutshell, temporal and spatial changes in species behaviour and distribution are not isolated processes. They are connected through interactions with other species at the same or adjacent trophic levels and beyond, through indirect effects of trophic cascades. In food webs, climate fluctuations influence the relative timing of food requirement and availability (Stenseth *et al.*, 2002). Climate change influences community composition and ecosystem processes not in isolation, but in concert with other drivers of global change like species invasions promoting arrival of nonnative species and reshuffling of communities, ultimately contributing to modified ecological networks and altering ecosystem processes (Walther 2004; 2011).

iii. Synergistic impacts on behaviour

Synergies with other environmental changes, such as habitat loss and fragmentation, will potentially have large effects on species' abilities to adapt to changing climate. Samways (2007) has cautioned that there is little point in monitoring individual species to assess the conservation status of specific ecosystems. There is a need for a holistic approach to explore how whole systems work, since conservation of species will ultimately depend on the conservation of the whole functional system (Ananthakrishnan, 2007). Samways (2007) also has expressed greatest concern that climate will be interactive and synergistic with other adverse factors, leading to multiple impacts on species. Travis (2003) has pertinently called the synergism between climate change and habitat loss a **"deadly anthropogenic cocktail"** for biodiversity. For instance, since the 1950s there has been a 70% decline in the larger British moths, probably due to agricultural intensification and widespread and intensive use of insecticides coupled with climate change (Conrad *et al.,* 2004). Furthermore, 30 to 35 species of British butterflies have not tracked recent climate change owing to suitable habitat. Indeed, for these butterflies the fragmentation and destruction of natural habitats resulted in isolation of suitable areas for colonization. Evidence has suggested that it is only the more mobile and generalist butterflies and dragonflies that could track climatic compatibility (Samways, 2007).

2. SPECIFIC BEHAVIOURAL PATTERNS

i. Foraging behaviour

Increase in carbon dioxide levels is expected to affect the biology of a number of living organisms including insects. Increases in atmospheric levels of carbon dioxide can cause increases in plant growth rates and changes in the physical and chemical defenses of plants and the chemical composition of their tissues. Phytophagous insects are indirectly affected by those changes in their host plants. Observed decreases in the protein content and increases of the C/N ratio in leaves under elevated carbon dioxide imply a reduction in food quality. In some cases, herbivorous insects compensate for that change through increased consumption (Subrahmanyam *et al.*, 2009). Many adult lepidopterans feed on nectar, and it has long been assumed that flowers are important sources of carbon dioxide. Production of nectar and carbon dioxide in the nocturnal, moth-pollinated flowers of *Datura wrightii* is positively correlated temporally (Guerenstein *et al.,* 2004), so that floral carbon dioxide emission may indicate food abundance for the moths. The idea that moths use carbon dioxide as a cue for foraging could imply that they can predict volumes before probing the flowers and would hover in front of only promising flowers, thus increasing their foraging efficiency.

ii. Pheromone-mediated behaviour

In contrast to plant-mediated effects on insect herbivores, there is paucity of information on natural enemy-mediated effects under changing atmospheric conditions (Hunter, 2001). As natural enemy efficacy is governed largely by behavioural mechanisms, altered prey finding and prey defence may change insect population dynamics. Mondor *et al.,* (2004) have shown that pheromone-mediated escape behaviours and hence the vulnerability of insects to natural enemies, are divergent under atmospheric conditions associated with global climate change. *Chaitophorus stevensi,* a common aphid on trembling aspen trees, *Populus tremuloides*, have diminished responses in enriched carbon dioxide environments, while those in enriched ozone have augmented escape responses, to alarm pheromone.

These results suggest that divergent pheromone-mediated behaviours could alter predator-prey interactions in future environments, in the context of carbon dioxide (CO_2) and tropospheric ozone (O_3) levels increasing 31% and 35% respectively, since mid-1800s.

iii. Oviposition behaviour

Adult Lepidoptera possess an organ in their labial palpi – the labial-palp pit organ (LPO), which is specialized for sensing CO_2 (Kent *et al.,* 1986). In the moth, *Cactoblastis cactorum,* this organ plays a role in the oviposition behaviour (Stange, 1997). *C. cactorum* is a nocturnal moth that oviposits on *Opuntia stricta.* These plants, found mainly in arid regions, assimilate CO_2 during the night and thus are sinks for CO_2. Plants that have more photosynthetic activity during the day receive more eggs during the night. It is thought that these moths sense the magnitude of the negative CO_2 gradients associated with a plant's metabolic activity and thus detect the healthiest plants, that is, the plants that assimilate more CO_2 during the night (Myers *et al.,* 1981).

iv. Social behaviour

Insect social behaviour includes the control of environmental variables inside nests. For example, excessively high temperatures or CO_2 levels within nests can have deleterious effects on the colonies. As part of the control of nest-climate conditions, some social insects measure and regulate the levels of CO_2. Regulation in honeybees involves collective ventilation by wing-fanning workers near the entrance, which actively drives air out of the hive. Thus, under natural conditions, there is a positive temporal correlation between CO_2 levels in the hive and the number of fanning bees (Seeley, 1974). Large colonies control CO_2 levels more precisely and at lower levels (mean CO_2 level of 0.0 ppm) than do small colonies (mean CO_2 level of 7800 ppm) (Subrahmanyam *et al.,* 2009).

3. CASE STUDIES

i. Insect pests, predators and parasitoids in agro ecosystems

Increasing temperatures may result in a greater ability to overwinter in insect species limited by low temperatures at higher latitudes, extending their geographical range (Elpginstone and Toth, 2008), and sudden outbreaks of insect pests can wipe out certain crop species, and also encourage the invasion by exotic species (Kannan and James, 2009). Global warming will lead to earlier infestation by *Helicoverpa* zea in North America, and *Helicoverpa armigera* in North India, resulting in increased crop loss. Rising temperatures are likely to result in availability of new niches for insect pests. Temperature has a strong influence on the viability and incubation period of *H. armigera* eggs. Egg incubation period can be predicted based on degree days required for egg hatching, which decreases with an increase in temperature from 10 to 27°C, and egg age from 0 to 3 days (Dhillon and Sharma, 2007).

Overwintering of insect pests will increase as a result of climate change, producing larger spring populations as a base for a build-up in numbers in the following season. These may be vulnerable to parasitoids and predators if the latter also overwinter more readily. Diamond black moth, *Plutella xylostella* overwintered in Alberta in 1994 (Dosdall, 1994), and if overwintering becomes common, the status of this insect as a pest in North America will increase dramatically. Many insects such as *Helicoverpa* spp. are migratory, and therefore, may be well adapted to exploit new opportunities by moving rapidly into new areas as a result of climate change (Sharma, 2005).

Relationships between insect pests and their natural enemies will change as a result of global warming, resulting in both increases and decreases in the status of individual pest species. The majority of insects are benign to agro-ecosystems, and there is considerable evidence to suggest that this is due to population control through interspecific interactions among insect pests and their natural enemies – pathogens, parasites and predators (Price, 1987).

Oriental armyworm, *Mythimna separata* populations increase during extended periods of drought (which is determined by natural enemies), followed by heavy rainfall because of the adverse effects of drought on the activity and abundance of the natural enemies of this pest (Sharma *et al.,* 2002). Temperature not only affects the rate of insect development, but also has a profound effect on fecundity and sex ratio of parasitoids (Dhillon and Sharma, 2008; 2009). Insects that spend important parts of their life histories in the soil may be more gradually affected by temperature changes than those that are above ground simply because soil provides an insulating medium that will tend to buffer temperature changes more than air.

There will be an increase in variability in insect damage as a result of climate change. Higher temperatures will make dry seasons drier, and conversely, may increase the amount and intensity of rainfall, making wet seasons wetter than at present. Natural plant products, entomopathogenic viruses, fungi, bacteria and nematodes and synthetic pesticides are highly sensitive to the environment. Increase in temperatures and UV radiation, and a decrease in relative humidity may render many of these control tactics to be less effective, and such effect will be more pronounced on natural plant products and the biopesticides (Isman, 1997). Rapid dissipation of insecticide residues due to increases in temperature and precipitation will require more frequent application of insecticides. Therefore, there is need to develop appropriate strategies for pest management that will be effective under situations of climate change and global warming in future.

ii. Bioacountic Ecology and entomogenic climate change in forest ecosystems

Forest ecosystems result from a dynamic balance of soil, plants, insects, animals and climate. This balance can be destabilized by outbreaks of tree-eating insects. These outbreaks, in turn, are sensitive to climate, which controls precipitation. Drought stresses trees, rendering them vulnerable to insect predation. The net result is increased deforestation driven by insects and modulated by climate (Dunn and Crutchfield, 2006). Classical example is the havoc done by bark beetles in causing rapid deforestation leading to increased fire risk and consequently resulting in climate change on the North American boreal forests. These bark beetles have remarkable bio acoustic abilities and they use a combination of chemical and acoustic signals to regulate aggression, attack on host trees, courtship, mating behaviour and population density. Thus bark beetle communication is largely multimodal and that both pheromone and mechanical signaling are interwoven. Dunn and Crutchfield (2006) have proved conclusively that current characterization of increasing insect populations instead of being merely symptomatic of global climate change point out some insects as significant generative agents of climate change. In other words, climate change in this instance is entomogenic.

iii. Behavioural patterns in aquatic insects

Insects are poikilothermic animals whose metabolism, rate and magnitude of growth, development and overall behavioural activities respond significantly to thermal change on a diel, seasonal, and annual basis (Ward and Stanford, 1982). Despite this sensitivity to temperature, most aquatic species can be found in aquatic habitats over a broad geographic area that includes a wide range of thermal regimes. Obviously, these aquatic insect species possess bioenergetic, developmental, and/ or behavioural mechanisms that enable conspecific populations to survive and reproduce in very different environmental conditions. The conspicuous behavioural impacts of climate change on aquatic insects are on adult flight and larval ventilation. For example, a species that normally swarms in the afternoon may find air temperatures at that time too warm for flight; this could require individuals to fly either earlier or later when temperature was cooler or, in extreme cases, flight might be inhibited completely. In contrast, flight periods (diel or seasonal) of some species that are often limited by cold temperatures (e.g., at night, in the early morning, or in early Spring) may expand as a result of warmer temperature. These changes in flight periodicity could affect many aspects of reproductive behaviour, including the temporal barriers that contribute to reproductive isolation of closely related, sympatric species. One potential effect of elevated water temperature on the behaviour of larval insects will be on ventilator behaviours and gas exchange. Unlike most substances, the solubility of gases (including oxygen) in water generally decreases as water temperature increases. Low oxygen concentrations in water do not affect air-breathing aquatic insects; but, they can be important to aquatic insects that acquire their oxygen directly from water. Larvae have been observed to exhibit a variety of behaviours in response to low oxygen concentrations. For example, some caddisfly larvae undulate their abdomens, some stonefly and damselfly larvae do "push ups," and some mayfly larvae beat their abdominal gills (Eriksen *et al.,* 1984). All these behaviours generate greater water movement over gas exchange surfaces on the body, but they are also energetically costly and may therefore actually increase oxygen demand.

Water temperature is crucial for aquatic insects and all life cycle changes respond to changes in temperature (e.g., Brittain, 1980; Ward and Stanford, 1982; Sweeney, 1984). This is particularly apparent in the egg stage, and in many mayflies there is a clear relationship between water temperature and the length of egg development, as well as distinct temperature limits for successful development (Elliott and Humpesch, 1980).

Many mayfly species, like for instance the widespread western Palaearctic species, *Baetis rhodani* display considerable life cycle plasticity, being able to change the number of generations per year in response to changes in temperature. The timing of mayfly emergence also frequently depends on temperature (Brittain, 1980) and earlier emergence is likely in many mayfly species in a warmer climate. Adult body size in many hemimetabolous insects depends largely on thermal conditions during development while adult size in mayflies is closely correlated with fecundity (Brittain, 1980; Elliott and Humpesch, 1980). Survival of the adult of aquatic insects is likely to be an extremely important factor regulating the size of aquatic insect populations. Yet very little is known about the impact of global warming/climate change causing mortality during the adult stage. Stress caused by high air temperatures may affect the ability of female insects to develop mature ovaries, potentially reducing population numbers particularly in mayfly, stonefly and caddisfly species inhabiting cool-water habitats. Hence, during this stage of their life cycle, a suitable riparian environment is potentially very important, especially in the context of global warming because riparian zones (stream-side vegetation) can provide a supply of food, sites for mating and completion of egg development and also refuge from predators.

Climate change scenarios predict rapid shifts in environmental conditions, not only air temperatures, but also in precipitation and runoff. This will change the frequency and magnitude of floods and droughts. Precipitation will also affect the extent and duration of ice cover in cold regions. Floods have a major structuring effect on aquatic communities, triggering species replacement and succession. In alpine streams in central southern Norway, an unusually high spring flood significantly decreased densities of the mayfly, *Baetis rohodani*, which was present as nymphs at that time. In contrast, densities of summer species, *Acentrella lapponica*, present in the egg stage down in the substrate, increased in the two years after the flood (Brittain, 2008). Floods are also important in terms of dispersal, opening up waterways and establishing connectivity between watercourse elements and permitting the establishment of new populations. Droughts and low flows will also impact mayfly communities. In regulated rivers for hydropower, flows are frequently reduced for all or part of the year. Such flow reductions favour genera characteristic of lentic habitats, such as *Baetis*, *Rhithrogena* and *Epeorus* (Brittain and Saltveit, 1989).

iv. Behavioural strategies of insect vectors of diseases

Nearly half of the world's population is infected by vector-borne diseases, resulting in high morbidity and mortality. Global climate change has impacted almost every aspect of infectious diseases, from their life cycles to humans' susceptibility to them, fostering their virulence and increased transmission. Weather affects vector population dynamics and disease transmission, with temperature and humidity considered key variables. Vector behaviour has many effects on transmission of vector-borne pathogens. The most notable impact of vector behaviour occurs during vector-host contact, which is the result of the vector searching, locating a host and feeding. This is influenced by elements of vector biology such as host preference (choice of one host over another) and distribution of vectors between hosts and dispersal. However, host behaviour also plays important roles. Modifications of host behaviour resulting from seasonal changes or vector density can be particularly important. Another consideration is individual variation in vector behaviour. Often, vectors are considered only at the population level, with most parameters reflecting averages. However, variation in vector behaviour may be of considerable importance and should not be overlooked at any cost.

Due to space constraints, behavioural diversities of insect vectors of diseases are briefly highlighted citing the impact of temperatures on the life stages, and the gonotrophic cycle of the malarial parasite and on the projection of increase/decrease in transmission windows of malaria in view of climate change, which aspects were excellently reviewed by Dhiman *et al.*, (2008). The distribution of different malaria vectors, diversity in geography and environmental conditions result in variable endemicity of malaria. It takes about 10 days for an egg to reach the adult stage of an anopheline mosquito at an optimum temperature of 28°C. At lower temperature, the duration gets prolonged while at increased temperature the duration is reduced. However, at more than 40°C, mortality occurs in adult mosquitoes. Reduction in the duration of gonotrophic cycle and in the extrinsic incubation period of malarial parasite are related to increasesd rate of transmission (Molineaux, 1988). These two entomological variables are especially sensitive to changes in environmental temperature, being reduced with increments in temperature (Martens *et al.,* 1999). Examination of the projection of rise in temperature vis-à-vis the distribution of malaria vectors reveals that increased temperature would affect the biology of *Anopheles culicifacies*, *Anopheles fluviatilis* and *Anopheles minimus* to a major extent. It is also possible that the adult anopheline mosquito species may find a micro niche for their survival e.g., *Anopheles stephensi* rests in tanks in India to overcome high temperatures. Vector species like *Anopheles minimus* and *A. fluviatilis* may also seek outdoor resting places in thick vegetation under a changed climatic scenario.

CONCLUSION

The wider implications of effects of future climate change on distribution patterns and survival of insects of biodiversity value should be borne in mind while planning management for effective conservation. For example, the synchrony in both space and time between insects and their food plants may become disrupted, so that insects no longer have easy access to critical resources, and in turn many plants may lose access to specific pollinators. Enforced adaptations to changed seasonal development may affect diapause regimes so that current 'temporal refuges' are no longer available or effective. Critical resources or critical sites may simply disappear, become unsuitable, or otherwise diminish in accessibility. In this context, optimal management can be planned emphasizing conservation measures along ecological gradients, and maintaining habitat heterogeneity to facilitate choices for local adaptations such as microclimate and host selections (New, 2009). Climate change will also increase the risk of infectious diseases through range extension of insect vector species, changing pathogen dynamics in environmental reservoirs and altering the pathogen transmission cycles. An understanding of the eco behavioural dynamics of disease transmitting insect vectors along with other aspects of control will go a long way to promote effective control strategies of insect vector species.

REFERENCES

Ananthakrishnan, T.N. 2007. Insects and climate EAI Base Paper No. 1. Entomology Academy of India, Madras Christian College, Chennai, India.

Bonam, G. 2002. Ecological Climatology: Concepts and Applications. Cambridge University Press, New York, USA.

Brittain, J.E. 2008. Mayflies, Biodiversity and Climate change, pp. 1-14. In: (eds.) Hauer, F.R., Stanford, J.A. and Newell, R.L. *International Advances in the Ecology, Zoogeography, and Systematics of Mayflies and Stoneflies.* University of California, Entomology, Vol. 128.

Brittain, J.E. 1980. The biology of mayflies. *Annual Review of Entomol.* 27: 119-147.

Brittain, J.E. and Saltveit, S.J. 1989. A review of the effect of river regulation on mayflies (Ephemeroptera). *Reg. Riv.: Res. Manage.* 3, 191–204.

Conrad, K.F., Woiwod, I.P., Parsons, M., Fox, R. and Warren, M.S. 2004. Long-term population trends in widespread British moths. *J. Insect Conserv.* 8: 119-136.

Debinski, D.M. and Cross, M.S. 2008. Conservation and Global Climate Change. Princeton University Press. http://press. princeton.edu/

Dhillon, M.K. and Sharma, H.C. 2007. Effect of storage temperature and duration on viability of eggs of *Helicoverpa armigera* (Lepidoptera: Noctuidae). *Bulletin of Entomological Research* 97: 55-59.

Dhillon, M.K. and Sharma, H.C. 2008. Temperature and *Helicoverpa armigera* food influence survival and development of the ichneumonid parasitoid, *Campoletis chlorideae. Indian Journal of Plant Protection* 36: 240-244.

Dhillon, M.K. and Sharma, H.C. 2009. Temperature influences the performance and effectiveness of field and laboratory strains of the ichneumonid parasitoid, *Campoletis chlorideae. BioControl* 54: 743-750.

Dhiman, R.C., Pahwa, S. and Dash, A.P. 2008. Climate change and malaria in India: Interplay between temperatures and mosquitoes. *Regional Health Forum,* Vol. 12 (1): 27-31.

Dosdall, L.M. 1994. Evidence for successful overwintering of diamondback moth, *Plutella xylostella* (L.) (Lepidoptera: Plutellidae), in Alberta. *Canadian Entomologist* 126: 183-185.

Dunn D.D. and J.P. Crutchfield. 2006. Insects, trees, and climate: The bio-acoustic ecology of deforestation and entomogenic climate change, santa Fe InstituteWorking Paper 06-12-055.

Elliott, J.M. and Humpesch, U.H. 1980. Eggs of Ephemeroptera. *Annual Report of the Freshwater Biological Association*, UK 48: 41-52.

Elphinstone, J. and Toth, I.K. 2008. *Erwinia chrysanthemi* (*Dikeya* spp.) - The facts. Oxford, U.K: Potato Council.

Eriksen, C.H., Resh, V.H., Balling, S.S. and Lamberti, G.A. 1984. Respiration in aquatic insects. In: (eds.) Merritt, R.W. and Cummins, K.W., An Introduction to the Aquatic Insects of North America. 2nd Ed. Kendall/Hunt Publ., Dubuque.

Guerenstein, P.G., Yepes, E.A., van Haren J, Williams, D.G. and Hildebrand, J.G. 2004. Floral CO_2 emission may indicate food abundance to nectar-feeding moths. *Naturwissenschaften*, 91: 329-333.

Hance, T., van Baaren, J., Vernon, P. and Boivin, G. 2007. Impact of extreme temperatures on parasitoids in a climate change perspective. *Annu. Rev. Entomol*. 52, 107-126.

Hegland, S.J., Nielsen, A., Lazaro, A., Bjerknes, A.L. and Totland, O. 2009. How does climate warming affect plant-pollinator interactions? *Ecol. Lett*. 12, 184-195.

Hunter, M.D. 2001. Effects of elevated atmospheric carbondioxide on insect-plant interactions. *Agricultural and Forest Entomology*, 3, 153-159.

IPCC 2001. Climate Change 2001: Synthesis Report. Third Assessment Report. Cambridge University Press, Cambridge, UK.

Isman, M.B. 1997. Neem and other botanical pesticides: Barriers to commercialization. *Phytoparasitica* 25: 339-344.

Kannan, R. and James, D.A. 2009. Effects of climate change on global diversity: a review of key literature. *Tropical Ecology* 50(1): 31-39.

Kent, K.S., Harrow, I.D., Quartararo, P., Hildebrand, J.G. 1986. An accessory olfactory pathway in Lepidoptera: the labial pit organ and its central projections in *Manduca sexta* and certain other sphinx moths and silk moths. *Cell and Tissue Research,* 245: 237-245.

Lamb, H.H. 1995. Climate, History and the Modern World. Routledge, London, UK.

Lewis, O.T. 2006. Climate change, species-area curves and the extinction crisis. *Phil. Trans. R. Soc*. B 361, 163-171.

Martens, P., Kovats, R.S., Nijhof, S., De Vries, P., Livermore, M.T.J., Bradley, D.J., Cox, J. and McMichael, A.J. 1999. Climate change and future populations at risk of malaria. *Glob Environ Change 9*: S89-S107.

Molineaux, L. 1988. The epidemiology of human malaria as an explanation of its distribution, including some implications for its control. In WH Wernsdorfer, SI McGregor (eds), *Malaria, Principles and Practice of Malariology*, Churchill Livingstone, London, p. 913-998.

Mondor, E.B., Tremblay, M.N., Awmack, C.S. and Lindroth, R.L. 2004. Divergent pheromone-mediated insect behaviour under global atmospheric change. *Global Change Biology* 10: 1820-1824.

Myers, J.H., Monro, J. and Murray, N. 1981. Egg clumping, host plant selection and population regulation in *Cactoblastis cactorum* (Lepidoptera). *Oecologia*, 51: 7-13.

New, T.R. 2009. Insect Species Conservation. Cambridge University Press, Cambridge, UK. 256 pp.

Parmesan, C. and Yohe, G. 2003. A globally coherent fingerprint of climate change impacts across natural sytems. *Nature* 421, 37-42.

Price, P.W. 1987. The role of natural enemies in insect populations. In: P. Barbosa and J.C. Schultz (Editors), *Insect Outbreaks*. San Diego, CA, USA: Academic Press. pp. 287-312.

Regniere, J. 2009. Predicting insect continental distributions from species physiology, Unasylva 231/232, Vol. 60.

Rubenstein, D.I. 1992. The Greenhouse Effect and Changes in Animal Behavior: Effects on Social Structure and Life-History Strategies. 180-192 pp.

Samways, M.J. 2007. Insect Conservation: A Synthetic Management Approach. *Annu. Rev. Entomol.* 52: 465-487.

Saseendran, S.A., Singh, K.K., Rathore, L.S. Singh, S.V. and Sinha, S.K. 2000. Effect of climate change on rice production in the tropical humid climate of Kerala, India. *Climatic Change*, 44 (4): 459-514.

Seeley, T.D. 1974. Atmospheric carbondioxide regulation in honeybee (*Apis mellifera*) colonies. *Journal of Insect Physiology*, 20: 2301-2305.

Sharma, H.C., Sullivan, D.J. and Bhatnagar, V.S. 2002. Population dynamics of the Oriental armyworm, *Mythimna separata* (Walker) (Lepidoptera: Noctuidae) in South-Central India. *Crop Protection* 21: 721-732.

Sharma, H.C. (ed.). 2005. Heliothis/Helicoverpa *Management: Emerging Trends and Strategies for Future Research*. New Delhi, India: Oxford & IBH, and Science Publishers, USA. 469 pp.

Stange, G. 1997. Effects of changes in atmospheric carbondioxide on the location of hosts by the moth, *Cactoblastis cactorum*. *Oecologia*, 110: 539-545.

Stenseth, N.C., Mysterud, A., Ottersen, G., Hurrell, J.W., Chan, K.S. and Lima, M. 2002. Ecological effects of climate fluctuations. *Science* 297, 1292-1296.

Stireman III, J.O. *et al.* 2005. Climatic unpredictability and parasitism of caterpillars: implications of global warming. *Proc. Natl. Acad. Sci. USA* 102, 17384-17387.

Subrahmanyam, B., Abd El Latif, A.O. and Kesava Kumar, H. 2009. Perspectives of Global Climate Change and Physiological Adaptations by Insects, pp. 140-162. In: (eds.) V.V. Ramamurthy, G.P. Gupta and S.N. Puri. *National Symposium on IPM Strategies to Combat Emerging Pests in the Current Scenario of Climate Change*, Central University, Pasighat, Arunachal Pradesh. Entomological Society of India, New Delhi.

Sweeney, B.W. 1984. Factors influencing life-history patterns of aquatic insects. P. 56-100. In: Resh, V.H. and Rosenberg, D.M. (eds.). The Ecology of Aquatic Insects. Praegar Publishers, New York, USA.

Travis, J.M.J. 2003. Climate change and habitat destruction: a deadly anthropogenic cocktail. *Proc. R. Soc. London* B 270: 467-473.

Tylianakis, J.M., Didham, R.K., Bascompte, J. and Wardle, D.A. 2008. Global change interactions in terrestrial ecosystems. *Ecol. Lett.* 11, 1351-1363.

Van der Putten, W.H., Macel, M. and Visser, M.E. 2010. Predicting species distribution and abundance responses to climate change: why it is essential to include biotic interactions across trophic levels. *Phil. Trans. R. Soc.* B 365, 2025-2034.

Visser, M.E. and Holleman, J.M. 2001. Warmer springs disrupt the synchrony of oak and winter moth phenology. *Proc. R. Soc. Lond.* B 268, 289-294.

Walther, G.R. 2011. Community and ecosystem responses to recent Climate change. *The Royal Society,* 1471-2970.

Walther, G.R., Post, E., Convey, P., Menzel, A., Parmesan, C., Beebee, T.J. C., Fromentin, J.M., Hoegh-Guldberg, O. and Bairlein, F. 2002. Ecological responses to recent climate change. *Nature* 416, 389-395.

Walther, R. 2004. Plants in a warmer world. *Perspect. Plant Ecol. Evol. Syst.* 6, 169-185.

Ward, J.V. and Stanford, J.A. 1982. Thermal responses in the evolutionary ecology of aquatic insects, Ann. Rev. Entomol., 21: 97-117.

Williams, S.E., Bolitho, E.E. and Fox, S. 2003. Climate change in Australian tropical rainforests: an impending environmental catastrophe. *Proceedings of the Royal Society of London Series B – Biological Sciences* 270, 1887-1892.

Dynamics of Insect Behaviour
T.N. Ananthakrishnan & K.G. Sivaramakrishnan
Scientific Publishers (India), Jodhpur, 2012, pp. 189-210

Chapter - 13

MULTIDIMENTIONAL BEHAVIOUR OF INSECT VECTORS OF PUBLIC HEALTH IMPORTANCE

B.K. Tyagi

INTRODUCTION

Arthropods, and particularly insects, by sheer virtue of their stupendous numerical supremacy and diverse adaptations, have often been rightfully judged as the mankind's most formidable competitors for survival on Earth (Goddard, 2000). They damage crops in field and go-downs, suck blood of humans and animals and cause annoyance in practically every part of the world. They are found in all zoogeographical regions, but more so in the tropical countries where a large number of insects and other arthropods, such as mosquitoes, flies, fleas, lice, mites and ticks infect humans and other animals directly or indirectly with the organisms of many dangerous and debilitating diseases that essentially impede national and individual development, make fertile land inhospitable, impair intellectual and physical growth, and demand huge costs in treatment and control programmes (Barraud, 1934; Christophers, 1933; Rao, 1984; Spielman, and D'Antonio, 2001; Tyagi, 1994). Vector-borne diseases are particularly rampant in tropical and subtropical countries (Table 1). List of these infections is pretty long, however, as far as India is concerned, malaria, dengue, filariasis, leishmaniasis and Japanese encephalitis are considered major public health problems. These diseases no longer confine to the developing nations, including India, where they were often credited in past for slowing the national economic advancement and human health development. Consequently, these insect- and other arthropod-borne infections, which were once considered diseases exclusively of the rural poor, have presently assumed enormous significance by becoming "diseases of development" associated with people's need, such as massive migration from rural to urban areas for improving economy, creation of new irrigational avenues, and mining projects attracting vast labour forces from areas known for endemic character of various different tropical diseases.

Table 1. Facts and figures about arthropod vectors and diseases they transmit in humans and animals

Human Disease	Pathogen(s) involved	Vector(s) involved	Vector incrim-ination	Country	Authority
Filariasis	*Wuchereria* and *Brugia* spp.	Mosquitoes	1886	China	Manson
Texas cattle fever	Babesia gemina	Tick	1891	N. America	Smith & Killbourne
Nagana disease in cattle		Tsetse flies	1895		Bruce
Malaria	*Plasmodium* spp.	Anopheline mosquitoes	1897	India	Ross
Avian malaria	*Plasmodium* ?	Mosquitoes	1898	India, UK	Ross
Plague	Bacterium (*Y. pestis*)	Fleas	1898		Simond
Human malaria cycle	*Plasmodium* sp.	Anopheline mosquito	1899	Italy	Grassi, Bignami and Bastianelli
Yellow fever	Togavirus	Mosquitoes (*Aedes aegypti*)	1990	Cuba	Finlay/Reed, Carroll, Lazear and Agramonte
Americal Sleeping sickness	*Trypanosoma* spp.	Tritomine bugs	1895 1908	S. America S. America	Bruce Chagas
African sleeping sickness	*Trypanosoma* spp.	Tsetse flies	1903	Africa	Bruce & Nabarrow
Dengue	Flavivirus	Mosquitoes	1906	Africa	Graham

Fowl spirochaetes	Spirochaetes (*Borrelica conserina*)	Tick (*Argus persicus*)	1903		Marchoux & Salimbeni
Rlelapsing fever	Lice		1907		Mackie
Onchocerciasis	*Onchocerca* spp.	Blackflies	1926	Sierra Leone	Blacklock
Encephalitides -WEE, EEE, SLE etc.	Virus/ Bacteria	Mosquitoes	1930-1990	Various countries	Various scientists
-Lyme borreliosis and Ehrlichiosis		Ticks			

MOSQUITO – THE MAN'S DEADLIEST FOE

From the public health viewpoint, mosquitoes (Family Culicidae, Order Diptera) have always attracted maximum attention due to their involvement in mediating transmission of some of the most deadly diseases. Mosquito-borne illnesses infect about 700 million people each year worldwide. Of all the predators that stalk the planet, mosquito has taken more lives than any other organism (Tyagi, 2004). Mosquitoes are vectors for many deadly and debilitating human diseases such as malaria, yellow fever, dengue, filariasis and various types of encephalitides – just to name a few important ones. Consequently, other than mosquito, no animal on earth has touched so directly and profoundly the lives of so many human beings with respect to their health, economy, social activities and tranquility. Each year some 500 million people are affected by mosquito-borne diseases and nearly 3 million die from these. With a tiny little bite this pest, with the skill of a phlebotomist, transmits more than a hundred diseases, from protozoan (*e.g.*, malaria) and helminthic (*e.g.*, lymphatic filariasis) to viral infections (dengue, yellow fever, several different kinds of encephalitides). There are over 3200 species of mosquitoes but only a small percentage of these phlebotomize and/or carry disease pathogens to humans.

A vector is defined as an arthropod or other animal that carries a parasite from one host to another host by allowing the parasite or pathogen to complete, in principle, a part of its life cycle in its body. However, at times a vector may not be essential for the completion of life cycle of the parasite and in such instances it is referred to as a mechanical vector. A pest can be defined within animal world as any insect or, for that matter, any animal which becomes a source of trouble or loss to humans. When the losses incurred are enormous or few, the causative insect is accordingly referred as a major or minor pest. Medically speaking, insects and other arthropods which feed on human blood and play a part in disease transmission are most important.

ECONOMIAL BEHAVIOUR OF VECTOR-BORNE DISEASES

Cumulatively the arthropod-borne diseases exert unfathomable economic losses to humans. These losses are not merely monetary but relate to health as well. Diseases like malaria, filariasis, dengue and Japanese encephalitis incur annual losses worth several trillions of US dollars all over the world, if only expressed in terms of Disability Adjusted Life Years (DALYs) (Table 2).

Table 2. Global public health importance of insect-borne diseases with their vectors and distribution (after Morner *et al.,* 2002)

Disease	Vector species	Burden of disease[1] estimated 2000 DALYs	Distribution
Malaria filariasis	*Anopheles* spp.	40,213,000	Tropics, main burden Sub-Saharan Africa
Lymphatic filariasis	*Culex* spp.	5,549,000	Tropical urban areas
Onchocersiasis	*Simulium* spp.	951,000	W.- Africa, C.- America
Leishmaniasis Old world New world	Sandflies (*Phlebotomus* spp.) (*Lutzomyia* spp.)	1,810,000	Patchy, Old World (semi)arid zones, New World humid forests
Trypanosomiasis African	Tsetseflies (*Glossina* spp.)	1,585,000	Patch, in W.- and southern Africa
American trypanosomiasis (Chagas' disease)	Triatomid bugs	680,000	S.- and C.- America linked to poor housing
Arboviral diseases			
Dengue	*Aedes* spp.	433,000	Urban tropics

Japanese encephalitis	*Culex* spp.	426,000	S. and SE Asia, linked to irrigated rice/pigs
Yellow fever	*Aedes aegypti*	na	Africa, South America

[1]Global estimate of Disability Adjusted Life Years (DALYs) in 2000.

In Sub-Saharan Africa, a major malaria infested region, only malaria-related annual economic burden in 1987 was to the tune of US$ 0.8 billion (0.6% of gross domestic product) which was expected to further hike to US$ 1.7 billion (1.0% of GDP) by 1995. In India losses due to only dengue is estimated to be about US $ 1.8 million per annum, which is a gross underestimate. Economic losses have a tendency to rise along with the national income and improved socioeconomic conditions. Also, there are a number of direct and indirect factors which are considerable when to calculate economics of malaria. Often, the related documentations are faulty and grossly incomplete. In spite of this, even if only the recorded facts are taken into consideration, the economic loss due to malaria in India comes somewhere between Rs. 15,777.85 and 29,934.85 million at the 1991 level.

MALARIA, THE *NUMERO UNO* VECTOR-BORNE DISEASE

This estimate obviously did not consider major areas like enormous losses in agriculture, industry, mining, railways and other important productive sectors in the country susceptible to malaria-related perils, in addition to indiscernible yet apparent losses to the physical and intellectual growth of the population living in endemic areas. Keeping in mind some of the factors like estimates of morbidity and mortality due to malaria expressed as DALYs, Sharma (1996) postulated an overall loss falling in the range of Rs. 16,192 – Rs. 20,192 million, provided prevalence figures of only 2 million malaria cases are considered (which, according to the WHO SEARO, is at least 6-7 times low than probably the actual incidence in the country!). The above information is a clear indication about the enormity of seriousness malaria disease needs to be dealt with in India. It is needless to re-emphasize that the management of various vector-borne diseases in India takes away more than half of the financial layout earmarked for the entire health sector!

Malaria offers perhaps the best example for determining the economic burden levied by any arthropod-borne disease in a country, since (i) among all the vector-borne diseases malaria continues to be one of the biggest contributors to disease burdens in terms of death and suffering, and from the public health point of view it is undoubtedly the most devastating parasitic disease in the tropics, and (ii) it is definitely the most extensively studied subject. By undermining the health and capacity to work of hundreds of millions of people, it is closely linked to poverty and contributes significantly to stunting social and economic development. Nearly four decades ago there were hopes that malaria could be eradicated in a fairly short time, mainly through the intensive use of insecticides. However, malaria still remains a serious threat in endemic countries. Those at greater risk of dying from the disease are children under age 5 in malaria-endemic areas, pregnant women, people moving from non-malarious to malarious zones for reasons of work, migration, refuge, war or tourism, and travelers who visit endemic countries and return home with the disease. In many parts of the world, malaria is becoming even a greater problem than it has been before. Epidemics are recurring in areas where transmission has been interrupted, and are generally associated with deteriorating social and economic conditions.

Throughout the world more than 40% of population is exposed to varying levels of risk for malaria in 91 countries. Of a total world population of approximately six billion, 1.7 billion (35%) live in areas where endemic malaria has been considerably reduced or even eradicated, but where transmission has begun again and in which the situation is now unstable or deteriorating. In some areas, severe malaria problems developed following major ecological or social changes, particularly the agricultural or the exploitation of forests or socio-political interests. Each year 300-500 million cases of malaria, 90% of them in Africa, and between 1.5 and 2.7 million deaths are reported worldwide. Severe malaria morbidity and mortality are linked with areas endemic for *Plasmodium falciparum*, the predominant species in Africa. According to a report malaria kills a child somewhere in the world every 30 seconds, whereas it has been estimated that 800,000 children die each year of malaria in Africa alone. Mortality is lower but nevertheless important in the Americas and Southeast Asia, especially where *P. falciparum* drug resistance is widespread. The Southeast Asia region embodying India is of particular interest to us as there occur approximately 3.54 million cases every year.

In India, before malaria control programme was launched in 1953, nearly 100 million people suffered from the disease annually and close to 0.1 million people died every year. The latter figure clearly appears to be a gross underestimate since during frequent epidemics so common in the country those days the overall number would have been much higher, as in 1931, the number of deaths indirectly or directly resulted from malaria had exceeded 2 millions. Besides, malaria affected the national as well as individual development as economic losses of the order of Rs. 2936 million per year were forced on India. Soon after the country-wide National Malaria Eradication Programme was launched in 1958 (in fact, the formal approval for the initiative was accorded by the Government of India as early as the late 1957), malaria cases were reduced to less than 0.1 million in 1965-66, without single death, and it appeared that malaria would be wiped off from India before very long! It is noteworthy here that decline in cases was in case of *P. vivax* and not in *P. falciparum*. Within the next ten years malaria epidemics were widespread and the disease, recording the highest ever number of malaria cases (6.47 million) by 1976, invaded even those areas which were earlier not known for malaria transmission or were once freed from its scourges. Subsequently, during 1980s and early 1990s, following a series of critical reviews of the malaria eradication programmes, the malaria incidence was brought down to less than 2 million, but soon increased to nearly 3 million in the mid-1990s, as reported by the National Malaria Eradication Programme (NMEP). Ironically, the malaria incidence and deaths reported by NMEP were based on the parasitologically proved cases, thus excluded a large number of asymptomatic carriers and/or those who approached other medicinal systems for cure of disease or even resorted to self-medication. As a consequence, both morbidity and mortality were apparently underestimated (NMEP, 1985). While the NMEP reported 2.5 to 3 million malaria cases and about 1000 malarial deaths annually, the WHO South East Asia Regional Office put the estimates to 15 million malaria cases (nearly 6 times) and 19,500 deaths due to malaria. Therefore, it appears safer to assume that the malaria incidence reported offers at most the disease trend and not its true incidence, knowledge of which, for logical reasons, is inevitably indispensable for effective implementation of malaria control in the country.

Malaria control in India is presently hampered mainly by two reasons: (i) Development of drug resistance in malarial parasites, and (ii) Development of insecticide resistance in malaria vectors. *Plasmodium vivax* and *P. falciparum* are dominant species in India, with the former responsible for 60-65% cases and the latter for 35-40% cases (Sharma, 1999). Among the six major vector species prevalent in India, *An. culicifacies*, the main transmitter of rural and peri-urban malaria in the peninsular India, is alone responsible for 65% malaria and 55% *P. falciparum* annually. This mosquito is resistant to DDT and HCH all over the country, and to malathion in Maharashtra, Gujarat and certain other States. *Anopheles stephensi* transmits nearly 12% malaria cases in the country, mostly in the urban and industrial areas. This species, too, has developed multiple resistance against various insecticides, many of which are however used as larvicides. Yet, another vector species, *An. fluviatilis,* which like above two species is common in Rajasthan, transmits 15% of total malaria cases and 30% *P. falciparum* in the country.

The strategy of malaria control in India has been to cause interruption in the disease transmission by spraying residual insecticides in the rural areas (351.8 million population) and by source reduction and larviciding in urban areas (62.1 million population). Because most of the primary malaria vectors in India are endophilic and/or endophagic in behaviour, indoor residual spraying of insecticides in the rural areas and anti-larval operations in the urban set-ups have been the major means to control malaria in the country. Led by widespread use of DDT in controlling vector population, now largely jeopardized owing to development of resistance in vector mosquitoes, other insecticides like HCH (later banned in 1997), malathion, and certain synthetic pyrethroids such as cyfluthrin, deltamethrin and lambdacyhalothrin have also been selectively employed in the control of malaria, although spray targets were rarely achieved due to logistic and operational constraints.

In recent times India has experienced major outbreaks of malaria, dengue, Japanese encephalitis and leishmaniasis, not to forget the once controlled flea-borne plague epidemics of 1994 (Maharashtra and southern Gujarat) and 2002 (Himachal Pradesh). Ironically, the Thar Desert which has until recently been widely acknowledged as an hypoendemic area for malaria with only sporadic outbreaks is presently transformed into a hotbed for *Plasmodium falciparum*-dominated malaria, particularly in the Indira Gandhi Nahar Pariyojana-irrigated north western desert areas. Likewise, dengue fever and tropical leishmaniasis, too, seem to establish itself in the Thar desert.

MOSQUITO'S IMPORTANCE IN MEDICAL ENTOMOLOGY

Insects playing a vectorial role for any disease pathogen are of two types. The first group includes blood sucking micro-predators which are organisms that take a little blood from many hosts. They do not stay on one host for their entire life (*e.g.*, mosquitoes, sand flies, tsetse flies, black flies, fleas, bed bugs etc.) (Fig. 1). In principle, they are characterized to find a blood capillary in the host's body, have specialized mouthparts to obtain fluids (blood or lymph), secrete anti-coagulants to prevent host blood from clotting, and many of such micro-predators secrete some anaesthetic to reduce pain. The second group comprises ectoparasites which live on the surface of the host, feeding by prolonged stay on that one host (*e.g.*, lice, ticks, mites etc.).

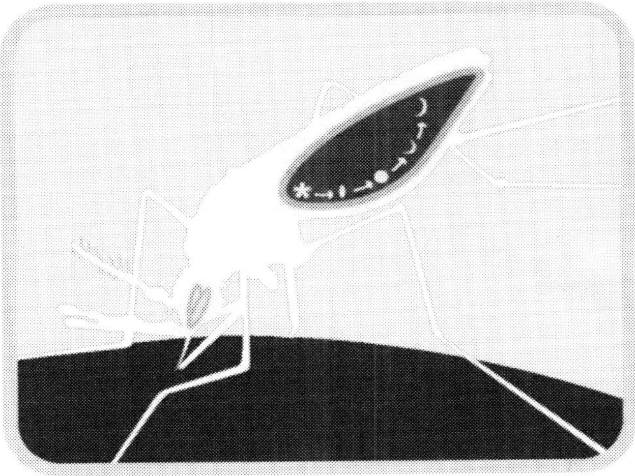

Figure 1. Diagrammatic presentation of a mosquito in the process of feeding, with particular exhibition of salivary glands and the engorged gut.

Insects and other arthropods derive further significance of great evolutionary importance by the fact that the disease organisms they naturally transmit are ordinarily helpless and inactive outside their hosts due to their inability to withstand exposure to light or dry air. The dependency of many pathogens on their insect hosts may be obligatory, and unless they can escape and find a sanctuary in another host, they are bound to perish with their victimized-host. The arthropod involved may be just an accidental carrier of the disease pathogen or an obligatory vector, one which helps the organism complete its development.

VECTOR-BORNE AND ZOONOTIC DISEASES

Arthropods, and particularly the insects, are responsible for transmitting a large variety of infections including protozoa (malaria and leishmaniasis), helminthes (filariasis, loaosis, onchocerciasis), bacteria (plague, flea-borne typhus, tularaemia etc.), and viruses (yellow fever, dengue, Japanese encephalitis). Malaria parasites are transmitted by certain mosquito species belonging only to the genus *Anopheles,* while leishmaniasis is spread by the sand flies. Mosquitoes belonging mainly to the genera *Culex* and *Mansonia* are incriminated for transmitting human lymphatic filariasis, whereas Japanese encephalitis is principally transmitted by certain *Culex* species only. The infamous Yellow Fever and dengue viruses are transmitted by *Aedes* mosquitoes. There is found a very strong and "specific" inter-relationship between vector species and those of the parasite/pathogen. However, as it is seen in the case of viruses and bacteria, the concerned parasite or pathogen may find more than one species or routes to complete its part of development in the vector or intermediate host's body.

Zoonosis is defined as a disease of animals that may be transmitted under natural conditions to man. Certain vector-borne diseases are, in fact, zoonoses for their effective accomplishment but humans are also not spared in the process of survival of the pathogen. Japanese encephalitis and the Khyasanur Forest Disease (KFD) are good examples of zoonoses caused by pathogens (viruses) having nonhuman animals as their natural host (e.g., pigs, goats and birds in case of Japanese encephalitis and the monkeys in case of KFD). Because they are not part of natural transmission cycle, humans are only incidentally infected. Zoonoses usually persist in nature in silent transmission cycles between vectors and nonhuman hosts, going undetected unless they spill over and infect the human population (e.g., plague, JE). In contrast, the anthropogenic vector-borne diseases, such as dengue fever and malaria, require no animal host and are transmitted from human to human by mosquito vectors, with the exception in Malaysia where the former disease is suspected to involve monkeys as zoonosis. In many of the advanced countries, although some of these disease-causing pathogens may now be rare, but due to constant presence of their mosquito vectors and the factor of extensive human mobility from hyper-endemic countries or regions the danger of re-entry of disease pathogens there is being constantly feared.

VECTOR PATHOLOGY AND ITS SIGNIFICANCE IN DISEASE TRANSMISSION

What happens after a parasite infests its invertebrate host, the vector – responsible for accomplishing the imperative other half of the development cycle of the parasite, as a necessity for the survival of the latter – is a billion dollar question, which has been baffling the scientists for decades! The subject of vector pathology is of immense interest, but also the one greatly neglected so far. This process unfailingly signifies the requirement of the parasite to complete its development within the intermediate host's body to reach the infective stage. It would generally appear true that the human pathogens requiring insects as their intermediate hosts would also adversely affect their 'health". This can be understood by the fact that the various pathogens, particularly; large-sized (*e.g.,* helminthes) penetrate the gut wall to cross over to haemoceal before journeying to the final destination of the salivary glands. Repeated perforation of the gut wall (*e.g.*, in case of mosquitoes) makes the insect weaker, on one hand, and more susceptible to future successful "cross-over journeys" through gut wall by the pathogen, imparting greater chances to the insect of becoming an efficient vector.

Flea offers a classical example as to how a serious pathogen of plague could be transmitted, by a simple "regurgitation" phenomenon. The aetiology of plague was first discovered during the third pandemic of the disease occurring in Hong Kong in 1894. After Alexandre Yersin had identified the causal agent *(Yersinia pestis)*, Paul-Louis Simond proved the flea's role as vector. These discoveries were of prime importance for the subsequent development of efficient means for fighting plague as well as for preventive and curative treatments and vaccines. Vaccination brought about a sharp decrease in plague mortality and morbidity. However, the disease has never been eradicated. It is still prevalent in various Asian, African and American countries and is among the re-emerging diseases at the present time. The genetic basis of transmission mechanisms and pathogenicity of the bacillus are only beginning to be understood. We now know that the attenuation of the EV76 strain used by Girard and Robic as an anti-plague vaccine in Madagascar is due to the spontaneous excision of a large chromosomal DNA fragment of 102 kb, a part of which contains a group of genes implicated in the pathogenicity and appropriately called high pathogenicity island. These mechanisms of flea - bacillus transmission are also beginning to be known. Two bacterial loci participating in the blocking of the ectoparasite's proventriculus have been identified. One is situated next to the high pathogenicity island on the unstable 102kb chromosomal fragment, the other on the large 95kb plasmid specific to *Y. pestis*. The molecular basis of the bacillus' acquisition of multi-resistance to antibiotics have likewise recently been characterized. However, although *Y. pestis* is one of the most pathogenic micro-organisms of the bacterial world, the mechanisms responsible for this high level of pathogenecity have still not been identified. This is well worth noting, since a certain number of genes acting as pathogenicity factors in other species are present but altered in *Y. pestis*. Plague still withholds many secrets.

Do evolutionary interactions between pathogen and its vector play any role in characterizing the pathogenicity of the vectoring insect. A close view on the malaria parasites' specificity to only anopheline mosquito will possibly help understand the phenomenon. Human malaria parasites are transmitted by a few dozen species of anopheline mosquitoes found throughout the tropics. Why is it, then, that more than 90% of the fatal cases of malaria (over two million deaths annually) are concentrated in tropical Africa alone or, in more exclamatory terms, more than 95% anophelines denying to transmit the malaria parasites? Although there is no one simple answer to the first querry, it is not coincidental that Africa is home to the two most significant malaria vectors, *Anopheles gambiae* and *An. funestus*. Despite the long-term prospect of vaccines and the promise of antimalarial drugs, the only current strategy for malaria control targets vector populations with indoor residual insecticide spraying and/or insecticide-treated bed nets. A future means of control may entail replacing vector populations with genetically engineered non-vector ones. For either of these vector-based strategies to be successful against counteracting evolutionary pressures (e.g., insecticide resist-ance), a detailed understanding of the genetic relationships and levels of gene flow within and between natural populations of these vector species is a prerequisite. Moreover, *An. gambiae* and *An. funestus* belong to cryptic species complexes, i.e., groups of morphologically indistinguishable and closely related species. The isomorphic components of these complexes nevertheless differ at both intraspecific and interspecific levels in vectorial capacity due to poorly understood behavioral and physiological differences. Uncovering the molecular basis for these differences may help us to understand the evolutionary dynamics of these species, and may offer new tools for malaria control. At a higher taxonomic level, it is essential to understand evolutionary relationships among members of the genus *Anopheles*, a group of mosquitoes distinguished from all others by their ability to vector human malaria parasites. Efforts are being made in this direction by using a variety of genetic markers such as the microsatellites, SNPs, maternal- (mitochondrial) and paternal- (Y chromosome) derived loci.

DOES CLIMATE PLAY ANY SIGNIFICANT ROLE IN FULMINATING VECTOR PREPONDERANCE AND DISEASE EPIDEMICS ?

Environment, particularly the climatic variability, plays an inevitably indispensable role in regulating the dynamics of a number of vector-borne diseases (Gubler *et al.,* 2002). Most vector-borne diseases exhibit a distinct seasonal pattern, which clearly suggest that they are weather sensitive. The significance of environment in shaping the biology of these diseases can be amply gauged from the fact that the entire gamut of development, behaviour and survival of arthropod vectors as well as the intensity of the diseases they transmit is strongly influenced by climatic factors. Temperature (in all seasons), rainfall, and humidity are especially important, but other factors such as wind and duration of daylight can also be significant. In particular, daily maximum and minimum temperatures affect the pathogen's rate of multiplication within the insect, which in turn affects the rate of salivary gland infection and hence the likelihood of successful transmission to the human host. If the development time of the pathogen exceeds the life span of the insect, transmission can not occur. Vector longevity is thus very important and can be shortened by elevated temperatures. The tremendous growth in international travel increases the risk of importation of vector-borne diseases, some of which can be transmitted locally under suitable circumstances at the right time of the year. But, demographic and sociologic factors, as well as land use, agricultural practices, residential patterns, human behaviour and vector preponderance also play a critical role in determining disease incidence. It is, in fact, a complex interplay of all these factors that determine the overall effect of climate on the dynamics of arthropod-borne diseases.

Table 3. Impact of weather and climate on vector- and rodent-borne infections

Environm-ental factor	Vector/pathogen/ vertebrate host	Impact
Temperature	Vector	Survival can increase or decrease depending on species.
		Some vectors have higher survival at higher latitudes and altitudes with higher temperatures.
		Changes in the susceptibility of vectors to some pathogens, e.g., higher temperatures reduce the size of some vectors but reduce the activity of others.
		Changes in the rate of vector population growth.
		Changes in feeding rate and host contact (which may alter the survival rate).
		Changes in the seasonality of populations.
	Pathogen	Decreased extrinsic incubation period of pathogen in vector at higher temperatures.
		Changes in the transmission season.
		Changes in distribution.
		Decreased viral replication.
Precipitation	Vector	Increased rain may increase larval habitat and vector population size by creating a new habitat.
		Excess rain or snowpack can eliminate habitat by flooding, thus decreasing the vector population size.
		Low rainfall can create habitat by causing rivers to dry into pools (e.g., dry season malaria).
		Decrease rain can increase container-breeding mosquitoes by forcing increased water storage.
		Epic rainfall events can synchronize vector host-seeking and virus transmission.

		Increased humidity increases vector survival; decreased humidity decreases vector survival.
	Pathogen	Few direct effects but some data on humidity affects on malarial parasite development in the anopheline mosquito host.
	Vertebrate host	Increased rain can increase vegetation, food availability, and population size.
		Increased rain can also cause flooding and decrease population size but increase contact with humans
		Decreased rain can eliminate food and force rodents into housing areas, increasing human contact, but it can also decrease population size.
Increased sea level	Vector	Altering estuary flow and changing salt marshes may affect drastically mosquito species by decreasing or eliminating selected mosquito breeding sites (e.g., reduced habitat for *Culiseta melanura*).

Vector-borne pathogens spend part of their life cycle in cold-blooded arthropods that are subject to many environmental factors. Changes in weather and climate that can affect transmission of vector-borne diseases include temperature, rainfall, wind, extreme flooding or drought, and sea level rise. Rodent-borne pathoges can be affected indirectly by ecologic determinants of food sources affecting rodent population size, and floods can displace and lead them to seek food and refuge. The relationship between ambient weather conditions and vector ecology is complicated by the natural tendency for insect vector to seek out the most suitable "microclimates" for their survival, e.g., resting under vegetation or pit latrines during dry or hot conditions or in culverts during cold conditions. Climate-related disasters may change the dynamics of human-mosquito contact; floods, may create conditions that allow mosquito proliferation and enhance mosquito-human contact (Table 3). The apparent increased instability of weather patterns in recent years, the new insights into recurring phenomenon, such as the *El Nino*, that influences world weather, and the global warming possibly owing to the greenhouse-gas accumulation in the troposphere, have re-emphasized to draw fresh attention on the actual and potential health consequences. The inter-relationship between climate, weather and health (in the present context it is to be understood as the state of well-being under the impact of vector-borne infections) is quite intricate. It is now realized the world over that climatic changes across time-scales influence ecological systems through direct and indirect events, in turn affecting disease conditions.

Climatic changes in time scales are of different kinds and intensities. The *El Nino* seasonal variations occupies a special place in the global weather formation, even though the ENSO (El Nino Southern Oscillations) is one of several natural variabilities. Its influence on global climate is second only to that of seasons, and it is predictable. An *El Nino* event is a major change in the global climate system associated with epidemic warming of the upper ocean layer in the eastern tropical Pacific Ocean, lasting for three or four months. The *El Nino* events are linked with rainfall and temperature extremes, and are a major cause of inter-annual climate variability. During the 1997 *El Nino* event a postulation was cast for increased risk of serious drought in India, Australia, Brazil and east Africa, on one hand, and heightened the danger of torrential rains and flood in American deserts from Peru to California, and of forest fire in Indonesia. Studies have already proved a strong correlation between outbreaks of malaria and the ENSO in the Thar Desert, in north western India (Bouma & van der Kaay, 1994, 1996; Bouma *et al.,* 1995, 1997; Tyagi, 1999). Even for certain other highly infectious diseases, like cholera, a definite understanding has now been developed over its epidemic occurrence and the rise in the local sea surface temperatures and growth of algal blooms. In similar fashion, therefore, knowledge is being strengthened on ascertaining perceptible correlation for eruptions of other equally devastating diseases like cryptosporidiosis, leptospirosis, giardiasis and Legionnaire's disease to the *El Nino* (warm phase) and *La Nina* (cold phase) situations, since the transmission of these diseases is affected by the quality and distribution mechanisms of drinking water, which in turn, is affected by rainfall patterns triggered by the *El Nino* patterns. In fine, therefore, it has become more than crystal clear that all the vector-borne diseases have a very strong correlation with climatic variability.

The possible pathway through which climate affects health, particularly the onset and/or intensities of vector-borne infections, vary widely. However, this highly complex web of events cannot be comprehended within a single-discipline framework and, perhaps to understand relationship of climatic variability with that of any weather- based discipline as well as dynamics of any vector-borne disease, the cross-pathways of inter-disciplinary or inter-events cohesiveness have to be essentially traced out and modeled up. One major point of discourse, notwithstanding, is that of all those health impacts working under climatic variability, the climate *per se* is most certainly not the only factor responsible for bringing about any catastrophic disease pandemic. In fact to build up a strong relationship between climate and health, it is inevitable to also consider a multitude of social, political and cultural factors. Economically and politically unstable populations are most vulnerable under extremes of climatic variabilities due to their inability to adapt or respond adequately and rapidly to counteract the devastating outcomes such as those during malaria and dengue epidemics, to example a few of the most virulent vector-borne diseases associated with high morbidity and mortality. According to a recent study, based on projected population and climate change worldwide, nearly 5-6 billion people (50-60% of the projected world population) will be at risk of dengue fever transmission compared with 3-5 billion people (35% of the population) if climate change did not happen. Based on a model developed by them, they have suggested that excessive resource consumption in rich countries should be addressed to avoid this increase in the generation of dengue fever.

Even though there is no panacea available to completely revert the consequences of the climatic variability such as the onset of vector-borne diseases like malaria, dengue, Japanese encephalitis and leishmaniasis, all of which prevail in India, nevertheless a complete scientific understanding of weather events and development of appropriate early warning tools may certainly still do a good help and save several lives. However, the use of climate forecast information in public health arena not only requires greater scientific understanding of climate-health symptoms, but also a paradigmatic shift in thinking from response to preparedness and mitigation. Therefore, to offer useful insight into causality and ultimately to integrate climate tools into public health policy and decision making, research must be strongly interdisciplinary amongst at least, so to say, climatology, disease epidemilogy, medical entomology, sociology and geography, but not to forget in the least the administration and political support for the cause in reference. On any level, be it regional (like that of the Thar Desert), national or international, a concerted research effort involving various different agencies, organizations and the private sectors is a must.

Some important vector-borne diseases like malaria, dengue and certain encephalitides are best demonstrable to have a working interaction with environment.

(1) Dengue fever

There are clear indications that an increase in ambient temperature in rather cooler areas, such as the outer lower scarps of Himalayas, has resulted in elevating the altitudenal distribution of the dengue vector, *Aedes aegypti*. Similarly, denudation of forest cover, comprising mainly rubber plantation harbouring another important dengue vector, *Ae. albopictus*, in the Western Ghat region in Kerala State has also resulted in an obvious increase in the average annual temperature and, consequently, the increased vector preponderance and distribution (Hiriyan & Tyagi, 2004; Hiriyan *et al.,* 2003; Tyagi *et al.,* 2003). As to the dengue virus, there is a proof of its proliferation in *Ae. aegypti* mosquitoes in the laboratory when temperature was increased. Appropriate models developed to prove this hypothesis suggest that relatively small increases in temperature in temperate regions, given viral introduction into a susceptible human population, should increase the potential for epidemics. However, epidemic potential depends on the interaction of many parameters including virus, host and environmental factors. To predict actual risk of human cases it is important to incorporate site-specific human, viral and environmental determinants of transmission (Gubler & Kuno, 1997; Guzman & Kouri, 2002).

(2) Malaria

More than anything temperature and rainfall, the former factor in particular, affect the bionomics including disease transmission potential of the vector species and the epidemiology of malarial disease. Most vector species can not survive temperatures above 40°C and below 8°C. While the temperature threshold for the development of *P. falciparum* in its anopheline hosts is 18°C, for *P. vivax* it is 15°C. The incidence and prevalence of malaria may be limited by altitude, a good proxy for temperature, but malaria transmission depends on many factors besides temperature. In the Thar Desert of western India, which until recently remained virtually a terra incognito for high intensity malaria, particularly *P. falciparum*-dominated, malaria epidemics were associated with the catastrophic transformation in the xeric physiography brought about by extensive canalized irrigation (Tyagi, 2002 Tyagi & Chaudhary, 1997). Although malaria is seasonal, there is substantial interannual heterogeneity of malarial incidence around the globe. Extremes of rainfall (both drought and floods) associated with El Nino events are linked to variability in malarial incidence in different regions (Bouma & van der Kaay, 1994; Bouma et al., 1997). Even though, there are clearly two different schools of thoughts on whether or not malaria endemicity will change in already high-risk areas, it is possible that small increases in minimum temperature in cooler regions may disproportionately increase malaria transmission, provided there are no socio-economic barriers to transmission.

In temperate countries, which were either never affected by malaria or where the disease once occurred but subsequently eradicated, malaria may well be there knocking on the door if climate conditions continue to warm This situation may be exacerbated by the ever increasing inflow of infected travelers and/or vectors, particularly the drug resistant parasites, from the endemic countries.

While in United States, increased frequency of local transmission as a result of increased imported disease due to migrant workers and immigrant populations living in substandard housing has been observed in the past decade, other non-endemic countries have also reported malaria recently. After 4 years of being malaria free, transmission of *P. vivax* malaria by local vectors has been reported in Italy. In Britain, indigenous cases have been recently detected, while in France, about 5,000 imported cases are reported every year, with seldom local transmission of the disease. However, a major resurgence of epidemic malaria has occurred in the southern republics of the former Soviet Union because of the breakdown of public health services. In India, cross-border infiltration of refugees from across the neigbouring countries like Mian Mar and Bangladesh have helped resurge the malarial disease with possibly more drug-resistant cases sneaked in the country.

(3) Japanese encephalitis and other encephalitides

Vectors of Japanese encephalitis, belonging mainly to *Culex vishnui* group (e.g., *Cx. tritaniorhynchus, Cx. pseudovishnui* etc.)

have a strong relationship to rainfall and particular agricultural practices pertaining to paddy cultivation, as observed in several south-east Asian countries including India. Warming of climate as well as precipitation both have been associated with outbreaks of SLE, transmitted by *Cx. tarsalis* and *Cx. nigripalpus*, in North America, where in certain parts it appears to be associated with the sequence of warm, wet winters, cold spings and hot dry summers. Mere 3-5 oC increase in average temperature may cause a northern shift in the distribution of both WEE and SLE viruses and a decreased range of WEE in southern California.

(4) Plague

Temperature and precipitation affect abundance of both fleas and rodents. The 1994 plague epidemic in western India followed a disastrous earthquake, implying that varied climatic events strongly affect rodent population dynamics, largely through effects on food availability. Flea-borne plague incidence has been found to rise in conjunction with increasing rodent populations following unseasonal winter/spring precipitation in United America.

(5) Tick-borne infections

Ticks transmit several infections some of which are of course affected by ambient situations of climate. Human contact with Kyasanur Forest Disease (KFD) in India was understood to be a result of man's deforestation-related activities where ticks like *Haemaphysalis* spp. and others were originally infecting monkeys but, with the human activities in the forest, accidentally inclined to infect the intruding humans. This implies that ticks and their mammalian hosts are influenced by land use, land cover, soil type, elevation and weather conditions.

REFERENCES

Barraud, 1934. *The fauna of British India. Diptera* III. *Family Culicidae,* IV. Taylor & Francis, London.

Bouma, M. J. and H.J. van der Kaay 1994. Epidemic malaria in India and El Nino southern oscillation. *The Lancet* 344: 1638-1639.

Bouma, M.J and H.J. van der Kaay 1996. The El Nino Southern Oscillation and the historic malaria epidemics in the Indian sub-Continent and Sri Lanka: An Early warning system for future epidemics. *Tropical Medicine & International Health* 1:86-96.

Bouma, M.J., H.E. Sondorp and H.J. van der Kay 1994. Health and climate change. *The Lancet* 343: 302.

Bouma, M.J., G. Poveda, W. Rojas, D. Chavasse, M. Quinones, J. Cox and J. Patz 1997. Predicting high risk years for malaria in Colombia using parameters of El Nino Southern Oscillations. *Tropical Medicine & International Health* 2(12): 1122-1127.

Christophers, S.R. 1933. *The fauna of British India. Diptera III. Family Culicidae, Tribe Anophelini.* Taylor and Francis, London, pp. 371.

Goddard, J. 2000. *A physician's guide to arthropods of medical importance.* CRC Press, London, pp. 422.

Gubler, D.J. and Kuno, G. 1997. *Dengue and dengue haemorrhagic fever.* CABI Publishing, New York.

Gubler, D.J., P. Reiter, K.L. Ebi, W. Yap, R. Nasci and J.A. Patz 2002. Climatic variability and change in the United States: Potential impacts on vector- and rodent-borne diseases. *Environment Health Perspectives* 109: 223-233.

Guzman, M.G. and G. Kouri 2002. Dengue: an update. *The Lance Infectious Disease* 2:33-39.

Hiriyan, J. and Tyagi, B.K. 2004. Cocoa pod (*Theobroma cacao*) – a potential breeding habitat of *Aedes albopictus* (Skuse) in dengue sensitive Kerala State, India. *J. Am. Mosquito Cont. Assoc.*: 20(3): 323-325.

Hiriyan, J., S.C. Tewai and B.K. Tyagi 2003. Plastic cups inveigling *Aedes albopictus* (Skuse) breeding around tea vendor spots in Ernakulam city, Kerala State. *Dengue Bulletin* 27: 195.

Rao, T.R. 1984. *The anophelines of India.* Malaria Research Centre, New Delhi. pp. 594.

Spielman, and D'Antonio 2001. *Mosquito: The man's deadliest foe.* Faber & Faber, London.

Tyagi, B.K. 1994. Alternative control strategies for mosquito vectors of human disases in India, pp. 359-372 (In: *Perspectives in Entomological Research*) Scientific Publishers, Jodhpur (India), pp. 406.

Tyagi, B.K. 1999. Vector-borne diseases of the Thar Desert. *Abstr. Pap. Man in Desert – Medical Problems. CME – 99,* Military Hospital, Jodhpur, p. 3.

Tyagi, B.K. 2002. Malaria in the Thar Desert: Facts, figures and future. Agrobios (India), Jodhpur. pp. 165.

Tyagi, B.K. 2003. *Medical entomology: A handbook of medically important insects and other arthropods.* Scientific Publishers (India), Jodhpur, pp. 265.

Tyagi, B.K. 2004. *The Invincible Deadly Mosquitoes.* Scientific Publishers (India), Jodhpur, pp. 265.

Tyagi, B.K. and Chaudhary, R.C. 1997. Outbreak of falciparum malaria in the Thar Desert (India), with particular emphasis on physiographic changes brought about by extensive canalization and their impact on vector density and dissemination. *Journal of Arid Environments* 36: 541.

Tyagi, B.K., Hiriyan, J. and Tewari, S.C. 2003. Dengue in India: a review, with special emphasis on the climate and environment variabilities. *Paper Abstract, First International Seminar of Medical Entomology,* Bhopal.

Dynamics of Insect Behaviour
T.N. Ananthakrishnan & K.G. Sivaramakrishnan
Scientific Publishers (India), Jodhpur, 2012, pp. 211-210

OVERVIEW

The ability of insects to modify, maintain and create habitats has been critically assessed and adequately justified by the contribution of several authors on diverse groups of insects. Both in ecological and evolutionary time, insects have been known to colonize large plants, those with a greater variety of resource types supporting more species. For an assessment of the dynamics of communities and constituent species, information presented on diverse aspects such as pollination, gall formation, predation and parasitism, forest and agricultural pests, besides the dynamics of aquatic insects, has been adequately discussed, leading to a better appreciation of the functioning of diverse communities. Curious ecological phenomenon like the massive infestation of the rubber beetle is an aspect deserving interest, curiosity and concern needing public awareness.

The need for an appreciation of the differentiation processes controlling growth and development of plant galls has been adequately emphasized, these galls representing highly regulated growth manifestations of plants, ensuring nutrition and shelter for the concerned insects. Specialized gall inducing behavior is host specific enabling efficient utilization of food resources, the concerned insects tending to adapt its behavior to plant reaction.

Ecological services tend to create a new range of habitat niches which can be exploited by a wide variety of insects. Organized complexity as evident in ants, which communicate with one another coordinating overall colony behavior is of interest. The knack for engineering and social coordination is unique in ants and notably worker ants are known to switch from nest building behavior to foraging. Being major constituent of terrestrial communities, they are ecologically and behaviorally different. Their role in regulating litter dynamics and soil organic matter are well known. Besides, the role of dung rolling beetles exhibiting diverse dung processing strategies, very beneficial to the soil, inducing organic matter, enabling increased soil aeration and percolation.

Besides terrestrial insects, critical assessment has been made of the behavioral diversity of aquatic insects, with larval behavioral trends as template, initiating several different kinds of ecobehavioral traits in adults. Behavioral strategies such as emergence, mating, swarming and oviposition are geared for maximum survival, protection from enemies and species propagation.

In short, the intricacies of insect-plant interactions involve ecological, biochemical and behavioral intricacies. In some insects with multiple species performing a function, a decrease in the density of one species may be compensated by an increase in the density of another, with no loss of ecosystem function.

T.N. ANANTHAKRISHNAN

Inside Indian Schools

Some Books from Social Science Press and Aurum Books

Public Policymaking in a Globalized World (Revised Edition) HB
Edited by ROBIN LEWIS

Society and Culture in India
Edited by SUBAS MOHAPATRA

Labour State and Society in Rural India: A Class-Relational Approach (HB)
JONATHAN PATTENDEN

Much Ado Over Coffee: Indian Coffee House Then and Now
BHASWATI BHATTACHARYA

Gitanjali Reborn: William Radice's Writings on Rabindranath Tagore
Edited by MARTIN KÄMPCHEN

Gandhi: Anti-biography of a Great Soul (HB)
MICHAEL DE SAINT-CHERON

Shades of Difference: Selected Writings of Rabindranath Tagore with Audio Visual Material, *Tagore and His World* (HB)
Edited by RADHA CHAKRAVARTY

Calcutta: The Stormy Decades (HB)
Edited by TANIKA SARKAR AND SEKHAR BANDYOPADHYAY

Dalit Women: Honour and Patriarchy in South India (HB)
CLARINDA STILL

Tamil Brahmans: The Creation of a Middle-Class Caste (HB)
CJ FULLER AND HARIPRIYA NARASIMHAN

Persistence of Poverty in India (HB)
Edited by NANDINI GOOPTU AND JONATHAN PARRY

Savage Attack: Tribal Insurgency in India (HB)
Edited by CRISPIN BATES AND ALPA SHAH

The Year of Blood: Essays on the Revolt of 1857 (HB)
RUDRANGSHU MUKHERJEE

The Autobiography of a Revolutionary in British India (HB)
KALI GHOSH

The Many Worlds of Sarala Devi: A Diary
Translated by SUKHENDU RAY and
The Tagores and Sartorial Styles: A Photo Essay (HB)
MALAVIKA KARLEKAR

Viramma: Life of a Dalit (PB)
VIRAMMA, JOSIANE RACINE, AND JEAN-LUC RACINE

Everyday Nationalism: Women of the Hindu Right (HB)
KALYANI DEVAKI MENON

Marriage, Love, Caste & Kinship Support: Lived Experiences of the Urban Poor in India (HB)
SHALINI GROVER

Religious Division and Social Conflict: The Emergence of Hindu Nationalism in Rural India (PB)
PEGGY FROERER

Literature and Nationalist Ideology: Writing Histories of Indian Languages (PB)
HANS HARDER

The Everyday State and Society in Modern India (PB)
Edited by CJ FULLER AND VERONIQUE BENEI

Economics and Its Stories (HB)
AMAL SANYAL

Telecommunications Industry in India: State Business and Labour in a Global Economy (HB)
DILIP SUBRAMANIAM

Cultural History of Modern India (PB)
Edited by MEENAKSHI KHANNA

Income Poverty and Beyond: Human Development in India
Edited by RAJA J CHELLIAH AND R SUDARSHAN

The Politics of the Pharmaceutical Industry and Access to Medicines: World Pharmacy and India (HB)
Edited by HANS LOGREN

SERIES: EVERYDAY LIFE OF POLITICAL STRUGGLES

Windows into a Revolution: Ethnographies of Maoism in India and Nepal (HB)
Edited by ALPA SHAH AND JUDITH PETTIGREW

Rebels from the Mud Houses: Dalits and the Making of the Maoist Revolution in Bihar (HB)
GEORGE KUNNATH

Unruly Hills: Nature and Nation in India's North-East (HB)
BENGT G. KARLSSON

SERIES: GERMAN WRITINGS ON INDIA AND SOUTH ASIA

'When the war began we heard of several kings': South Asian Prisoners in World War I Germany (HB)
Edited by FRANSIZKA ROY, HEIKE LIEBAU AND RAVI AHUJA

Cultural Encounters in India: The Local Co-Workers of the Tranquebar Mission, 18th-19th Centuries (HB)
HEIKE LIEBAU

'Everwhere is Becoming the Same'? Regulating IT-work between India and Germany (HB)
NICOLE MAYER-AHUJA

SERIES: FRENCH WRITINGS ON INDIA AND SOUTH ASIA

Scholars and Prophets: Sociology of India from France in the 19th and 20th Centuries (HB)
ROLAND LARDINOIS

Tocqueville in India (HB)
JEAN ALPHONSE BERNARD

AURUM BOOKS

Sabotage (PB)
ANITA AGNIHOTRI
Translated by ARUNAVA SINHA

Opium Poppy (PB)
HUBERT HADDAD
Translated by RENUKA GEORGE

Inside Indian Schools
The Enigma of Equity and Quality

Vimala Ramachandran

Published by
Esha Béteille
Social Science Press
69 Jor Bagh, New Delhi 110003

Social Science Press gratefully acknowledges the generous grant given by Azim Premji Foundation towards the publication of this book.

ISBN 978-93-83166-25-1

Set in Plantin Std
Typeset by Manmohan Kumar, Delhi 110035
Printed by Vee Enn Print-O-Pac
A-5, Flatted Factory Complex, Okhla Industrial Estate,
New Delhi 110 020

This book is dedicated to

Anita Kaul (1954–2016), a dear friend and co-traveller in my education journey. An administrator from the Karnataka cadre of the IAS, she was a steady and silent force behind the Total Literacy Campaign of 1989, the Nali Kali (Activity Based Learning Programme in Primary Schools) of Karnataka in 1996, the National Curriculum Framework of NCERT 2005, and the Right to Education Act of 2009. Her love for children and her faith in teachers was legendry – she believed 'when you are in doubt, when you are apprehensive about initiating change, go to the schools and spend time with teachers and children... then all your doubts will melt away'.... Her discussions and debates and frank criticism enriched my understanding.

Contents

List of Tables

List of Figures

Acronyms

ANM – Auxiliary Nurse Midwife
APPEP – Andhra Pradesh Primary Education Programme
AS – Alternative School(s)
ASAR – Age Specific Attendance Ratio
ASER – Annual Status of Education Report
BC – Backward Caste
BPL – Below Poverty Level
BRC – Block Resource Centre
BSS – Balika Shikshan Shivir
CBSE – Central Board of Secondary Education
CCE – Continuous and Comprehensive Evaluation
CRC – cluster resource centre
CSSM – Child Survival and Safe Motherhood
CWSN – children with special needs
DFID – British Department for International Development
DI – District Inspector
DIETs – District Institutes of Education and Training
DISE – District Information System for Education
DM – District Magistrate
DPEP – District Primary Education Programme
DPSC – District Primary School Council
DSE – Directorate of School Education
EBB – Educational Backward Blocks
EGS – Education Guarantee Scheme
EVS – Environmental Studies

FC	–	Forward caste
FGD	–	focus group discussion
FP	–	Family Planning
GER	–	Gross Enrolment Ratio
GOI	–	Government of India
GPI	–	Gender Parity Index
GPS	–	Government Primary School(s)
ICDS	–	Integrated Child Development Services
ICSE	–	Indian Certificate of Secondary Education
IEC	–	Information, Education and Communication
IED	–	Integrated Education
IFA	–	Iron/Folic Acid
IUD	–	Intrauterine Device
IIM	–	Indian Institute of Management
ITI	–	Industrial Training Institute
JRM	–	Joint Review Mission
KGBV	–	Kasturba Gandhi Balika Vidhyalaya
KSSP	–	Kerala Sastra Sahitya Parishad
LSA	–	Lok Sampark Abhiyan
MCH	–	Maternal and Child Health
MDM	–	midday meal
MHRD	–	Ministry of Human Resource Development
MoTA	–	Ministry of Tribal Affairs
MPCE	–	Monthly Per Capita Expenditure
MS	–	Mahila Samakhya
MSK	–	Mahila Shikshan Kendra
MTA	–	Mother-Teacher Association
NAR	–	Net Attendance Ratio
NCF	–	National Curriculum Framework
NER	–	Net Enrolment Ratio
NFHS	–	National Family Health Survey
NIEPA	–	National Institute of Educational Planning and Administration
NPE	–	National Policy on Education
NRBC	–	Non-residential Bridge Course
NSS	–	National Sample Survey
NSSO	–	National Sample Survey Organisation

NUEPA	–	National University of Educational Planning and Administration
OBC	–	Other Backwards Caste
ODA	–	Official Development Assistance
OOSC	–	out-of-school children
PIRLS	–	Progress in Reading Literacy Study
PRIs	–	Panchayati Raj Institutions
PTR	–	pupil-teacher ratio
PVTG	–	Particularly Vulnerable Tribal Groups
RCH	–	Reproductive and Child Health
RMSA	–	Rashtriya Madhyamik Shiksha Abhiyan/Secondary Education Mission
RPM	–	Review-Planning-Monitoring
RTE	–	Right to Education Act
SC	–	Scheduled Caste/Dalit
SDMC	–	School Development and Monitoring Committee
SHG	–	self-help groups
SI	–	School Inspector
SIDA	–	Swedish International Development Agency
SKP	–	Shiksha Karmi Project
SKS	–	Shiksha Karmi Sahyogi
SKs	–	Shiksha Karmis
SMC	–	school management committee
SSA	–	Sarva Shiksha Abhiyan
SSE	–	Statistics of School Education
ST	–	Scheduled Tribe/Adivasi
SWRC	–	Social Work and Research Centre
TBA	–	traditional birth attendant
TIMSS	–	Trends in International Maths and Science Study
TLM	–	teaching-learning materials
TPR	–	Teacher-pupil ratio
TSG	–	Technical Support Group
VEC	–	Village Education Committee
VECC	–	Village Education Construction Committee
WBBPE	–	West Bengal Board of Primary Education

Preface and Acknowledgements

This compilation of essays was written at different points of time, starting in 2011 when I was awarded a National Fellowship at the National University for Educational Planning and Administration (NUEPA) in New Delhi. Drawing upon over two decades of my research, I have tried to explore the contours of equity and quality issues in India's educational system. Starting with an overarching statement on the relationship between equity and quality, I then explore different dimensions of the equity-quality divide. I have acknowledged the contributions of researchers who worked with me at different points of time – not only to acknowledge their contributions, but also to beat a path of ethical research whereby all contributors, no matter how young they are or in which capacity they assisted me, are given their due.

I do not have a formal degree in education. I developed a passion for it, and especially for girls' and women's education, when I was asked to be part of a team to conceptualize and design a women's education programme for the Government of India in 1987. Along with Srilatha Batliwala, I travelled across the country trying to understand the educational status of women. I met and interacted with the academic community engaged in research and with the grassroots organizations working with women. Later, in 1988, I joined the government as the National Programme Director of Mahila Samakhya (MS) – education for women's equality. I was an active participant in planning the Bihar Education Project of 1991 and the Lok Jumbish Project in 1992. When the government planned the District Primary Education Project (DPEP) in 1993, I was part of the group developing the project design. Later, in 2004, I was actively involved in designing and launching the Kasturba Gandhi Balika Vidhyalaya (KGBV) scheme. Involvement in mainstream government elementary education programmes provided me with the opportunity to travel across the country, and to engage intensively with schools, teachers, and administrators. Needless to say, launching and managing the MS programme involved intensive interactions with rural communities, especially Dalit and Tribal women and girls. I got an opportunity to visit government schools in many districts. This opened my eyes to the dismal situation of our schooling system and the helplessness of families that have no option but to send their children to dysfunctional government schools. I also noticed that when rural women earned a little surplus (above their survival expenses), the first decision they took was to pull their children out of government schools and send them to private ones. I became painfully aware of the dire predicament of the very poor who are left with little choice.[1] When the MS programme started residential programmes for out-of-school adolescent girls (Mahila Shikshan Kendra, MSK), women brought their daughters and cajoled the organizers to admit the children. This was the situation across the country. In 2003, I was asked to evaluate the Balika Shikshan Shivir (Residential Girls Education Programme, BSS) programme of Lok Jumbish, Rajasthan (Ramachandran, Pal, and Mahajan 2003). This seven-month programme for rural out-of-school girls was an amazing experience: poor Dalit and tribal families were eager to enrol their daughters, exploding the myth of a lack of interest in girls' education in rural Rajasthan.

There is little doubt that MSK and BSS made a difference in the lives of the girls who had the opportunity to participate and also furthered the cause of girls' education in general. At that point, I became a firm advocate of fixed-duration residential education camps for girls that could help break the inter-generational cycle of

[1] In November 1995, I was invited by the Government of Andhra Pradesh to document the experience of the Podupulakshmi (self-help) groups of Anantapur and Nellore District. It was here that I saw first-hand that the first major economic decision they took was to send their children to private schools (Ramachandran 1996).

illiteracy and the poor educational levels among the most deprived communities. Ensuring girls study up to the age of 16 or even 18 is the surest way to increase the age of marriage, improve their self-esteem and confidence, and promote role models in society. I argued that this model would be of immense value in remote and sparsely populated regions, and among the poorest of the poor in the state. Educating adolescent girls may perhaps be the only way to put an end to the seemingly never-ending cycle of illiteracy, low age of marriage, and low status of women.

Based on the experience of MSK and BSS, the Government of India introduced the Kasturba Gandhi Balika Vidhyalaya (KGBV) programme in 2004. I was closely involved in communicating the programme's design through workshops across the country. In 2008 and again in 2013, I was asked to lead the national evaluation of the KGBV programme. In the 24 states that we covered, we saw that the programme had the potential to respond to the educational needs of out-of-school girls in the 11+ age group – especially those from socially and economically disadvantaged groups. In particular, the programme was of immense value in areas where girls drop out after primary school because of distance or difficult terrain. The ability of states to reach out to girls who have dropped out of school depends a great deal on the proactive interest taken by the state government. Where the leadership at the state level has not given the scheme the priority it merits, implementation down the line has also not adhered to the schematic pattern. Conversely, where the state government leadership has accorded priority to the scheme and appreciated the spirit behind it, implementation has been good and it has been able to reach out-of-school girls (MHRD, GOI 2008b).

It was an amazing journey, one that forced me to question fundamental beliefs about equity, gender, and education – and about educational administration. Everywhere I saw and heard how much poor rural women value literacy and education, especially for their children. Dalit and tribal women talked about their exasperation with government schools, the experiences of their children, and the systemic exclusion of the poor and marginalized from good quality schooling. I searched for published research and papers that analysed the situation that I saw with my own eyes. What shocked me was the absence of good, field-based research that teased out the contours of equity and quality in India's schools. The government essentially relied on the data gathered by the system, though this often does not reveal the true picture of exclusion. The larger education community questioned the veracity of the official data and debated the nature of the real situation; discussions of what needs to be done were often polarized on ideological lines.

It is this divergence between what I saw and what I read that compelled me to question the value of exclusively quantitative research and convinced me of the importance of weaving in qualitative insights in a rigorous manner. This collection of essays is a product of my work over the last sixteen years.

A substantial part of this book was written when I was a National Fellow at National University for Educational Planning and Administration from 2011–14. I would like to thank NUEPA for the generous fellowship and an ideal working environment.

The opportunity provided by Nehru Memorial Museum and Library to give a public lecture on 5 September 2013 enabled me to sharpen my thoughts and get critical feedback from peers.

Some of the chapters in this book have grown out of special articles written in the *Economic and Political Weekly*. I gratefully acknowledge the huge visibility and feedback that I got as an EPW author.

This book was made possible by the generous publication grant given by Azim Premji University, Bangalore. My thanks to S. Giridhar, Dileep Ranjekar and Anurag Behar for reposing confidence in me and supporting this venture.

Special thanks are due to Archana Iyer, Anugula N Reddy, Kameshwari Jandhyala, Niti Saxena and Prerna Goel Chatterji for their support for this book project. This book would not have been possible without their timely support, feedback and critical comments.

PART I

THE BIG PICTURE

1 Introduction

As early as the pre-Independence period, the education policies of the Indian government recognized the existence of wide social and gender disparities in children's enrolment in school and years of schooling completed. In the pre-1947 period, social reformers argued that there should be no difference in the educational content provided to boys and girls, at least until the end of the primary stage. It is noteworthy that in the immediate post-Independence period the idea that education was a 'universal good' for men and women of all social groups and classes was formally accepted. However, as the decades rolled by, subtle changes have crept in. In the 1970s, during the peak of the population-control era, women's literacy and education were linked with having smaller families and with issues of maternal and child health and nutrition. Later, in the 1990s and 2000s, programmes concerning reproductive and sexual health and HIV/AIDS again turned the spotlight on adolescent girls' education, which was considered a be the panacea for all health issues. For both the government and the international development community, the notion of 'functional literacy' – as an instrument or tool by which women could access developmental resources, keep accounts, and read information about contraceptives, childcare, safe motherhood, nutrition, and so on – became the peg on which to hang women's education. Official policy documents positioned literacy as the critical, determining variable that would have a positive impact on other social and demographic indicators.

By the end of the 1980s, a rights-based approach to education gained greater acceptance in the government, especially in the renewed interest in the 1974 Status of Women report and in the discussions leading to the reformulation of the 1986 National Policy on Education. After many years of debate and struggle, the Right to Education Act was passed in 2003 and notified in 2009. However, one should hasten to add that the international development community has continued to adopt an instrumentalist approach to the education of girls and women, and that the debate between the two approaches continues even today.

UNESCO's *World Declaration on Education For All*, ratified in Jomtien, Thailand in 1990, was an important milestone in the march towards universal elementary education. Forty years after the 1948 adoption of the *Universal Declaration of Human Rights*, this conference reminded the world that everyone has a right to education and that basic education is more than an end in itself: it is the foundation for lifelong learning and human development. This 'rights' perspective on educational access and equality received a tremendous boost during the International Conference on Population and Development, held in Cairo, Egypt in 1994, which declared: 'Everyone has a right to education, which shall be directed to the full development of human resources, and human dignity and potential, with particular attention to women and the girl child. Education should be designed to strengthen respect for human rights and fundamental freedoms, including those related to population and development' (UNPF 2014: 14). The very fact that a conference on population compelled the international community to look into the inter-linkages between population, poverty, gender, social inequality, and development was a major departure from the past. Shortly thereafter, in the 1995 Copenhagen Declaration on Social Development, education was placed centre stage in the struggle for human development and social justice: 'We affirm that, in both economic and social terms, the most productive policies and investments are those that empower people to maximize their capacities, resources and opportunities' (Paragraph 7).[1]

[1] The text of this declaration is available from the United Nations website at http://www.un-documents.net/cope-dec.htm.

The Fourth World Conference on Women, held in Beijing, China in 1995, also reaffirmed government commitments and called upon the international community to intensify efforts to enhance women's access to education. Four important world conferences, one after another, bombarded the world with similar messages of equality, empowerment, human rights, and universal access to the tools of knowledge: education.

In India, the mid-1980s ushered in another important change to the educational landscape: for the first time, the government agreed to accept the assistance of international donors for primary and elementary education. This was a direct outcome of the global activism advocating 'Education for All'. 1987 to 1994 was a very creative period in the history of basic education in India. Six new initiatives were launched by the government, namely: the British ODA (Official Development Assistance) – supported Andhra Pradesh Primary Education Project and Swedish-aided Rajasthan Shiksha Karmi project in 1987; the Netherlands-supported Mahila Samakhya – Education for Women's Equality project in Karnataka, Uttar Pradesh, and Gujarat in 1988; the UNICEF-funded Bihar Education Project and World Bank-supported Uttar Pradesh Basic Education Project in 1990; and the Swedish-aided Rajasthan Lok Jumbish in 1992. In 1994, the District Primary Education Programme (DPEP) was conceived and launched as an umbrella programme through which support could be channelled from various funding agencies.[2] DPEP, which covered nearly 60% of the country, aimed at increasing enrolment, retention, and achievement at the primary stage of education, as well as reducing gender and social disparities in districts with low female literacy rates.

In 2001, the central government launched the Sarva Shiksha Abhiyan (SSA; a flagship programme of the government to universalize elementary education) across the country, covering both upper primary and primary education with the goal of providing universal elementary education by 2010. While Mahila Samakhya

[2] Programme funds were obtained from the World Bank, the European Commission, and the British Department for International Development (DFID). State governments contributed 15% of the total cost.

was a women's education and empowerment programme, the other basic education projects expressly stated 'bridging gender and social equity gaps' as one of their main goals.

With international interest in and support for these new educational programmes, the demand for data grew – particularly data that could help assess the progress made towards bridging the gender and social equity gaps in access to elementary education. As a result, the government agreed to introduce periodic evaluations of the projects. Joint government–donor review missions visited the project areas and interacted with project functionaries. The DPEP supported the collection, compilation, and publication of this primary education data. Together, the development partners (or donors) and the Indian government evolved mechanisms to monitor programme progress. These included:

(1) The government-initiated annual gathering and compilation of school-based information known as the District Information System for Education (DISE). This was set up to capture key educational indicators – essentially enrolment and other inputs. Within a few years, the government realized that school-based data did not capture the full picture. As a result, some states started organizing household surveys to ascertain the number of out-of-school children, starting with the Lok Sampark Abhiyan in Madhya Pradesh in 1996. In 2001, household surveys were made an essential part of educational planning. Later, when researchers and the media questioned the veracity of this household survey data, the Government of India commissioned a sample survey to estimate the number of out-of-school children. The first was done in 2005, and the third in 2014.

(2) Six-monthly Joint Review Missions (popularly known as JRMs) were introduced to review the collected reports and data, and also to visit project areas. These missions involved state-wise six-monthly reports as well as national-level documentation on select issues, sectors, innovations, and challenges. The documentation of good practices was also supported through the technical assistance fund. This provided an opportunity for researchers and educational institutions to align their research interests with those of the government.

Starting in 1996, the Indian government in collaboration with these donor agencies gave many researchers the opportunity to conduct qualitative research on gender and social equity issues. However, the fact that the government and its partner donor agencies were commissioning research studies became a contentious issue: several independent academic institutions argued that commissioned research was only directed toward the areas that the government thought were important, while decreased funding was available for non-commissioned independent research. There was also a lot of debate about whether the findings of the commissioned research were actually used. Not only did it take up to five years before the results were published and made available to policy makers and the larger education community, but the government itself rarely used the findings to inform policy.

Another issue was the use of the data. While data was gathered by social group (SC, ST, Muslim) and gender, its tabulation was done in silos. For example, while the data on enrolment or dropouts were tabulated by social group and gender, in the reports the data on retention, out-of-school children, or children with disabilities were categorized separately by social group and by gender – the reports did not indicate how many ST girls were out-of-school, for example.

Apart from this quantitative data, there were few field-based, qualitative studies on gender or social equity issues. As one of the goals of the donor-assisted education projects was to bridge the gender gap in educational participation, almost every JRM demanded information on the participation of boys and girls. The donors asked for research-based analyses of barriers to the educational participation of girls, of specific social groups, and in specific geographic areas. This greater demand for and interest in gender issues spurred greater engagement with methodologies meant to capture gender data and gain better insight into social relations and education.

The Technical Support Group (TSG) attached to the Department of Elementary Education, GOI financed and commissioned research on a range of issues, including issues of social and gender equity and the inclusion or exclusion of children within schools. These studies provided an opportunity to critique existing methodologies and approaches and to demand better techniques for capturing gender and social equity issues of educational access, participation, and outcomes. For example, in the first few years most of the 'evidence' collected by these research projects was in the form of enrolment and dropout data, disaggregated by gender. Gradually, the review missions asked for data that showed gender categorizations within each social group to understand who the non-enrolled and dropout girls were: were they from Scheduled Castes or Tribes; were they primarily rural; were they the children of illiterate parents? When micro-studies revealed that more boys were attending private schools and more girls were in government schools, there arose a greater demand for data from both government and private schools, disaggregated by social group and gender.

As the JRMs and the larger educational research community started analysing the newly generated information, the finding that students from socially disadvantaged groups – especially girls – could be 'pushed out' of schools (as opposed to dropping out of them) demanded new approaches to understanding gender- and caste/community-based practices inside the classroom. This led to the commissioning of classroom process studies designed to explore gender and social discrimination within the classroom: how and why some children were being pushed out of school. At the same time, researchers from universities and academic institutions also started to engage with issues of caste, class, and gender in studies of educational participation.

This was also when the gender review of textbooks was initiated: in many states, researchers and practitioners of school education were asked to review the existing school textbooks for gender bias or stereotypes. By the end of the DPEP in 2003, government-funded inquiries into gender and equity in education had become more varied. It was officially recognized that social equity and gender issues intersect and overlap, and that it was therefore necessary to look at gender relations within social groups and within different situations and locations.

However, after the end of the DPEP project there was a lull in the research. Taking stock of DPEP, the government and the donor community realized that, while enrolment and retention rates had gone up, there was very little information on what and how much the children were learning in school. DPEP claimed to have significantly bridged gender and social equity gaps, but when Sarva Shiksha Abhiyan was formulated in 2001, its goals[3] told a different story. The sad reality was that the DPEP had not made a significant dent on the equity and quality front. By 2005, the donor community and education-related NGOs, including the global community, had shifted the focus to learning outcomes. While this is indeed an important issue and merits focused attention, the tragedy was that the emergence of new challenges meant that old challenges were left behind. In the post-2005 period, the focus again shifted to enrolment and retention; gender and social equity were put on the back burner.

Issues in Quality and Equity

Research and evaluations that I have participated in since the mid-1990s have highlighted a range of issues that continue to plague the Indian educational system. For example, while children belonging to migrating families and the urban poor have been found to be the hardest to reach in educational interventions, the government continues to prioritize rural areas. Similarly, the steady increase in private school enrolment and corresponding decline in government school enrolment – even in educationally backward areas! – has not received the attention it merits. While this phenomenon has been highlighted by independent surveys and the media, it has not been seriously addressed by the government. Equally worrying is the evidence of an increasing number of single-teacher, and hence multi-grade, schools in rural and remote areas, which urban-oriented

[3] The goals of the Sarva Shiksha Abhiyan were:

1. All children in school. Education Guarantee Centre, Alternate School, 'Back-to-School' camp by 2003.
2. All children will complete five years of primary schooling by 2007.
3. All children will complete elementary schooling by 2010.
4. Focus on providing an elementary education of satisfactory quality for all children, with an emphasis on education for life.
5. Bridge all gender and social category gaps at the primary stage by 2007, and at the elementary level by 2010.
6. Universal retention by 2010.

teacher training, curriculum development, and teaching-learning resources continue to ignore entirely. Another significant finding is the continued use of traditional methods of teaching, with an emphasis on memorization and rote learning. These issues have rarely grabbed the attention of policy makers and administrators. Reviews of all of the evaluations I have done over the years have thrown up many questions, that remain unanswered even today (Ramachandran et al. 2004; Ramachandran and Goel Chatterjee 2014):

What are the reasons for low levels of learning in different circumstances and different kinds of schools?
What systems are in place to measure teacher accountability and teacher attitudes/practices in the classroom?
What is the behaviour of teachers towards students from various social and economic backgrounds and children with special needs (CWSN)?
How are exclusionary or discriminatory practices towards children and women teachers being addressed by schools?
And, within all of the above, how do gender relations, stereotypes, and prejudices play out?

What is undeniably clear is that a child is at a greater disadvantage if he/she lives in a rural area or poor urban slum and comes from a poor family. This situation becomes worse if the child is a girl – especially an older girl – from one of the socially disadvantaged groups such as Dalits, Scheduled Tribes, or Muslims. Therefore, addressing gender and social equity issues in education requires a framework that can capture these heterogeneous gendered realities and disadvantages, especially the following factors: location (rural, urban, remote, tribal, etc.); identity (caste, cultural, religious, occupational); socio-economic status of the family; type of school (government or private); ability and disability; and, within all of these factors, gender relations. These factors intermesh and influence not only the formal access to schools but also, more importantly, how children are treated inside the school, their ability to actively participate in school activities, their ability to learn in school, and the kind of support they get or do not get at home and within their community. Consequently, it would be fair to say that achieving the equity goals for education requires simultaneous work on several fronts and cannot be one-dimensional.

An important theme running through this collection of essays is that the government has inadvertently reproduced existing social and economic inequalities in its effort to expand educational opportunities. Government schools across the country primarily cater to the poor, thereby reaffirming their social segregation from the economically better off, who opt for fee-paying private schools. Further, the gradual segregation of educational facilities by economic status and by social group feed off of each other and are also influenced by the phenomenon of identity politics.

Equally significant is that procedural accountability has taken precedence over outcome accountability in the drive to collect and compile data for monitoring educational progress. In other words, student learning is not on the government's radar. While it may not be prudent to argue that learning levels have come down in the last twenty years, the fact remains that they continue to be low – as borne out by the successive learning-assessment surveys conducted by the both the government and non-governmental organizations.

Part I: The Big Picture

This book is divided into two parts, the first drawing from macro data and the second from qualitative microstudies. The first part argues that equity and quality are two sides of the same coin. Gender, caste, class, and location influence the ability of children to access quality education: it depends on the child's caste, her family's economic status, where she is located, the kind of school available to her, and most importantly the number of teachers available in her school. One of the challenges that researchers face is to separate the school-specific from the family/community-specific issues that influence children's ability to pursue education. This demands the disaggregation and classification of quantitative data by social status (caste and religion), economic status, and location; when this kind of highly stratified data is not available, performing a more nuanced analysis using quantitative data becomes challenging.

Chapter 2 provides a historical analysis of equity and quality and how they have become disjointed in India's march towards universal elementary education. It sets the stage for the arguments presented in the rest of the book and argues that quality education is the heart of equity.

Chapter 3 attempts to capture the revelations of different data sources about the educational landscape in India. Interestingly, different data sources capture different dimensions of reality; it is only by triangulating them, juxtaposing one against the other, that researchers can get a nuanced picture of India's educational situation.

It is well known that Scheduled Tribes are the most educationally disadvantaged groups in India. Drawing upon secondary sources, Chapter 4 attempts to capture the realities facing tribal communities in different regions of India and argues that the situation on the ground is varied and that it is not possible to look at tribal communities as one homogeneous group.

The final chapter of Part I (Chapter 5) explores the complex web of gender and education. As enrolment in primary and upper primary schools has increased, many more children from diverse backgrounds have entered schools – thereby increasing the diversity within the classrooms. As the government has opened more schools to reach out to children in rural and remote areas, there has also been an exponential increase in the number of small, multi-grade schools. As a result, today there are many more first-generation learners and children from disadvantaged groups in schools. Equally significant is that, since the early 2000s, there has been an increase of children with special needs in elementary schools. The classroom is not only diverse in terms of children from different social and economic groups, but also in terms of children of different ages who study in the same class. The enrolment of underage children in Class I, for example, has compelled policy makers and administrators to think about preschools as an adjunct to primary schools (SSA 2002). The web of gender and equity in our schools has therefore become complex; this chapter attempts to define a framework that will allow a deeper understanding.

Since the mid-1960s, Indian educationists have argued for a common school system in which the economic status, caste, location, or gender of a child would not determine what kind of education he or she will get. Despite the compelling arguments made over the last five decades – and despite policy documents that outline the importance of equitable access to quality education – there is little sign of change on the ground.

Part II: Behind the Smokescreen of Data

One would imagine that the ground-level situation would be transformed after almost 30 years of targeted school education programmes and projects. Unfortunately, this is not the case. The tragedy is that the last ten-to-fifteen years have instead seen the exacerbation of social and economic inequality. It may be prudent to ask why inequalities have been exacerbated despite government programmes meant to universalize elementary education. In a recent publication, Jos Mooij (2014: 165) throws light on this subject by characterizing the Indian welfare state. While the rights-based approach that framed the 2009 Right to Education Act (RTE) was a result of the social mobilization and international commitments that also propelled DPEP (1990s), SSA (2001), and RMSA (2009) to provide a school to all children, contradictory social and economic processes have effectively negated any positive impact of these programmes. As the elite and the middle classes have marched out of government schools in favour of private ones, those without a voice were left to deal with the government. As a result, the quality of India's public educational institutions have gone from bad to worse. It is also distressing that with every passing decade the rights-based discourse is becoming more pronounced in policy documents and at the same time 'shape[ing] the way in which people think about rights and entitlements and the way they seek to secure better access or better quality – individually or collectively' (Mooij 2014: 167). Resentment against affirmative action is also growing: there is an increasing backlash against reservation policies and related initiatives meant to protect the entitlements of the disadvantaged. In the last two years, these very forces have been trying to dilute the RTE Act with a view to protecting their turf in private schools. Caste and religious identities have become far more pronounced, especially in the political arena.

In Part II of this book, I revisit qualitative studies that I participated in between 2002 and 2015. As I moved from one study to another during this time, my own understanding of the equity and quality issues that frame school education improved and I was able to develop a far more nuanced understanding.

One of the early qualitative studies (discussed in Chapter 6) was commissioned by the government in 2002–03, in an attempt to examine how gender and social equity issues were addressed in the flagship District Primary Education Programme (DPEP) of the Government of India (Ramachandran 2004). This study involved micro studies in schools in six states: Madhya Pradesh, Chhattisgarh, Haryana, Karnataka, Andhra Pradesh, and Tamil Nadu. However, the government almost rejected the entire study because it argued that the school system was reinforcing existing social divisions and that a new kind of segregation was emerging. The findings of this study were widely debated, and gradually the education community not only accepted them but also encouraged others to inquire into the equity dimensions of our schooling system.

Soon after the results of the 2002–03 study were shared with the Government in 2003 (and before it was published in 2004), the World Bank commissioned another qualitative study to explore the factors that facilitate or impede successful completion of primary school (Ramachandran et al. 2004). This World Bank study was done in 2003 in Andhra Pradesh, Karnataka, and Uttar Pradesh, and explored the school-, family- and community-level issues that enable children to enter school and complete at least six years of primary education. Like the previous study (Ramachandran 2004), this one also attracted the attention of both the education community and the government because the findings were so startling. Three issues hit home: the persistence of short-term hunger among children attending school; the work done by children before and after school (questioning the notion that once a child in enrolled in school she/he does not work); and the very limited amount that children in school are actually learning. Before this study, there had been no large-scale surveys on learning, and the government had essentially relied on assessments done under the DPEP; the World Bank report indicated that change was needed.

Chapter 7 summarizes the key insights from this landmark piece of qualitative research.

Several years later, the government was forced to acknowledge the unfortunate reality of social exclusion inside schools. While journalists like P. Sainath and researchers like Geetha Nambissan and Manabi Mazumdar had all argued that both subtle and blatant forms of exclusion were pushing children out of school, field-based qualitative studies about the schooling experience of Dalit and Tribal children were rare. In 2009, UNICEF's regional office in charge of South Asia commissioned me and others to conduct a qualitative study to explore how school water and sanitation programmes affect the inclusion of children from specific social groups (UNICEF 2009).[4] It was well known that hierarchies of caste/community manifest through social practices related to eating, drinking, and the use of common facilities. Traditionally, certain kinds of work, especially cleaning toilets and the disposal of dead animals, have been done by the most disadvantaged groups in the social hierarchy, particularly Dalits (SC). Though social reform movements have made a big difference in many parts of the country, caste discrimination continues to be a reality, and there was a lot of anecdotal evidence pointing to the persistence of this kind of discrimination within schools. Exclusion could also be subtle, and not discernible in routine classroom observations or during school-level interactions.

UNICEF's study confirmed the worst fears of the education community (UNICEF 2009): exclusion manifests in different forms, and children were discriminated against based on five intersecting identities:

Social group and community (Dalit, Sahariya, ST, and OBC);
Economic status (not well dressed, dirty clothes, do not bathe);
Performance in school (good in studies/not good in studies; regular/irregular);
Visible disabilities, infectious diseases (scabies, persistent cough); and
Gender (differential duties assigned to boys and girls).

[4]While this study does not form a chapter of this book, I have included a brief discussion here to provide context.

Children were assigned duties according to their age, gender, and (in many schools) the social standing of their family. Teachers reinforced dominant social attitudes and caste prejudices. This report was received with alarm from both the government and civil society. Dalit organizations circulated the report widely; we were informed that Mayawati Prabhu Das, the first Dalit Chief Minister of Uttar Pradesh, used it to demonstrate the existence of caste-based exclusion to others in her government. However, there were others who vehemently denied this reality.

As this debate raged on, the Ministry of Human Resource Development (MHRD), GOI commissioned a national study (Ramachandran and Naorem 2012) to explore the issue of inclusion and exclusion in 2011–12 (Chapter 8), especially what it means to be a Dalit or Tribal child in our schools in six states (Andhra Pradesh, Assam, Bihar, Orissa, Madhya Pradesh, and Rajasthan). One of the overarching insights of this study was the need to view inclusion and exclusion from different vantage points: from the outside (who goes to what kind of school); from the inside (what happens inside the school); and from society (who is visible and who is not visible; for example, seasonal or new migrants are often invisible in data on out-of-school children). Equally significant is the influence of the larger society and social norms on what happens in school, the attitude and behaviour of the teachers, and the involvement or lack of involvement on the part of parents and community leaders. The study also found that political and social assertion of the rights of Dalit and Adivasi communities in the larger society also influences practices and attitudes in schools.

Part II concludes with an essay on teachers (Chapter 9) that draws upon three qualitative studies done between 2004 and 2015 (Ramachandran et al. 2005; Ramachandran, Bhattacharjea and Sheshagiri 2008; and Jandhyala and Ramachandran 2015). Starting with the question of why our teachers are demotivated and disheartened, this chapter explores the problems that teachers themselves have highlighted and the factors that contribute to the erosion of their accountability. The essay then moves on to look at the working conditions for schoolteachers and whether they make a difference to the overall schooling experience of children.

As the chapters in Part II reveal, not much seems to have changed in the last three decades. If anything, caste, class, and religious identities have become *more* pronounced, and this is only exacerbated by the current political environment. Across this diverse country, one thing is common: the predicament of very poor, Dalit, and Tribal boys and girls has not changed much. Children who attend government primary/elementary schools learn little. They move from one class to another and reach Class VIII. The cumulative burden of poor learning then hits them like lightning, and they are not able to cope with secondary school. They try hard, even seeking private tuition, but continue to fail. As a result, they drop out before the Class X board examinations. Ten years of schooling has little impact on their learning level...and they are back at square one. They enter the workforce as unskilled or semi-skilled workers, eking out a living as their parents did – thereby perpetuating the inter-generational cycle of poverty.

2 Equity and Quality

Two Sides of the Same Coin

"A student of Class VIII of today cannot read what Class IV children could thirty or forty years ago. Our children enrol in secondary school and drop out because they are unable to cope. Children today do not have rudimentary skills in language, mathematics, or science. What is the use of such schooling?" This was a remark made by a parent in Rajasthan in 2014.

The Indian government is in the process of revising its education policy; in June 2016, an expert committee (popularly known as the Subramanian Committee) brought out a report for the consideration of civil society and the government. This has led to considerable discussion and debate about where we are with respect to education. The government does not tire of announcing that the enrolment in primary schools in India showed impressive growth in the 1990s and 2000s. Government and non-government surveys both confirm that close to 97% of children between 6 and 11 years old are enrolled in primary school (MHRD, GOI 2014; ASER Centre and PRATHAM 2015). More girls are in school, as are more children from socially and economically deprived communities. Large numbers of teachers have been appointed in the last ten-to-fifteen years (MHRD, GOI 2014). The Right to Education Act was notified in 2009. In fact, if we go by the numbers published by the government, there is much to be upbeat about in the primary education scenario in India.

Yet, despite the impressive data generated every year, there is a sense of disquiet when we talk of education. The report of the committee set up to draft a new education policy notes: 'Today, we find that as a result of efforts made during the last few decades, while accessibility, infrastructure and literacy levels have improved significantly, there remains much to be disturbed about when one reflects on the continuing inequity, and the poor quality of education' (MHRD, GOI 2016: 3). In July 2014, my colleagues and I conducted several focus group discussions with parents of children who had dropped out after Class VIII. What we heard from them was distressing: in almost all of the villages in Rajasthan, parents said that their children had learnt little. Parents said that students of Class VIII could not read a Class IV text, and did not have rudimentary skills in mathematics and science. This was not the case thirty years ago. As a result, many children drop out because they are unable to cope. This experience compels parents to question the value of this kind of education.

Almost everyone will agree that mere access to schooling, without a commensurate focus on learning, is meaningless. This is one of the unfathomable dilemmas in India's educational planning – how could our policy makers decide to fix access first, and then worry about quality? The essence of education is learning. Only a mindless bureaucracy could have thought of separating quality from access.

Even the issue of access has been addressed in fits and starts. Access to primary education has improved significantly in the last ten years, but the same cannot be said for upper primary and secondary education. For example, in 2010–11, the Gross Enrolment Ratio (GER) at the primary level was 116 with little gender gap, but at the upper primary level it drops to 85.5 (boys 87.7 and girls 83.1).[1] It drops even further to 65 (boys 69 and girls 60.8) in Classes IX and X, and to 39.3 (boys 42.2 and girls 36.1) in Classes XI and XII. The situation in higher education remains quite worrisome, with a GER of 16%, far below the global average of 27% (Planning Commission, GOI 2013). The situation had not

[1] The calculations in this paragraph are my own, based on data from NUEPA 2012a, 2012b.

significantly improved by 2015; while the GER for primary and upper primary remained the same, at 116 (girls 116.3) and 85.2 (girls 82.9) respectively, the GER for Classes IX and X drops to 65.2 (girls 60.9), and for Classes XI and XII it drops further to 39.4 (girls 36.2). Many children reach Class IX despite learning little, as they are mechanically promoted; the real picture only emerges from the enrolment rate in Class XI. As we go further up the educational ladder, many more children drop out – some because they do not have the adequate knowledge and skills to cope with higher levels of schooling, others because there are no secondary schools within reach. What is disturbing is that the dropout rates are much higher among disadvantaged social groups – among the rural and urban poor, and among children of migrant families. At all levels, there are huge rural-urban, male-female, inter-state, and inter-community differences. The biggest marker is the kind of school a child attends, alongside a combination of poverty, social status, type of school and location.

For many decades, we have known that the social and economic status of a child's family is one of the strongest determinants of the child's educational attainment. The last 60 years of educational data and other socio-economic surveys (like the National Sample Survey and National Family Health Survey) have demonstrated this link. This is not only the case for India; rather, it is a global phenomenon. There is considerable evidence to show that the educational mortality of different social classes is different, and that the cumulative burden of poor learning and exclusion is most visible in higher education (Deshpande 2013). Pierre Bourdieu and Jean-Claude Passeron also point out that

> The combination of the educational chance of the different classes and the chances of subsequent success attached to the different sections and types of schools constitutes a mechanism of deferred selection which transmutes a social inequality into a specifically educational inequality, i.e., an inequality of 'level' or success, concealing and academically consecrating an inequality of changes of access to higher levels of education (1977: 158).

With the cumulative burden of non-learning pushing children out at different stages, the gap in achievement levels continues to grow while children move from primary to high school to university. As Sean F. Reardon has argued, 'The relationship between family socioeconomic characteristics and student achievement is one of the most robust patterns in educational scholarship, yet the causes and mechanisms of this relationship have been the subject of considerable disagreement and debate' (2011: 4). In India, the mix of family poverty, location (rural, tribal, urban slum, etc.), caste/community identity, gender, and parental occupation is known to be a significant predictor of educational achievement (Ramachandran 2009).

The fight for equity is ultimately a struggle for quality education.

Why and How did Equity and Quality become Disjointed?

It is commonly believed that parents who are well off devote a larger proportion of their financial and social resources to their children's cognitive development. It is also argued that the rich have greater access to educational resources. The question that then arises is whether differentiation in the kind of schools that children attend, the quality of the schooling, and the opportunities that are available there also reflect differences in income. Similarly, is it fair to assume that more educated parents (especially mothers) devote more time and energy to the overall development of their children (Reardon 2011)?

At another level, is there so much differentiation of the kind of schools children attend that it makes a significant difference to their overall development and learning? Until the 1970s, Indian educationists had heard about and even met professionally successful people who had studied in modest rural government schools or urban municipal schools. We also heard of numerous instances where a good school or motivated teacher was able to neutralize the odds against very poor children and enable them to learn. Equally, we heard about headmasters and teachers who had made a big difference in the lives of very poor children through encouragement and, most importantly, active academic support and guidance. Sadly, such cases are becoming increasingly rare; in the last two-to-three decades, the difference in the ability of a child attending a rural primary school compared with children attending a private school or a government school with a lot of resources has only increased.

The more worrying trend is that a higher number of children who are poor or from a disadvantaged community (SC, ST, new migrants, Muslims) end up attending schools that are more likely to have very poor facilities, indifferent or poorly qualified teachers, or multiple grades sitting together and taught by one teacher (i.e., a multi-grade school). Such schools are now referred to as 'high-poverty schools' and exist in both the government and private sectors.[2] Further, income inequality has led to the increasing spatial segregation of high-income families from middle- and low-income families, with the poorest often relegated to the outskirts of cities, inner city slums, or specific habitations in rural areas. This spatial segregation of the rich from the middle class and the very poor essentially implies that children go to different kinds of, and differentially endowed, schools.

According to data from the 66th round of the National Sample Survey (NSS) (NSSO 2011), a startling 14.25 million children in the 6–14–year age group were estimated to be 'not attending'. While more and more children are moving from primary to upper-primary, the numbers drop sharply in secondary schools, and the Age Specific Attendance Ratio (ASAR) decreases sharply.[3] The 64th Round of the

[2] The term 'high poverty schools' was first used in the US in the mid-1960s to describe schools that were poorly resourced in terms of infrastructure, learning resources, and teacher quality. Typically such schools were located in neighbourhoods where poor African Americans or migrant communities lived.

[3] The Age Specific Attendance Ratio (ASAR) is 88% (all) and 87% (Rural) for age 6–10; 86% and 85% for age 11–13; 64% and 61% for age 14–17; and 18% and 15% for age 18–24. The NAR for Classes I to V is 88% for all (87% rural, 91% urban); for Classes VI to VIII, it comes down to 59% (57% rural and 65% urban); decreases sharply to 41% in Classes IX and X (38% rural and 51% urban); and to a low of 27% in Classes XI and XII (23% rural and 40% urban) (NSSO 2010: 62). Age Specific Attendance Ratio (ASAR): For each age group, this measure gives an idea of the proportion of persons of a particular age group who are currently

National Sample Survey revealed that close to 84% of children from the poorest economic quartile attended government and government-aided schools (NSSO 2010). As we go up the economic ladder, the proportion of children attending government schools decreases and that of children attending private school increases. If we look deeper, even those poor children who are able to attend private schools go to the 'low cost' private schools, where the quality of instruction is, at best, marginally better than that of government schools and the infrastructure is far worse.

It is indeed noteworthy that more than 90% of both rural and urban households reported having a school with primary classes within 1 km. But the availability of schools with middle or secondary-level classes differed considerably between rural and urban sectors. Only 61.6% of rural households, compared to 82.5% of urban households, had a middle school within 1 km. For secondary-level classes, the proportion was 30.7% for rural households, compared to 68.6% for urban households. Further, about 33% of rural households did not have a secondary school within a distance of 3 km (NSSO 2010). It is fair to say that access is not equal and that the ability of children to continue their education is largely determined by their location and the availability of public transportation.

Crossing the secondary education barrier is a major hurdle before young people can hope to enrol in colleges and universities. The quality of secondary education exerts a huge influence on students' opportunities after school. Thomas Weisskopf points out:

---

attending educational institutions, irrespective of the level or class in which they are studying. For example, for age group 6–10 years, this attendance ratio = Number of persons in age group 6–10 currently attending educational institutions x 100 / Estimated population in the age group 6–10 years (NSSO 2010).

Net Attendance Ratio: For each class group, this is the ratio of the number of persons in the official age group attending a particular class group to the total number persons in the age group. For example, for class group I-V, the ratio (in %) = Number of persons of age 6–10 years currently attending Classes I-V x 100 / Estimated population in the age group 6–10 years. This number can also be obtained for the remaining class groups, i.e. VI-VIII, IX-X, and XI-XII, with the corresponding official age groups of 11–13, 14–15, and 16–17 years, respectively (NSSO 2010).

> In India roughly 40% of higher education students are enrolled in relatively low prestige arts programmes/bachelors programmes. This percentage is 60 for SC and 75 for ST [...] The question is what kind of higher education do they have access to? [Their] academic performance and how many finish their degree in the stipulated time [...] [there are] varying estimates, but in professional courses the percentages are far lower than the general category. Why do SC/ST students fall behind in the course? (Weisskopf 2013)

In fact, this is not a new revelation. Way back in 1972, Suma Chitnis's work in Maharashtra revealed that the overwhelming majority of SC students (in the sample she was tracking) were studying in grade D (poor quality) colleges, their performance in the course was poor, and their dropout rates were high. She also found that a higher percentage of SC students were enrolled in non-professional and non-science courses (Chitnis 2013 [1972]). In reviewing the higher education terrain, Jandhyala B. G. Tilak notes, 'In India the emphasis is laid either on quantity or quality or equity, not on all three at the same time and we end up with a scalene, a triangle of three unequal sides' (Tilak 2013: 47).

Having overcome the first barrier of access to primary schooling, there is no guarantee that attending school will bring learning. ASER 2011[4] revealed that nationally only 48.2% of Class V children could read simple Class II-level texts, and only 29% of Class V children could solve a two-digit subtraction problem. These findings reflect only the tip of the iceberg; the reality on the ground is quite sombre. While more children may be enrolling in school, the big question is whether they are learning, and to what extent does the kind of school they attend influence this learning. PISA test results (2011) have raised similar issues.

[4] ASER: The Annual Status of Education Report (ASER) is a national survey that collects information regarding children's schooling status and basic reading and math abilities. Conducted every year since 2005, ASER is facilitated by PRATHAM and conducted by local institutions in every rural district of the country, reaching about 3 lakh households and 7 lakh children each year. ASER results show that at the national level, there has been no improvement in children's ability to read a simple text or do basic arithmetic in the seven years for which data are available.

When most administrators, educationists (even those working in government institutions), and the larger education community agree that measuring inputs and enrolment is inadequate, the question that arises is: Why does government data continue to focus primarily on enrolment and retention/transition, and to a far lesser extent on learning? Why is it that we, as a nation, shrug at the thought of millions of our children attending school and emerging barely literate? What does it say about our education system? And how did we reach this point?

Quality: The Heart of Equity

Before India attained independence in 1947, Mahatma Gandhi dreamed of a school system rooted in India's social and cultural context, while at the same time ensuring that good quality, secular education would be accessible to all of the children in the country. But as the first government assumed office, 'education for all' was quickly forgotten. We plugged along with the colonial system, partly because there were more pressing economic and political issues to tackle. Gandhiji's plea for a *Nai Taleem* (fresh approach) to education was lost in the din of building a command economy, which meant focusing on heavy/basic industries, institutions of higher technical education, and so on.

It is not that education was neglected. In addition to regular schools, the government set up a chain of 'central schools' (known as *Kendriya Vidhyalaya*) that were well endowed and of good quality. But the real attention was reserved for higher education: the ruling elite was committed to building quality institutions for professional and technical education. National institutes of technology, science, medicine, and management produced graduates comparable to the best in the world. It is no surprise, then, that India managed to ride the global information technology wave in the 1990s with confidence. In this rarefied world, excellence and quality were key.

However, the euphoria of India being a knowledge-driven economy was short-lived. In 2005, the independent non-governmental organization PRATHAM facilitated a nation-wide sample survey of children's learning outcomes. The results were shocking. Close to 35% of children in the 7–14 age group could not read a simple paragraph (Class I level), and almost 60% of children could not read a simple story (Class II level). 65.5% of children in the 7–14 age group could not tackle simple arithmetic problems and, more worryingly, 47% of children in the 11–14 age group could not solve Class II-level arithmetic problems (ASER Centre and PRATHAM 2005). That is not all. In the same year, the government admitted that, out of all children who enter Class I, 52.8% drop out before they reach Class VIII (MHRD, GOI 2005). Among them, children from the most deprived communities and from rural and remote areas constitute an overwhelming majority of dropouts. Education meant little to these children, since they acquired few skills and little confidence. Regardless of the official claims, close to half of all children in the appropriate age group (6–14 years) receive less than the constitutionally mandated eight years of schooling.

How Can Two Entirely Different Systems Coexist?

In the early 1950s, education and quality were seen as coterminous, in part because education was a privilege of the few who could afford it. Most children from middle-class families studied in government schools. With modernization and the escalating demand for education, pressure for the rapid expansion of the public education system increased. By the mid-1960s, as increasing numbers of children started enrolling in school, the country also witnessed a gradual increase in the number of private schools, both those that accepted government aid and those that did not. As the 1960s rolled by, people with means moved out of public (government) schools and the feeling developed that quality education was, perhaps, better pursued in private schools. This movement of the children from economically better-off families from government to private schools has continued to gain momentum through the subsequent decades. The ASER reports reveal an increasing trend of private school enrolment, from 18.7% in 2006 to 30.5% in 2015 (ASER Centre and PRATHAM 2007, 2016). As of now, the socially and politically significant minority, i.e. the middle and upper classes, have more or less fully moved out of government schools. Interestingly, even those who do remain in government schools go to the better-endowed ones. In recent years, for example, the senior civil servant fraternity (IAS) created an exclusive school for the children of Class I officers in New Delhi known as the Sanskriti School. Now efforts are underway to replicate this in all of the state capitals.

In 2015, 61% of schools were run by the education department of the government, 17% were run by local bodies (*Panchayats/* municipalities), 3% were private aided, and 17% were private unaided. This distribution changes quite dramatically when we look at secondary and higher secondary schools, of which 37% were run by the government, 8.5% by local bodies, 17% were private aided and 41% private unaided (NUEPA 2015c). If we look at this distribution of schools alongside the dropout rates (calculated by cohort dropout rates in 2011), it is noteworthy that 27% of children drop out by the end of Class V (28.7% boys and 25.1% girls). Further, 40.6% drop out by Class VIII (40.3% boys, 41% girls). However, the annual average dropout rates at the primary, upper primary, and secondary levels were 4.13%, 4.03%, and 17.06%, respectively, in 2014–15 (NUEPA 2016 & 2017). It is worrying that government data does not tell us who (i.e., by location, income group, or family education status) drops out, even though micro or field-based studies point to the very poor, among them socially disadvantaged groups like SCs, STs, and Muslims, and within them first-generation school-goers.

In the 1990s, data on enrolment and retention took precedence over the painstaking process of monitoring the functioning of schools. The inspectorate system gave way to data-based monitoring. As a result, as long as the local education authorities could show high enrolment rates, the work was considered done. Even though the dropout rates were alarming, the ability of the system to retain children was not seen as an indicator of achievement until almost the middle of the first decade of the 21st century.

With growing pressure to meet globally accepted levels of school enrolment, India joined the great numbers race. Issues of access and quality were analytically separated, and the focus shifted to somehow getting children into schools, with the belief that the government could worry about quality later. Statistics acquired an unprecedented status and we started searching for simple measures like enrolment, transition, and completion rates to show progress – forgetting that highly complex webs of issues are responsible for the poor quality of government schools.

The complex web of issues framing poor quality education

1) The school system is not accountable to anyone. When the government is satisfied with monitoring numbers (of students enrolled, teachers appointed, textbooks distributed, midday meals supplied, buildings constructed, toilets built, etc.), there is very little motivation to do or look at anything else. Academic support institutions at the cluster and block levels[5] are preoccupied with gathering data from schools. Old-time inspectors no longer monitor what happens inside the schools or, for that matter, what the children are learning. Therefore, according to the old maxim 'what is monitored is what gets done', our school system has perfected the skill of enrolling children and showing them on the rolls.

2) Given the interpretation of the no-detention policy (whereby children are not detained in a class even if they have not achieved grade-specific competence), schools promote children from one grade to the next without bothering to find out if they have indeed acquired grade-specific knowledge and skills. As a result, we now have a situation where many children reach Class V without learning to read or write. Another danger looms large: the RTE Act has extended this no-detention policy right up to Class VIII. This means that we may have children who have been in school for eight years without learning language, mathematics, or science.

3) The numbers of days that schools actually work varies significantly across the country. Student attendance has been falling. For example, the average attendance of students in Bihar has declined from 59% in 2007 to 50% in 2011, whereas in Uttar Pradesh it fell from 67% to 57% over the same period (ASER Centre and PRATHAM 2012, 2015). Several studies have also shown high

[5] 'Cluster' is a group of 10–15 schools and 'block' is an administrative unit. Both are units used in the compilation of data and monitoring of schools.

teacher absence rates (Michael Kremer et al. 2005; Muralidharan 2013). In my own research, I found that teachers in rural schools keep an undated leave application ready and take turns attending class. If there are five teachers appointed to a primary school, on any given day only two or maybe three are likely to be present (Ramachandran 2004; Sharma and Ramachandran 2009). Because they rotate this 'French-leave', each teacher ends up managing more than one class. Over the last ten years I have found that the actual time spent on teaching-learning in any one grade/class can be as little as 25 minutes a day. A recent study commissioned by the government's Sarva Shiksha Abhiyan also found that the actual teaching-learning time (time-on-task) was low, and student attendance was also a matter of concern (Research, Evaluation and Studies Unit, Technical Support Group 2010). Why then are we surprised that our children do not learn?

4) The RTE has stipulated norms for basic infrastructure and facilities, for teacher-pupil ratios, and other academic matters that determine quality. The Sarva Shiksha Abhiyan of today, the District Primary Education Programme of the 1990s, Operation Black Board of the 1980s, and so on, also developed elaborate guidelines and notified norms. The unfortunate reality is that many of these plans remain on paper. Lofty intentions in policy announcements are not followed by concrete action. As a result, even after three decades of government-sponsored, central-sector schemes for universal elementary education, a majority of government schools in both rural and urban areas (the abysmal state of the Municipal Corporation of Delhi-run primary schools is a case in point) have crumbling infrastructure and poorly maintained facilities. If there are toilets, there is no water; if there are toilet and water facilities, cleanliness is an issue. Apart from these problems of basic maintenance, the quality of their construction is also poor. As a result, our schools are neither attractive nor safe.

The situation is not very different at the secondary level, despite the secondary education mission. In many educationally backward states, government secondary schools have poor infrastructure, inadequate teachers for some subjects, extremely poor science labs, no library, etc. This list is fairly long. In addition, a large number of girls-only schools in rural and remote areas do not offer Science or Commerce. Students are left with few options and end up focusing on Language or Arts. In a study in 2008–09 in New Delhi (Ramachandran 2009), I found that the girls' schools in some of the poorer areas did not offer science. When we asked the students in Classes IX and X what they would like to do, many wanted to enrol in Industrial Training Institutes (ITI), polytechnics, or nursing schools, but were unable to do so because their school did not offer science. This finding was confirmed by a 2015 study done in three districts of Rajasthan, in which we found that

> Girls, especially from marginalised communities, have limited access to science and maths education because most government girls' schools to which most of them go do not offer these subjects [...] This deficit in maths and science girl students continues into the college and BEd level and, as a result, very few women science and maths teachers are available. (Jandhyala and Ramachandran 2015: 50)

The problem is not a lack of financial resources or of good policies or strategies. The basic issue is that no one really cares whether government schools function. Our political leaders, administrators, media, and middle classes have all abandoned government schools for the education of their own children. In one study (Ramachandran, Bhattacharjea, and Sheshagiri 2008), we had an opportunity to interact with teachers across eight states of India. I was shocked to note that not one of them sent their children or grandchildren to government schools. All of them said that they wanted their children to go to a functioning school, with a teacher for every class. They also demanded that the school maintain discipline. Almost all political leaders want English medium for their own families, even though they enforce vernacular medium for the government schools. They want to see their own children go to schools that function, with proper classrooms and desks and trained teachers, while at the same time they look the other way as government schoolchildren are packed into small, windowless spaces, made to sit on the floor, have no lights or fans and, worse, are unsure about whether they will see the teacher every day.

This is a warped system! Official numbers tell us that all is well with our education, while our eyes and ears tell a different story. India has created parallel structures of education governed by different norms and playing by different rules. In the process, the educational system has become more differentiated: public, aided, and private; formal and alternative; permanent and transitional; pedestrian coexisting with institutions of excellence. The poorer and more disadvantaged are pushed into schools with poorer facilities, teachers, and overall learning environment. This has given rise to a new trend of 'hierarchies of access', whereby, paradoxically, the democratization of access to schools is accompanied by the reaffirmation of a child's caste, community, and gender, which decide the school she or he will attend (Ramachandran 2004). Having pushed the socially excluded into educational ghettos, the government then wants to neutralize the inequalities of past centuries through affirmative action in the form of reservations for a certain percentage of admissions to higher and technical education institutions for specific social groups. It is cruel to offer reservations at the higher levels while paying little attention to educational quality and learning in school. Are we not perpetuating or even deepening inequalities by not giving our children equal opportunities?

The economic cake is expanding fast for some people in India. New opportunities are available to those who have been able to access quality education. There is a clamour for better schools, English-medium education, and new technologies. Quality is the new battleground.

And yet we are at a loss to define what we mean by 'quality'. At the heart of this concern are questions related to the goal of education. Is education meant to enhance students' ability to negotiate an increasingly unequal, divisive, and polarized world from a position of strength? Is it meant to enable people to reflect critically on their life situation and understand the world they live in, and therefore to make informed choices? Good quality education, in essence, involves creating a system that enables children to learn to know, learn to do, learn to live with others, learn to appreciate interdependence and diversity, and above all learn to 'act with ever greater autonomy, judgment and personal responsibility' (Delors et al. 1998).

This segregation of education, with different social classes enjoying differential access to different kinds of schools, cuts at the root of and poses a real threat to democracy. As Joel E. Cohen points out,

> little has been done to address questions about the universality of educational content, which varies across countries and regions. Universal could mean that all children get the same education. Or it could mean that all children get some education, but the content differs by place, type of school, or technology of instruction. Or it could mean that there are universal principles that apply everywhere but that the implementation and specification of these principles in practice depend on the local context. If content differs, there may be some or no common elements for all children. Or it could mean that education is the same for everybody except for those children who are disabled, remote, rural, minority, or otherwise handicapped. (Cohen 2010: 32)

In an increasingly polarized society, caste, religious, and linguistic identities are reinforced by politics: political parties mobilize along caste and community lines, pitting one group against the other to capture votes. Schools no longer provide a common shared space for children of different backgrounds and communities. Children today are growing up without getting an opportunity to mix with children from other social groups. Unlike children from middle-class and affluent families, whose greater access to the world media may potentially expose them to different viewpoints, the majority of poor children not only attend schools where they mix only with their own kind, but also have little access to the media (either print or visual). They are thus doubly disadvantaged: by poverty and by poor quality education.

Given this segregation, and given that those who can access reasonable quality education opt for higher-end careers, teaching itself is increasingly becoming a career of last resort. Given the nature of teacher deployment – regular teachers are sent to urban and well-located rural schools, and contract or ad hoc teachers are sent to more rural/remote areas – the overall ability of teachers deployed in rural/remote areas is noticeably weaker. With the base of the educational pyramid remaining poor and our system's inability to attract competent teachers, we are only reinforcing the quality divide.

As we peel away the layers of the education system, more contradictions reveal themselves. Larger governance issues like corruption, rent seeking, and patronage networks make an already difficult situation even more resistant to reform. Even well-meaning reformers are at a loss about where to begin. Lofty goals are set every few years, new documents are released (the most recent being the 2005 nation-wide effort of NCERT to draft a national curriculum framework) and institutions such as the National Council for Teacher Education (NCTE) are revamped to regulate quality. Yet, institutions that are expected to set standards and provide academic leadership at different levels (national, state, and district) have been reduced to petty fiefdoms that are least concerned with larger issues, such as the purpose of education. And, unfortunately, the larger political debate on education continues to centre primarily on writing and re-writing history and the peculiar manifestations of affirmative action, which only reinforce social divisions and existing inequalities.

The crisis faced by Indian education reflects a global concern. Be it the rioting youth of migrant communities in Europe, the exasperated students in South and Southeast Asia and the Middle East, or the palpable unrest in the Americas and Africa, all are telling us something. The educational system that emerged in the early industrial period is no longer able to meet the growing aspirations and needs of people across the world. The 'best' are happy with the Ivy League, Oxbridge, Sorbonne, and the Indian Institutes of Technology and of Management, which are their stepping-stones to money, fame, and power. Simultaneously, the media has equalized people's aspirations and raised hopes that education could indeed be the great equalizer. Yet, the education that the poor receive adds little real value to their lives. The real world is harsh, and those on the margins are being pushed further out into the wild.

Quality is the heart of the struggle for equality and justice. Our unwillingness or inability to define what we really mean by 'quality' and the institutional mechanisms that are necessary to realize standards of quality is ultimately a political question. Unfortunately, the communist parties, the social democrats, the centrist liberals, and the right-wing ideologues all are on the same side in this education quagmire, together contributing to the growing schism between rhetoric and reality.

The situation may seem grim. But as we look closer and listen carefully, young people across the world are not in the mood to take this lying down. Maybe there is a silver lining. There is an urgent need to re-imagine education, overhaul the system, and link education to life, livelihood, peace, and social justice. While striving for equality, the education system should be able to respond to people's aspirations and opportunities while enhancing their ability to make choices. A farmer should be able to enhance her productivity, weigh the pros and cons of traditional and modern technologies, and make informed choices. Children living in a multicultural environment should have an opportunity to learn about each other while creating bonds from shared experiences and, at the same time, learning from the catastrophic consequences of racism, parochialism, and communalism. Skilled and unskilled workers should be able to negotiate just wages and resist gender and age inequalities. The education system should have the depth as well as the range to span the different worlds where people live. A child born into a community that is rooted in a traditional occupation must have the opportunity to move out of that occupation if she so wishes, or to infuse modern technologies or marketing opportunities into it. We may have to reconstruct education in such a way that it provides multiple points of entry and exit, accommodates varying paces of learning, and enables people to both stand tall where they are and reach out to the world with confidence. Young people who do not want to pursue formal schooling after Class VIII or X should have the opportunity to pursue learning in a different stream. Equally, a young person who drops out at age 14 should be able to pick up the threads at a later stage, if she so desires. There is a need to rethink the value of linear progression from primary to higher education. This can happen only if people who are committed to an alternative vision have the courage to end the mindless pursuit of numbers (also known misleadingly as 'goals') and call the bluff. This is the only way we can strengthen the voices clamouring for a meaningful education.

Is There a Way Forward?

It is universally accepted that three factors influence the performance of a school system, namely: inputs (facilities, teacher-pupil ratios, student enrolment, no-detention policy, incentives, midday meals), the pedagogical process employed in classrooms, and the overall governance of the school system. We know that expenditure on inputs has been going up steadily, and (especially under the DPEP and SSA programmes) teacher training has also been given importance. However, research done in the last ten years does not show any positive correlation between these inputs and student learning (Muralidharan 2013). Since 1993–94, educational planners and administrators have been pushing for more of the same, and many of them refuse to acknowledge reality. It is time the education community started thinking differently and preparing for sweeping reforms – however painful it may be.

Maybe we need to start with teachers, initiate a dialogue with them, and take them on board to reform the recruitment, academic support, monitoring, and teacher management systems. There is something fundamentally wrong with the way we have managed teachers as a cadre, teachers as professionals, teachers as individuals, teachers as experts, and teachers as government servants. We have swung like a pendulum from one strategy to another. Some state governments and administrators have thought that hiring local contract teachers was the way to manage absenteeism, others thought that this was a bad idea and that protecting teacher tenure was the key, yet others thought mandatory training would do the trick . . . we in India have tried many things at different points of time. Given that teachers have been seen as a powerful constituency, however, accountability systems have never been designed or implemented seriously.

Maybe the best way to move forward is to announce that learning outcomes will be the primary and most important goal for elementary education. This has to be made into the non-negotiable ground rule that informs the entire education system. I am not arguing for mechanical testing of all children and adding to the burden of tests, but I am arguing for an effective system that will rescue children from the cumulative burden of non-learning. Let us not see the entire school system through the eyes of the elite schools of Delhi, Mumbai, and Chennai; the majority of schools in the country have no idea what or how much their students are learning. The Right to Education bestows the right to *learn* – not just to attend school and be promoted from one grade to the next without mastering the basics of language, science, or arithmetic.

We need to evolve a system in which the learning levels of children are ascertained at periodic intervals and presented to parents and other stakeholders in the immediate environment of the school. Though we still may not know how, it is high time that we actively involved the parents, the education community, civil society organizations, local leaders, and the media in revealing what and how much children are learning. Yes, as many educational researchers would caution, testing cannot be an end in itself, and there is no guarantee that measuring and presenting learning outcomes would automatically lead to improvements. Active interactions between teachers, parents, and academic support structures are the way to arrive at a balance between these two extremes.

One of the fundamental issues is that teachers are expected to stick to the syllabus and go through the motions of completing it, regardless of whether the children are learning. Planners and administrators assume that all children in a class are of the same level and that teachers have to adhere to a specific timeframe in which to complete the syllabus. The teachers are accountable for completing the prescribed textbook, not for ascertaining what their students know. Should we not turn the system on its head and start with what children know and who is at what level? Should not our schools have adequate autonomy, with the power to work out what is best for the students delegated to the head teachers? The dividing line between syllabi-oriented teaching and remedial teaching needs to go. The two need to become a seamless process that allows children to learn at their own pace and enables teachers to work with groups of children at different levels, perhaps even providing evening classes for children who may not get support at home. This means that we have to reduce the pupil-teacher ratios at the primary level to around 1:20, so that from Classes I to V teachers have the space to innovate and work with every child. At the same time, teachers need to be made accountable to their children. Headmasters need to have the authority (power) to manage teachers, including sanctioning leave, determining their annual pay raise, withholding the salaries of those who are absent or do not teach, plan for teachers' academic improvement, and so on. They should also have an adequate budget for hiring additional local teachers to provide remedial instruction after school hours and to support slow learners and children with disabilities.

It is essential to build a strong foundation at the primary stage, since it is difficult for children to catch up later. There is a lot of evidence from around the world to show that children who learn to read well in childhood have the potential to become autonomous learners at a later stage. Building a strong foundation is critical at the pre-primary and primary levels.

We need to make sure every single school in the country comes up to a basic, minimally acceptable level: one teacher for each class should be non-negotiable; the pupil-teacher ratio has to be calculated for every school, not every district; and a robust, independent mechanism should be in place to audit the functioning of schools (through toll-free helplines for students and parents to lodge complaints, for example).

Further, teachers should be appointed to a school, rather than to a state or district cadre, and they should be made accountable to the children. Maybe all teachers should work in a single school for a period of five to eight years before they can ask for a change. Many countries require teachers to renew their 'licence' periodically to prove their knowledge and skill of teaching. I do to know if such a system can be implemented in India, but it is worth exploring the ways and means of enabling teachers to update their knowledge and skills, to show that they have the aptitude to continue as teachers, and to demonstrate their commitment to learning and growing. Maybe promotions and raises could be linked to a mechanism that acknowledges good teachers and discourages those who do not have the aptitude or interest. I know these proposals are controversial in India, but nevertheless we have to search for ways to overhaul our school system.

Transfers, promotions and other career opportunities could be linked to the overall learning level of the children a teacher is responsible for. Equally, teacher training should be based on the abilities and subject knowledge of teachers. Training-related decisions need to be taken at the school level and supported by cluster- or block-level institutions. This will happen only if the block- and cluster-level academic bodies created by the government are made accountable for the learning levels of children in their respective areas, and are made to take cues from the Headmasters, rather than from some official sitting in the state capital. In a nutshell, educational administrators have to take responsibility for students' learning.

We are what we think and what we believe. Teachers are no different from the rest of society. Attitudes and prejudices influence the way teachers teach, how much interest they take in their students, and most importantly, how much effort they put into reaching out to and touching every child. This issue is rarely discussed and, even worse, it is not formally acknowledged. Given the growing social distance between the middle-class, well paid teachers and the children attending government schools, it is important to turn the spotlight on teachers' beliefs and attitudes. If teachers believe that some children 'cannot learn', or that education is not of much value to some children, then they are not likely to make an effort. Equally, if teachers have pre-conceived notions of the ability of Dalit or Tribal children, or of girls, to understand mathematics or science, chances are that the teacher will just ignore them. The tragedy is that – at least in India – pre-service and in-service teacher training rarely addresses these attitudes and prejudices. While this issue is discussed in Chapter 9 (The Teacher Conundrum) of this book, highlighting and underscoring the importance of attitude changes for school transformation is important.

The bottom line is that the entire education system has to make sure children realize their right to be taught and their right to learn, even if it means flagging uncomfortable questions and repeating unanswered ones.

3 The Narrative Emerging from Different Data Sources

Educational inequalities have gradually become sharper since India's independence, and by the 1990s there was recognition of the gravity of the problem. Responding to global trends, the government initiated several educational programmes to realize the goal of education for all. The first, the District Primary Education Programme (DPEP) initiated in 1994, focused on providing primary education to all, especially girls and children from marginalized social and economic groups. Sarva Shiksha Abhiyan (SSA), initiated in 2001, focused on opening more elementary schools and strengthening the existing school infrastructure and academic support structures at the cluster, block, and district levels. In 2009, the government enacted the landmark Right to Education Act to guarantee all children the right to education. And now, in 2017, there are again moves to redraft the national education policy.

These initiatives have led to a steady increase in the number of children participating in primary education. This is more visible among girls and children from rural areas, educationally backward states, and socially and economically disadvantaged groups. However, while significant economic and employment growth has taken place in the skilled services (information technology, financial services, telecommunications, tourism, and retail) and skill-intensive manufacturing, the hard fact is that all of these jobs require, at minimum, a secondary education. 'Employer surveys done in the recent past indicate that shortages of skilled workers constrain new private sector investment and growth' (World Bank 2009: 9).

As discussed in the preceding chapter, the Indian government's focus has been on primary education since 1950. It was only in 2003 that a decision was taken to extend the focus to age 14 to cover the whole elementary cycle. Equally noteworthy is that the government only woke up to the importance of secondary education in 2010, with the launching of the Rashtriya Madhyamik Shiksha Abhiyan (RMSA, popularly called the Secondary Education Mission).

While it is indeed significant that, according to the 2011 Census (Office of the Registrar General & Census Commissioner, GOI 2011), the literacy rate in India is 82.14% for men and 65.46% for women, with the lowest gender literacy gap (16.68%) since 1951, it is also important to understand that literacy rates are a very crude indicator of the educational status of a country. When we look deeper and examine the data on Gross Enrolment Ratios (GER) at different stages, dropout rates, transition rates, and successful secondary school completion rates, it becomes apparent that quality education has remained elusive and that disparities across genders, socio-economic classes, and geographical areas persist.

In this chapter, I attempt to draw a picture of this situation using secondary data from government sources, such as the decennial censuses of population, National Sample Survey (NSS), National Family Health Survey (NFHS), District Information System for Education (DISE), Statistics of School Education (SSE), and a few independent surveys like the Annual Status of Education Report (ASER).

Needless to say, these data sources each have different infirmities, and are often not comparable (i.e., sample surveys and population surveys provide different kinds of data). Often the surveys use different age categories; some of the surveys collect data from schools, while others collect it directly from the families. Hence it is difficult to create disaggregated data tables from these different sources that reflect clear trends across genders and social groups for consistent or even seamless time periods.

Enrolment and Attendance

Nearly universal enrolment has been achieved at the elementary level, with almost 96.9% of children enrolled in either a government or private school – although there is a steady decline in enrolment as one goes higher in the age groups, with the 15–16 years cohort having only 84.7% enrolment and a slight skew in favour of boys (ASER Centre and PRATHAM 2017). This is corroborated by data from different government sources. For instance, the Gross Enrolment Ratio (GER) at the primary level was 100% in 2007–08 (NUEPA 2015a), but the GER at the lower secondary level (enrolment in Class IX–X as a percentage of 14–16-year-old adolescents) was 78.5% and at the higher secondary level (Class XI–XII enrolment as a percentage of 16–18-year-old adolescents) was 54.21% (NUEPA 2015b).

The DPEP and SSA programmes brought improvements in the number of primary and upper primary schools, their infrastructure

Figure 3.1: Gross Enrolment Rates for selected years

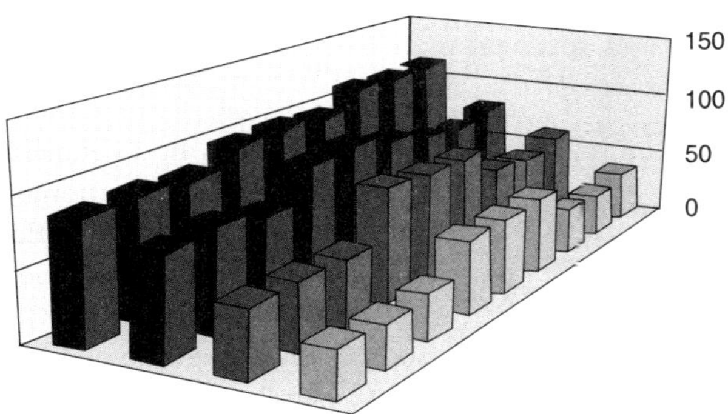

	2007–08			2014–15			2014–15 NER		
	Boys	Girls	Total	Boys	Girls	Total	Boys	Girls	Total
Primary (I–V)	115.26	112.58	113.97	98.9	101.4	100.1	86.3	88.7	87.4
Upper Primary (VI–VIII)	81.48	74.36	78.06	87.7	95.3	91.3	69.7	75.7	72.5
Secondary (IX–X)	62.62	53.23	58.15	78.13	78.94	78.51	48.1	48.9	48.5
Higher Secondary (XI–XII)	35.26	30.4	33.48	54.57	53.81	54.21	32.8	30.4	32.7

■ Primary (I–V) ■ Upper Primary (VI–VIII)
■ Secondary (IX–X) □ Higher Secondary (XI–XII)

Source: Select Education Statistics, MHRD; DISE 2007–08 and 2014–15.

and facilities, and their proximity to students. The impact of this is reflected in the enrolment data, which shows that primary enrolment increased from nearly 105 million in 1994–95 to 131 million in 2004–05 (MHRD, GOI n.d. [2008]). This growth incorporates the growth of girls' enrolment from 45 million in 1994–95 to 61 million in 2004–05. Since then, further improvements have been seen in girls' enrolment at the primary and upper primary levels.

This pattern seems to be continuing, even though the intensity has decreased. For example, GER at the primary stage is a little over 100%,[1] but at the upper primary level it is 91.3%, and at the secondary and higher secondary levels it is 78.5% and 54.2%, respectively (NUEPA 2015a, 2015b). This indicates that there is still nearly a 10% gap in GER between the primary and upper primary levels, a gap of nearly 13% between the upper primary and secondary levels, and a huge gap of 24% between the secondary and higher secondary levels. The Net Enrolment Ratio (NER) continues to be low, not only at the secondary level of education but also at the primary and upper primary levels. In 2014–15, the NER at the primary stage was 87.4%, at the upper primary stage[2] it was 72.5%, at the secondary stage it was 48.5%, and at the higher secondary stage it was 32.7% (NUEPA 2015a, 2015b). This contradicts the claim of near-universal enrolment at the primary and upper primary levels. Further, large differences in the GER at different levels confirm the persistence of high dropout rates.

This clearly shows that, while there has been improvement in overall enrolment across India, including that of girls and children from marginalized social groups, we are still far behind the goal of universal access and participation.

However, enrolment figures are frequently seen with suspicion, because teachers are expected to enrol all children between age 6

[1] Indicates that the cohort includes children who are older and younger than the accepted ages for each class.

[2] The DISE reports an Age Specific Enrolment Ratio of around 92% for age groups 6+ to 10+, 11+ to 13+, and 6+ to 13+ in 2014–15 (NUEPA 2015a, 2015b). This means children of age 6+ to 13+ are in schools. But one needs to look more carefully into both the data and the calculation methodology of the indicators to understand these inconsistent results.

and 14 in the schools. In a study done by the government in 2009, it was found that the average gap in attendance between the first and the last hour of school hours was 2.7% at the primary stage and 2.1% at the upper primary stage (Research, Evaluation and Studies Unit, Technical Support Group 2009). The gap between first and last hour attendance rates at the primary stage is even larger in the case of Bihar (3.6%), Rajasthan (7.2%), and Uttar Pradesh (4.4%). At the upper primary level, the difference between the first and last hour attendance rates is larger in Haryana (5.8%) and Rajasthan (4.5%). When we further break down the data by class, significant variation in attendance rates becomes visible. It is lowest in Class I (65.6%), and increases to 72.0% in Class V and to 78.8% in Class VIII. This can be attributed to the fact that students in Class I may be dropped at school by their parents or older members of the family; domestic chores or care for other members of the family may take precedence over dropping a child at school. Students in the higher classes do not abstain from going to school as frequently as those of lower classes because they are more independent by virtue of their age, and maybe also because they are more immune to infectious diseases. Lower attendance in the second half of the school day may be due to student fatigue, loss of interest and unattractive classwork in the afternoons, or being required to leave early because of work at home (Research, Evaluation and Studies Unit, Technical Support Group 2009).

Children's attendance as a percent of enrolment shows a decline from 73.4% in 2007 to 71.4% in 2014 for rural primary schools at the All-India level (ASER Centre and PRATHAM 2015). This decline is slightly steeper for the upper primary schools, where it decreased from 75.6% in 2007 to 71.1% in 2014. Although the attendance rates of children vary across states and through time, by and large, attendance rates were higher in 2009 than in 2014 (ASER Centre and PRATHAM 2015: 82). In some states like Bihar, Madhya Pradesh, and Uttar Pradesh, the attendance levels of children were much lower in 2014 – ranging from 50% to 60%.

As we move from elementary to secondary school, there is a steady increase in attendance rates; this trend is visible across gender and level of education. However, it is noteworthy that a ten-point

Figure 3.2: Gross Attendance Ratios by State for 2005–06 and 2014–15

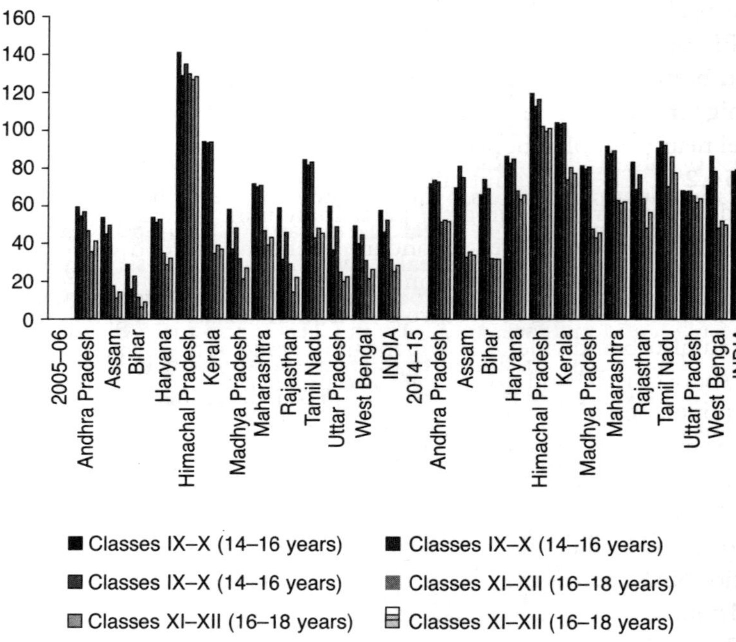

Source: MHRD, GOI 2005 and NUEPA 2015a, 2015b.

difference between boys and girls persists right up to the secondary level (World Bank 2009).

There has been significant progress in secondary education participation across all states. At the national level, secondary-level GER has increased from 57.6% in 2005–06 to 78.5% in 2014–15: a 26% jump in one decade. The progress achieved by some of the educationally backward states is also commendable. For example, the secondary level GER increased by 46% in Bihar, and by over 30% in Haryana, Madhya Pradesh, Rajasthan, and West Bengal, between 2005–06 and 2014–16. The progress in girls' enrolment is even more interesting. The GER of girls in secondary education improved by 32.7% at the national level, by more than 40% in West Bengal and Madhya Pradesh, and by nearly

59% in Bihar over the same period. In other states, the progress of girls is also remarkable. In many states, the Gender Parity Index (GPI) based on GER is more than 1, indicating that more girls than boys are enrolled in school. Similar trends can be observed in higher secondary education. The GER at the higher secondary level nearly doubled between 2005–06 and 2014–15, from 28.5% to 54.2%. The educationally backward states have made giant strides in general, and in girls' participation in particular, at both the secondary and higher-secondary levels. Interstate divergences and gender disparities have considerably narrowed.

However, looking at the Gross Attendance Ratio (GAR) across the rural-urban divide, it is evident that, while at the primary level the share of rural and urban areas is representative of their share of the population of children in the official age range, at higher levels the rural share of attendance falls below the rural share of population – leading to a 5% under-representation in secondary attendance and a 10% under-representation in higher secondary attendance (Bhog, Ghosh, and Mullick 2011). If we look at the primary Net Attendance Ratio (NAR) across the parameters of urban-rural, richest-poorest, and both sexes, the biggest male-female disparity is among the poorest (7.3%), followed by rural students (5.1%).

Compared to GER, data on the Net Enrolment Ratio (NER) reveals a less upbeat scenario. In 2014, NER was 48.4% at the secondary level and 32.7% at the higher secondary level (NUEPA 2015c). This indicates that a large number of children enrolled in secondary schools are either under- or (more likely) over-aged. Further, many children of the age that is supposed to correspond to the secondary level of education are not, in fact, in secondary education. It may be that a large number of these children are in the upper primary levels, and a few may have already crossed secondary education. The timely progress of children (according to their age) is critical for improving participation levels in secondary education.

This data gives us a broad understanding of the urban- rural divide at the primary and upper primary levels, across both sexes. But what we miss is the granularity of information on economically and socially backward groups, such as STs. In the later part of this chapter, I will examine the data that is available on these disadvantaged groups.

First, however, we need to understand the learning that is happening in our classrooms if we are to comprehend why a child would not like going to school.

Learning

With school enrolment and attendance, the first step towards learning is passed. However, it is important to underscore that, by itself, enrolment may not translate into learning. What students learn in school depends on several factors, many of which are school- and teacher-related. These factors affect not only students' learning in academic subjects, but also their personality development and achievement in various co-curricular areas. A lot depends on how teachers and students spend their time on teaching-learning activities in school. Apart from the time spent on active teaching by teachers and how they teach, another important dimension is 'what the students do and how much time they spend on different types of learning and other related or unrelated activities in school' (MHRD, GOI 2012).

In 2007, a study was conducted under SSA to estimate the time that teachers and students spent on learning and other activities in school, covering government and local body primary and upper primary schools in Assam, Haryana, Karnataka, Maharashtra, and Orissa (Research, Evaluation and Studies Unit, Technical Support Group 2010). This study revealed interesting insights into what really is happening with in the classrooms. For instance, teachers' activities in the classroom can be broadly classified as: (1) 'Student-centric activities', in which students are supposed to be mentally involved, alert, and active participants; (2) 'Teacher-centric activities', in which teachers play a dominant role and there is less interaction between teachers and students, and as a result students' initiative and mental processes are at a lower level; and (3) 'Supportive Instructional activities', which that are necessary but not explicitly part of teaching. The remaining two non-teaching activities are 'Class management' and 'Off-task'. On average, 83.8% of total classroom time is spent on student-centric (29.0%) and teacher-centric (54.8%) activities. 'Class management' and 'Off-task' consumed 1.5% and 2.1% of teachers' time, respectively.

Similarly, students' activities can be broadly classified as: (a) Active learning; (b) Passive learning; (c) Mechanical tasks; (d) Class management; and (e) Off-task. Of the total time spent on classroom activities, the percentage of time that students spent on each of these activities was 24.7%, 46.9%, 15.4%, 5.1%, and 7.8%, respectively. Passive learning time gradually increases from 41.2% in Class II to 53.7% in Class VI, whereas active learning time decreases from 26.4% in Class II to 22.0% in Class VI. This could be because as the grade levels go up the teacher-centric activities increase, with teachers mainly following the 'talk and chalk' way of teaching; hence, passive learning that includes listening attentively to the teacher and taking dictation increases (Research, Evaluation and Studies Unit, Technical Support Group 2010). Manabi Majumdar and Jos Mooij rightly point out,

> There is no doubt that the educational vision promoted by SSA is, indeed based on the idea that education should be enjoyable, creativity-enhancing and promoting the all-round development of the child. Many teachers however, live with other notions of education quality. At best, they believe that these new models are appropriate for young children, but from grade 4 onwards they feel children need to be subjected to a more rigid regime of rote learning. (Majumdar and Mooij 2011: 15)

Ten years of ASER reports show a startling decline in students' learning levels. Though the numbers are worrisome, it is important to note that the situation we see on the ground is the cumulative impact of a year-by-year drop in learning. For instance, around 55.5% of Class III children in Bihar could do subtraction and more in 2007, but by 2014 this number had shrunk to a mere 24.2% (ASER Centre and PRATHAM 2007b, 2015)! Bihar is not alone. A similar decline in reading and mathematics was also reported from Uttar Pradesh, West Bengal, Rajasthan, and Haryana. The survey found that falling attendance in the rural government schools in Bihar, Uttar Pradesh, and Rajasthan was a clear reason for this decline in learning levels. The average attendance of students in primary classes in Bihar declined from 59% in 2007 to 50% in 2011, but subsequently improved to 58.2% in 2014. In Uttar Pradesh, attendance fell from

64.5% in 2007 to 57% in 2011, and further to 55% in 2014 (ASER Centre and PRATHAM 2007b, 2012, 2015). It is important to note that student attendance is also tied to teacher attendance: without teachers present in the classroom, learning cannot happen. Teacher absenteeism and lack of accountability are prevalent issues across all states. Another reason for lower learning levels could be the increase of multi-grade classrooms in these states.

On the positive side, ASER noted that learning levels in Punjab and Tamil Nadu, where the state governments ran special programmes to improve reading ability and numeracy under the government's Sarva Shiksha Abhiyan (SSA) Programme, witnessed the most improvement. This indicates that a focused approach could lead to better learning outcomes. For instance, the focus of the Government of India (since 1991) has primarily been on enrolment, and not on the learning happening in the classroom. The outcome of this approach (as discussed in Chapter 2) is near-universal enrolment in primary schools. In the same vein, the focus of the Punjab and Tamil Nadu governments on reading is perhaps reflected in the recent National Assessment Surveys (NAS) (NCERT 2015). I must hasten to add that the jury is still out on whether short-term accelerated learning programmes can indeed have a long-term impact. Unfortunately, there are no studies that would help us understand the long-term impact of such interventions.

According to the most recent ASER report (ASER Centre and PRATHAM 2017: 43),

> Nationally, the proportion of children in Std III who are able to read at least Std I level text has gone up slightly, from 40.2% in 2014 to 42.5% in 2016. This proportion shows substantial increases among children in government schools in many states: Punjab, Uttarakhand, Haryana, Chhattisgarh, Gujarat, Maharashtra and Telangana. All these states show an improvement of more than 7 percentage points since 2014.
>
> Overall reading levels in Std V are almost the same year on year from 2011 to 2016. However, the proportion of children in Std V who could read a Std II level text improved by more than 5 percentage points from 2014 to 2016 in Gujarat, Maharashtra,

Tripura, Nagaland and Rajasthan. This improvement is driven by gains in learning levels in government schools in these states. Nationally, reading levels in Std VIII show a slight decline since 2014 (from 74.7% to 73.1%). Then and now, three out of every four children enrolled in Std VIII can read at least Std II level (the highest level assessed in the ASER survey). The state-wise picture for Std VIII reading levels does not show much improvement except for government schools in Manipur, Rajasthan, Maharashtra and Tamil Nadu.

It is interesting that, while similar proportions of children go to tutors, their contribution to the child's learning level may not be significant. Perhaps this is because private tuition is complementary: if school functioning declines, the effectiveness of tutoring is also low. Whether it is in the health sector, education, or even public services, the quality of government services sets the minimum expectations for private players. The private providers – especially those providing for the poor – will be only marginally better than the government schools, and this could also be the case for tuition centres. What is really worrying is that the quality of education – be it government or private – for the very poor continues to be abysmal.

Do Private Schools do Better than Government Ones?

The Quality Education Study was commissioned by Wipro Foundation and Education Initiatives in 2006 and 2011 (Educational Initiatives, Pvt. Ltd. 2006, [2011]). Conducted in urban schools that were considered good-quality educational institutions, the study focused on student achievement in subjects of English, Maths, Science, and Social Studies, along with co-scholastic achievements.

The main findings for the year 2011 indicate that students in the top schools of India perform lower than the international average on questions from studies such as 'Trends in International Maths and Science Study' (TIMSS) and 'Progress in Reading Literacy Study' (PIRLS) at the Class IV level, while they performed on par at the Class VIII level. The improvement in Class VIII levels was due to higher performance on procedural questions: students exhibited rote learning and performed comparatively better on questions that were procedural or did not involve a deeper understanding or application of concepts. There were significant differences between schools affiliated with different boards in different cities. This again brings forth comparisons between the different state boards and the Central Board of Secondary Education (CBSE) and Indian Certificate of Secondary Education (ICSE) boards. The differences between these boards themselves lend to inequity: children from CBSE or ICSE board schools performed better on the test compared with those from other boards.

In terms of the values of equity, diversity, and sensitivity, it was observed that there is a deep-rooted bias against the girl child, even in students from families that probably belong to the educated, higher socioeconomic strata of society. This was exhibited by nearly 43% of children in elementary school, who expressed that, if the choice had to be made, it would be better to educate a boy rather than a girl because educating the girl would only be a waste of resources in the long run. Also, 15% of girls in Class VIII felt that 'girls are burdensome to their parents'. Though 15% may seem like a small number, we need to remember that this survey was conducted in the better-quality urban schools, identified for their better and more sensitive learning environments. This can only lend perspective to how deeply this bias is ingrained in our social structure itself. The study also revealed that, while the principals and teachers said that co-scholastic activities were very important for the holistic development of the child, in practice most principals accepted not dedicating more than 9–10% of school time to physical education, sports, art, music, etc. (Educational Initiatives Pvt. Ltd. [2011]).

It can be seen that, while the perceptions of 'top schools' vary significantly from those of rural elementary schools, there are commonalities – especially when it comes to learning levels and achievements. While one may argue about the extent and severity of this problem, along with the veracity of the sources and studies conducted, it is important to see the message in the numbers and try to improve teacher actions and learning outcomes across geographies. It is imperative to look beyond the numbers of schools and enrolment or literacy measures and seek to incorporate quality of education as the focus of the education-delivery system.

Dropout and Out-of-school Children

A decline in dropout rates is an important indicator of quality education. While the government has been focusing on universal enrolment and access, focus is also needed on retaining children within the system so that they can complete their formal education.

Table 3.1: Dropout rates in Classes I-V, I-VIII, and I-X

Year	Class I-V			Class I-VIII			Class I-IX		
	Boys	Girls	Total	Boys	Girls	Total	Boys	Girls	Total
1999–2000	38.7	42.3	40.3	52	58.0	54.5	66.6	70.6	68.3
2009–2010	30.25	27.25	28.86	40.59	44.39	42.39	53.38	51.97	52.76
2011–2012	23.4	21.0	22.3	41.5	40.0	40.8	48.6	52.2	50.3
2013–2014	4.53	4.14	4.34	3.09	4.49	3.77	17.93	17.79	17.86

Source: MHRD, GOI 2014b, 2014c; NUEPA 2014a, 2014b; excel files from MHRD for 1999–2000.

As seen in Table 3.1, dropout rates have been declining since 1999–2000 across all levels, but the gaps between dropout rates as we progress from the primary to upper primary to secondary levels are glaring. Also, it is very interesting to note that, though in 1999–2000 the dropout rates of girls were higher than those of boys, by 2009–10 the trend was reversed, with the dropout rates of girls being lower than those of boys in many states.

This can be attributed to the special programmes and incentives provided by the government to keep girls in school. For instance, serious improvements to the facilities and infrastructure of schools under Sarva Shiksha Abhiyan could be an important contributor. For example, today more than 90% of schools have a functioning girl's toilet (NUEPA 2015c). The Kasturba Gandhi Balika Vidhyalaya (KGBV) Scheme was launched in 2004 to set up residential schools that would specifically target hard-to-reach girls from marginalized social groups like SC, ST, OBC, and (in limited areas) Muslims. As of December 2013, there were 3564 KGBV schools spread across the most educationally backward blocks of India. According to the national evaluation done in 2013, the total number of girls enrolled was 3,43,309; the community-wise break-up of the enrolled girls was:

SC (30.57%); ST (24.96%); OBC (30.50%); Muslims (7.48%); and Below Poverty Line (BPL) (6.49%) (MHRD, GOI 2013).

Five hundred and forty-three KGBVs function in 71 Muslim-concentrated Special Focus Districts. At 25%, the percentage of Muslim girls at these schools is high and, given that many of these districts also have significant SC or ST populations, the enrolment of SC and ST girls is also substantial. It is also significant that 319 KGBV have been established in 41 SC Special Focus Districts. These KGBVs have 46.24% SC, 8.04% ST, 30.16% OBC, 7.55% BPL, and 8.10% Muslim girls. In the 508 KGBVs established in 84 ST Special Focus Districts, ST girls form 67.95% of the total enrolment, while SC girls form 13.09%, OBC form 14.33%, BPL form 2.03%, and Muslims form 2.60%. The success of the KGBV programme partly accounts for the significant increase in the participation of girls at the upper primary level (MHRD, GOI 2013).

It is believed that the Balika Cycle Yojana, introduced in Bihar in 2007 to provide girls in secondary school with a bicycle for their commute, could also have made some difference for that state. Similarly, schemes like free bus passes for girls (in Tamil Nadu and Haryana) and similar initiatives could have made some difference. For example, the dropout rate for Class I-X came down from 85.4% in 2005–06 to 71.9% in 2014 (MHRD, GOI 2005, 2014). Government initiatives such as these, combined with higher economic growth, exposure to media, better telecommunications connectivity, and improved infrastructure, could also have made the difference. The overwhelming importance of nurturing and promoting the social norm of sending children to school acts as a propelling force; this, coupled with the presence of a functioning school, seems to make a significant difference.

The interplay of gender with social and economic status adds another important dimension. Gender relations in the family, community, and society significantly influence the ability of girl children to access services, nutrition, immunizations, healthcare (especially during bouts of illness), and schooling. The good news, however, is that if girls do reach school and are able to cross the initial barrier, they are highly motivated and struggle against all odds to remain in school (Ramachandran 2004). This trend was confirmed in a recent study on the transition from primary to secondary and higher secondary school (Ramachandran, Saxena, and ERU Consultants Pvt. Ltd. 2017). We found that once girls cross the Class VIII barrier they are highly motivated to complete secondary and higher secondary school.

Reasons for Non-attendance

The National Family Health Survey (NFHS) Round III (IIPS 2006) highlights the reasons for not attending school. This data is divided by urban and rural areas for both sexes. It can be observed that the reasons for non-attendance are different for boys and girls. A small example is the citing of 'School too far away' as the main reason for 7.0% of rural female children, but for only 1.4% of rural male children. From this we can infer that, while schools at one kilometre proximity are becoming a reality for primary schooling, for secondary schooling children need to travel a longer distance – maybe to the outskirts of the village – and parents may not be comfortable sending adolescent girls that far from home and thereby discontinue school.

NFHS Round II (IIPS 1999) revealed that the most common reason for school dropout was 'not interested in studies'. This reason is given for 36% of boys and 21% of girls between 6–17 years old. This reason was the most prominent in NFHS I (IIPS 1993) as well, and could be attributed to poor teaching-learning, indifferent learning outcomes, schools that lack facilities, teachers being irregular, corporal punishment and harassment in the classroom, discrimination and/or exclusionary practices or attitudes towards children from some communities, etc. Poor teaching at the elementary stage influences the level of preparedness of 14–17 year olds, leading to students dropping out after the elementary cycle.

The next most commonly reported reason for dropping out is 'costs too much' – for both boys and girls. There was an increase in this reason in NFHS Round II (IIPS 1999) when compared to NFHS I (IIPS 1993). This again needs to be seen in a wider context: while schooling is free, especially in rural areas, there are always hidden costs like study materials or attending private tuitions outside of school. Attending tuitions outside of school to supplement learning is no longer an elitist urban phenomenon. With teacher absenteeism and multi-grade classrooms leading to almost no individual attention in class, a vast number of students from both government and private schools are attending tuitions. According to ASER 2016, 24.5% of students in primary (Class I–V) and 27.3% of students in upper-primary (Class VI–VIII) from both government and private schools across India attend paid tuitions outside of school (ASER Centre and PRATHAM 2017). The incidence of paid tuition is higher in the eastern states. For example, nearly 54.2% and 65.5% of students in primary and upper primary schools in Bihar and 69.1% and 79.3% of students in primary and upper primary schools in West Bengal attend tuitions (ASER Centre and PRATHAM 2017). This obviously is an added expense, especially at the secondary level. In addition to gender-based preferences in household resource allocation, it is possible that this extra expense causes families to prioritize the education of boys – and so the axe falls on girl children's education.

The fourth most common reason given was that boys were 'required for outside work for payment in cash or kind' and girls were 'required for household work'. While household sizes are shrinking, about 1/5th of rural households still consist of seven or more members (IIPS 2006). This could be contributing to the dropout rates, since children (especially girls) are needed to contribute to the household income as well as to help with household chores.

Number and Proximity of Schools

With the DPEP and SSA programmes, there has been both a tremendous growth in the number of primary schools and improvement to the infrastructure of existing schools. During the last few years (i.e., between 1999–2000 and 2014–15),[3] the number of primary schools increased from 6.42 lakh to 8.47 lakh nationwide; upper primary schools increased from 1.98 lakh to 4.25 lakh. Between 1999–2000 and 2014–15, enrolment in elementary education also increased substantially, especially at the upper primary stage. Whereas the annual increase in enrolment was 1.1% at the primary

[3] The data for this discussion is taken from NUEPA 2015c.

level,[4] for upper primary it was 3.2%. In both the primary and upper primary levels, the proportionate increase in girls' enrolment was higher than that of boys: in primary classes, the annual growth rate for boys was 0%, whereas for girls it was 1.78%. Similarly, for upper primary, boys' enrolment increased at a rate of 2.1% per year while girls' enrolment increased by 4.5%. Now, with an average promotion rate of more than 90% at the primary and upper primary levels (NUEPA 2015b) and with the RTE Act stating clearly that no child should be held back, it is clear that the pressure on secondary and senior secondary schooling is going increase manifold.

To get a better understanding of this, let us refer to Figure 3.3, which shows the number of schools at each level of education. In 2014, there were 8.47 lakh recognized primary schools, followed by 4.25 lakh upper primary schools. These schools feed into 2.36 lakh secondary and senior secondary schools. This implies that there is one upper primary school for about every two primary schools, and one high school for approximately every 3.6 primary schools.

Figure 3.3: Number of educational institutions at each level

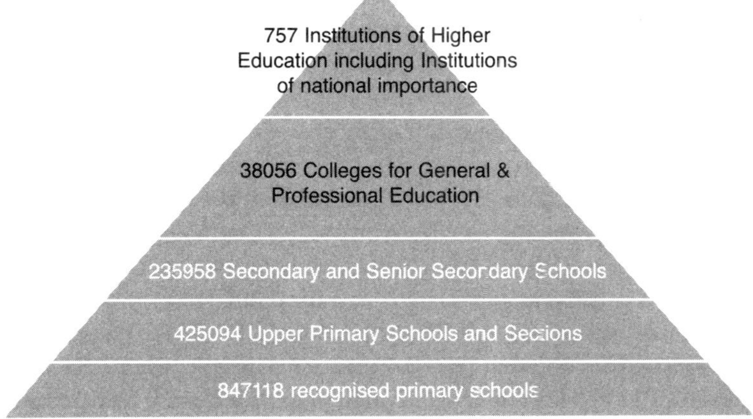

Source: MHRD, GOI 2016b and NUEPA 2015c.

[4] The annual growth rate was calculated with a semi-log equation using data from MHRD and NUEPA.

As we progress higher up the pyramid, we see that there are more than 38,000 colleges for general and professional education and 757 institutes of national importance.

In this situation, quality emerges as an important marker of equity, especially when it comes to availing the decreasing pool of educational facilities. Students from poorly functioning schools learn little; given that close to 50% of children in Class V cannot read a Class II-level text (ASER Centre and PRATHAM 2015), the implication is that many children just fall off the education ladder. Since the RTE now extends the no-detention policy up to Class VIII, chances are that children may just keep moving until Class VIII and then drop out.

The sheer numbers can act as a barrier to equity: because of the competition to enrol in higher education, children from government or alternative forms of schooling have no chance of competing with those who have access to the finances and resources required to prepare them for the pursuit of higher studies. Even for those who choose to pursue vocational training rather than a high-end professional education, there are not many options available. We have a large section of the population that really has nowhere to go once they finish their elementary school education.

Another problem is that, while elementary education prepares a student for reading, writing, and arithmetic, it is secondary schooling that actually makes an important contribution to his or her ability to read and write with comprehension, retain literacy skills, and acquire other skills, such as democratic citizenship and social cohesion. This simply cannot be done adequately at the elementary level – but the vast majority of India's youth have already left the education system before reaching the higher levels.

Another issue with these skewed numbers is the physical accessibility of these schools for the students. Whereas there are 61,152 secondary schools in urban areas to accept the students graduating from 70,996 upper primary schools (roughly 1.2:1 ratio), there are 174,806 secondary schools in rural areas to absorb graduates from 354,098 upper primary schools (a 2.3:1 ratio) (NUEPA 2015c). In addition, secondary schools in rural areas are typically located further from upper primary schools than in urban areas, making accessibility a real issue (NUEPA 2015c; World Bank 2009).

With the MHRD mandating a primary school within 1 km walking distance, an upper primary school within 3 km walking distance, and a secondary school within 5 km walking distance, access does not seem like it should be much of an issue. However, observations tell us that, in reality, only 85% of habitations have a secondary school within 5 km walking distance (NCERT 2016). According to the 71st Round of NSS (NSSO 2016), nearly 67% of rural households and 83% of urban households reported upper primary schools within 1 km from the house while only 37% of rural households compared to 73% of urban households reported secondary schools within such a distance. Within cities as well, while secondary schools seem to be available according to the numbers, their spatial concentration is again limited to the more 'well off' pockets, rather than the poorer areas or slums. For example, in a recent study (Jandhyala and Ramachandran 2015) we found that there was no secondary school for girls in the Muslim-concentrated area of Ajmer city.

The distance between schools and villages and the inequitable distribution of their locations becomes a deterrent for sending children – especially girls – to school. If we correlate this information with the NFHS III data (IIPS 2006), which shows that 5.8% of girl children stated 'school too far away' as their reason for non-attendance, we can understand the implications of proximity for their enrolment and attendance.

Pupil-teacher Ratios and Multi-grade Classrooms

While there is an issue with the number of schools catering to the needs of this country, there is also an issue in terms of students' access to quality education once in school. The Right to Education Act mandates a pupil-teacher ratio (PTR) of 30:1 and the presence of at least one teacher in each classroom to ensure that children learn better in primary school. Notwithstanding this regulation, in many schools across the country the PTR remains high. At the All-India level, there has been a marginal improvement in the proportion of schools complying with RTE norms for pupil-teacher ratios, from 38.9% in 2010 to 49.3% in 2014 (ASER Centre and

PRATHAM 2011, 2015). In 2014, Kerala stood out with 96.6% of schools in compliance; in Jammu & Kashmir, Nagaland, and Manipur, around 9% of schools are in compliance with these norms (ASER Centre and PRATHAM 2015). This is a quality issue because it has been proven that learning is correlated to PTR: classes with higher PTR do not achieve a lot in terms of learning because the teacher's attention is spread thin over many students, and he or she is therefore unable to provide the differentiated support for students from disadvantaged groups that is needed in the classroom.

While PTR is one side of the coin, the other side is the issue of single-teacher schools. In 2014–15, nearly 11.1% of the primary schools in India were single-teacher schools (NUEPA 2015a). The issue with single-teacher schools is that one teacher has to manage at least three to four classrooms, alone. This results in either students being kept in their respective grade-level classrooms without a teacher, or multi-grade teaching with several grade levels in the same classroom. Nationally, for rural primary schools, more than half of all classes visited by the team of ASER 2014 were multi-grade (ASER Centre and PRATHAM 2015). For example, Class II was sitting with one or more other classes in 62.8% of primary schools and in 59.8% of schools with upper primary sections. In Jammu and Kashmir and Jharkhand, more than 80% of observed Class II and IV classrooms in primary schools were multi-grade. Ideally, multi-grade teaching would not be a problem; it is better to group kids on the basis of accrued skills, rather than based on class levels defined by age. The issue is that our teachers' training caters to the norm of a single-grade classroom, rather than to the reality of multi-grade classrooms. This leads to lower learning levels; it has been observed that in classrooms with multiple grades, teachers tend to focus on the highest grade-level children while the others are simply not engaged. This can also be correlated to declines in student attendance (as shown in the ASER data from 2006 to 2015): children would rather be out of school playing or at home instead of sitting in a classroom where they are not engaged.

Teacher Absenteeism

While multi-grade classrooms do seem to impact student learning, we also need to consider the level of teacher absenteeism. Teacher facilitation is necessary for learning to happen in classrooms. A study (Research, Evaluation and Studies Unit, Technical Support Group 2009b) was conducted in Andhra Pradesh, Madhya Pradesh, and Uttar Pradesh to assess: (i) the percentage of teaching days lost due to teachers remaining absent from school for different reasons; (ii) the factors that contribute to teacher absences; and (iii) the effect of teacher absences on the attendance and achievement of students.

This study first looked at the number of teaching days lost per year compared to the total number of working days. As shown in Figure 3.4, Andhra Pradesh had both the least number of instructional days and the highest number of teaching days lost, compared to Madhya Pradesh and Uttar Pradesh. If we look at the state-level reports from the 10th Joint Review Mission of the Sarva Shiksha Abhiyan (SSA 2009), Bihar reported student attendance as low as 56% at both the primary and upper primary levels, while Jharkhand, Orissa, and West Bengal reported student attendance ranging from 71–75%. In 14 states, teacher attendance rates were between 75% and 85%. This

Figure 3.4: Teaching Days Lost

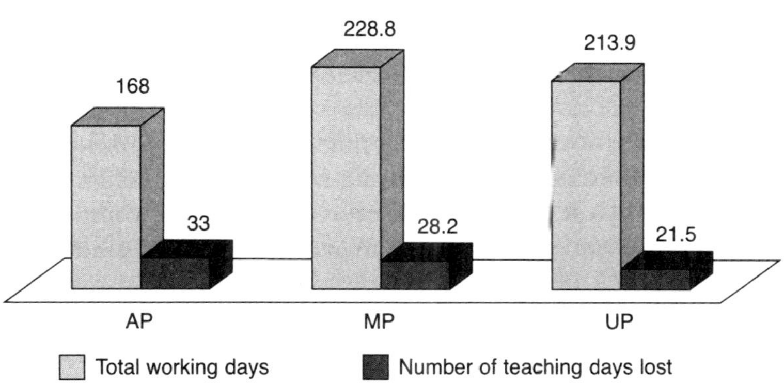

Source: Research, Evaluation and Studies Unit, Technical Support Group 2009b.

is corroborated by the ASER 2014 report, which pegs the average teacher attendance across India at 87% for the primary and upper primary levels (ASER Centre and PRATHAM 2015). This is cause for concern, as the MHRD has set a target of 90% teacher attendance across the country.

A recent study examined a sample of 619 schools and 2861 teachers across six states:

> The findings of the study show that teacher absenteeism, defined as 'absence without reason', is 2.5%. Although our sample is not statistically representative of the whole of the country, this number is roughly in the same order of magnitude as in other studies. [...] The different reasons that comprise teachers' absence in schools; reasons, given the realities of the government school system, range from official duties (academic and administrative) and official other departmental work to legitimate leaves that teachers are entitled to under their service conditions. Instead, teacher absence, at least in the popular narrative, is equated with teacher absenteeism. (Azim Premji Foundation Research Group 2017: 1–2)

This recent study is significant because it breaks down the reasons for teachers being absent from school. While the overall 'absence' rate was 18.9%, 3.8% were absent due to official academic duties; 2.1% for administrative duties; 0.9% for 'other department work'; 9.1% had taken authorized leave; and only 2.5% were absent on the day of the team's visit without prior intimation or a valid reason.

It is important to understand the reasons behind teachers' absence if we are to address the issue appropriately. According to a wide range of studies, it emerged that women teachers spent fewer days on duty away from school but took more leave for personal reasons. The number of days of medical leave taken by women was more than double that of men. However, it is not clear whether maternity leave was included in 'medical leave'. When the members of the VEC (Village Education Committee) and school leaders were asked about the reasons for absence, they highlighted 'family problems' and 'health of the teacher'; 'family problems' were reported by a majority of the respondents in Uttar Pradesh (Research, Evaluation and Studies Unit, Technical Support Group 2009b).

During focus group discussions in the community, the distance of teachers' residences from school was pointed out as a major factor that contributed to teachers not coming to school on time, leaving early, or even missing school in all three states. This was also expressed as transport not being available at the appropriate times, or teachers not staying in the villages where they were posted. Time spent on other income-generating activities, including private tuition, the pursuit of higher studies to improve their academic qualifications, and participation in political activities, was also given as a reason for teacher non-engagement and absenteeism, especially in Uttar Pradesh (Research, Evaluation and Studies Unit, Technical Support Group 2009b).

While a lack of infrastructure and facilities is surely one aspect of teacher absenteeism, the observed levels of disinvestment in the profession compelled us to delve deeper and seek out answers. The PROBE Report (PROBE Team 1999: 63) recognized this, and partly linked teacher absenteeism and shirking to the disempowering environment in which teachers have to work in India. While the vision of a teacher's role is defined grandly in policy documents, their lack of motivation and declining social status causes absenteeism. Currently, the autonomy of teachers in both teaching and learning is limited, leading to frustration (Majumdar and Mooij 2011).

Women Teachers

The lack of female teachers is another constraint on the delivery of equitable education for all. Though social norms are instrumental in encouraging women to become teachers, low female literacy and patriarchal values have had a negative impact in some states. In 2012–13, on average, for every 100 male teachers there are 79 female teachers in primary schooling, 76 female teachers in middle school, and a mere 66 female teachers in secondary school. This clearly shows that there is vertical segregation: as the education level goes up, the representation of female teachers comes down. This data varies significantly across states, as shown in Table 3.2.

Table 3.2: Number of women teachers per 100 male teachers in select states

States	Higher secondary schools	High school (post-Basic)	Middle (Senior Basic Schools)	Pre-primary & Junior Basic
	Number of female teachers (regular) per 100 male teachers			
All India	66	66	76	79
Assam	37	46	31	57
Bihar	27	22	64	68
Kerala	246	321	235	313
Madhya Pradesh	76	78	44	45
Maharashtra	44	58	92	81
Rajasthan	42	40	47	47
Tamil Nadu	123	130	200	288
Uttar Pradesh	19	26	47	60
West Bengal	58	57	35	81

Source: MHRD, GOI 2014c.

We can clearly see that in Tamil Nadu and Kerala the representation of female teachers is high across all levels. These states also have higher female literacy rates compared to most states in India. In states like Bihar, Uttar Pradesh, and Rajasthan, the numbers are low across all levels. However, analyses of the issue of gender in education need to go beyond just mechanically looking at the numbers. They need to focus on the everyday manifestations of gender in both learning and teaching.

We need to look at the attitudes and prejudices that act as a barrier to creating a more equally gendered environment in schools. Female teachers in rural India face multiple issues that hinder the growth of women's participation. It is tough for women to find accommodation and live alone in rural areas. Commuting is a possibility but has its own drawbacks, such as inconvenience, time wasted, and physical insecurity. Patriarchal norms enhance this problem. The family of a young, unmarried, female teacher in North India would hesitate to let her take a rural posting. If the teacher were married, then she would be expected to live where her husband finds employment, rather than the other way around.

Married teachers also have to face the double burden of teaching and household work.

The stress of dealing with patriarchal norms extends even to the school premises. Aside from sexual harassment, female teachers felt that they were treated like second-class employees by their male colleagues (PROBE Team 1999). Also, looking at the subjects men and women teach at the secondary levels, it can be observed that male teachers teach hard sciences such as physics, chemistry, or mathematics while female teachers teach arts, social sciences, and commerce-related subjects. This is because female teachers themselves do not have meaningful access to the 'hard sciences', since access to secondary education continues to be a challenge and teacher training courses are expensive.

States with lower percentages of female teachers in secondary schools, such as Rajasthan (28.2%), Orissa (25.1%), and Uttar Pradesh (25.6%), also have lower numbers of girls enrolled in secondary schooling (NUEPA 2015b). Female teachers are needed not only to help improve the enrolment of girl children in secondary education, but also to create a more gender-sensitive space in our schools. Currently, school teams and leadership are dominated by men, as are most teacher unions and associations.

School Stratification

The Indian educational system, like Indian society, is heavily stratified. The first level of stratification is between the government and private schools. Within government schools, there are different categories and types, from the well-funded Navodaya Vidhyalaya to one-teacher rural schools and the neglected Ashram Shalas for tribal children. Similarly, there is huge variation in the private sector, not only between aided and unaided schools, but between high- and low-cost schools, recognized and unrecognized schools, and finally, the tuition centres that pass themselves off as schools.

Although the number of and enrolment in private unaided schools is increasing, it is difficult to find a single data source to study these schools that can be deemed authentic, as both NCERT and DISE cover only recognized schools. While ASER, as a household survey, captures enrolment in unrecognized schools, it focuses mainly on rural areas. According to one national survey of 20 states, about half of all private rural primary schools are (perhaps) unrecognized (Muralidharan and Kremer 2006; Majumdar and Mooij 2011).

One of a parent's biggest expectations for a private school (irrespective of the category) is that it will perform better and guarantee better opportunities than government schools. This negative view of government schools could perhaps be attributed

Figure 3.5: Participation by different managements, 2014

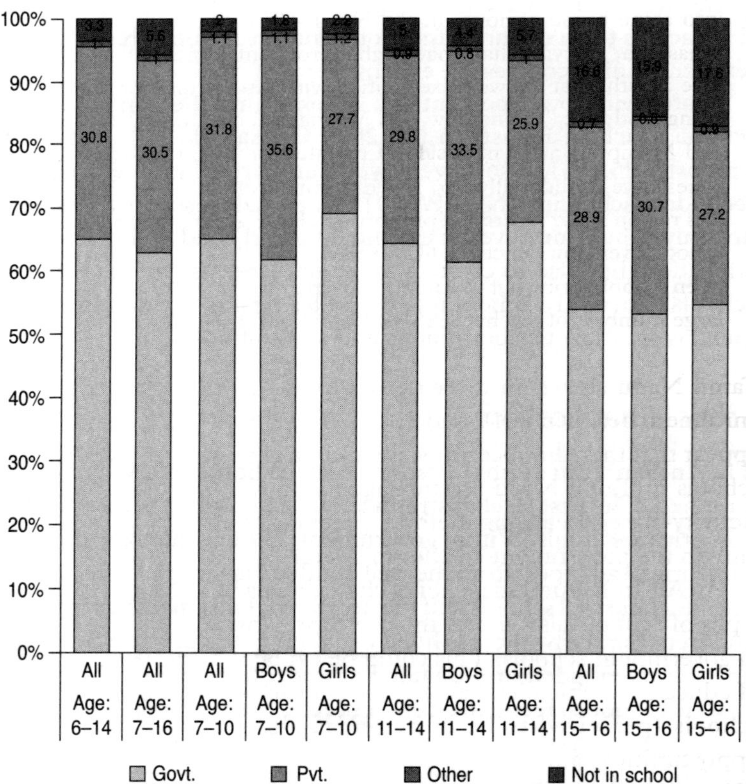

Note: 'Other' includes children going to Madrasa and EGS. Not in school = dropped out + never enrolled.
Source: ASER Centre and PRATHAM 2015.

to teacher absenteeism, the lack of accountability for teachers, and the poor infrastructure at government schools. Parents across urban and rural India choose to send their children to private schools (recognized or unrecognized) at considerable financial cost, with the expectation that private schools will impart a better education than government ones.

ASER captures this trend toward private enrolment and highlights the states with higher enrolment figures for private schools:

> As far as private school Enrolment is concerned, India can be divided into some broad regions. In the North-West, states like Punjab, Rajasthan, Haryana have had high Enrolment in private schools. The North-East shows mixed ratios with Assam and Arunachal being moderate, Tripura low, and Meghalaya, Mizoram, Nagaland and Manipur on the high side of private Enrolment. In the East, states have traditionally had low Enrolment in private schools, In this region Bihar where the proportion children enrolled in private school is very low which is a likely reflection of the massive efforts to open schools, bring out of school children into school and appointing large numbers of teacher. (ASER Centre and PRATHAM 2012)

Tamil Nadu shows an increase of around 10% in private school enrolment between 2007 and 2014 (NUEPA 2015c). While it may appear that there has been a positive change in the way government schools in Tamil Nadu are managed with the introduction of the Activity-Based Learning method, parents still do not seem to have faith in the government education system.

An All-India rural snapshot of children's participation in different types of schools also shows the overrepresentation of girl children in government schools. This could be explained by the rising costs of education, which lead parents to choose to send boys to private schools (perceived to be of better quality and providing better opportunities) and girls to government schools (perceived to be of lesser quality). This trend is evident in Figure 3.5 above, which shows the enrolment of different age groups in different types of schools, broken down by gender. In the 7–10 year-old group, 69% of girls are enrolled in government schools, versus 62% of boys.

In the 11–14 age group, girls' enrolment in government schools is 68%, compared to 61% for boys. There is not much difference in the 15–16 age group, because there is significant re-enrolment into government secondary schools so that children will qualify to take the Class X board examination.

If we look at the types of schools attended by secondary level students disaggregated by Monthly Per Capita Expenditure (MPCE) quintiles, it can be seen that there is a correlation between the expenditure capability of a household and the choice of school. As we move up each quintile, we observe a steady decrease in percentage of government school enrolment and an increase in private unaided school enrolment. This goes to show that, while private schools are mushrooming across the country, the lowest group in terms of economic prowess is still relying on the government's educational delivery system.

Data on school enrolment for 2010–11 shows that, for the first time, the number of children enrolled in English-medium schools from Classes I to VIII had crossed the 2 Crore mark: a 274% rise since 2003–04 (NUEPA Countrywide Enrolment Report). With more private schools than government ones offering English medium instruction from the beginning of elementary school, the choice of picking a private school becomes easier for parents who see learning

Figure 3.6: Types of schools attended by secondary level students as per MPCE quintiles

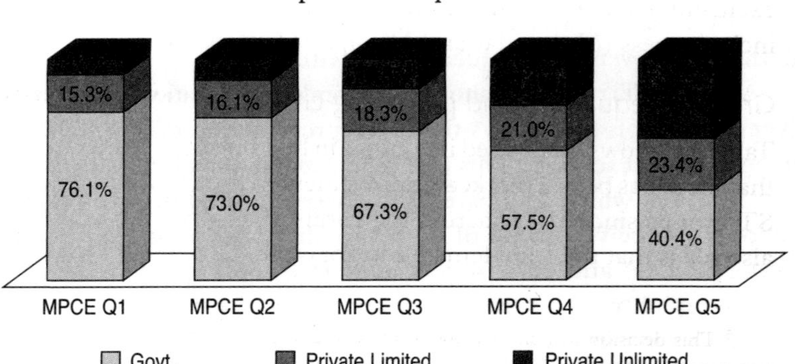

Source: NSSO 2010.

English as a ticket to upward mobility. However, this figure cannot be strictly interpreted as a marker of private school enrolment, since states like Punjab and Jammu and Kashmir have adopted English as a medium of instruction in government schools.[5]

Apart from low-cost private schooling, there is also a middle level of private schools that conduct entry exams, interview parents to see if they meet the school's unspoken criteria for social level, and charge fees around 2000–8000 rupees a month. There are also the additional costs of school uniforms, trips, coaching classes, or individual tuitions. At the top end of the spectrum is a small section of elite, global Indians whose children go to private schools with PTRs closer to 15:1 instead of 30:1, swimming pools and horseback riding facilities, international field trips, personal laptops for each student, and teachers with training from across the world.

The systemic class- and caste-wise sorting and streaming of children into different categories of schools has become so clear-cut and solidified that schools themselves appear to be socially stratified. Schools are branded as 'Dalit schools', 'elite schools', and so on, as opposed to socially mixed schools. Importantly, having more school options effectively means that the elite opts out of government-run schools, at a time when children from less privileged backgrounds have begun to enrol in such schools in greater numbers. This phenomenon of 'elite flight' reflects much more than just frustration with the quality of government schools: it indicates an ambition of exclusivity. It is only a short step from exclusivity to becoming exclusionary and fleeing from even the ideals of integration and inclusiveness (Majumdar and Mooij 2011; Ramachandran 2004).

Graded inequalities and persisting challenges

Table 3.3 shows the graded inequities in literacy rates, demonstrating that there has been a marked improvement in the literacy rates of SC/ST groups since 1991, across the rural-urban divide. But what we also see is that the rural female category in both of these groups lags

[5] This decision has, in fact, had terrifying consequences, as many teachers are not trained to teach in English and suddenly feel pressured to transact teaching in a foreign language.

quite far behind their respective male counterparts. Similar gaps are also seen between the urban female categories in both of these groups and their male counterparts. This indicates that, while it is heartening to see literacy rates improve, the struggle of a girl child belonging to a marginalized group to access quality education continues.

Table 3.3: Graded inequalities in literacy rates (Census 1991–2011)

Category	Census 2011	Census 2001	Census 1991
Rural Female ST	46.94	32.4	16
Rural Female SC	52.56	37.6	19.5
Rural Female Non-SC/ST	61.08	50.2	35.4
Rural Male ST	66.80	57.4	38.5
Urban Female SC	68.64	57.5	42.3
Urban Female ST	70.32	59.9	45.7
Rural Male SC	72.58	53.7	46
Rural Male Non-SC/ST	79.92	74.3	63.4
Urban Female Non-SC/ST	80.96	75.5	67.5
Urban Male ST	83.16	77.8	66.6
Urban Male SC	83.32	77.9	66.5
Urban Male Non-SC/ST	89.73	87.6	83.4

Source: Adapted from Lall and Rao 2011.

Looking at the caste and religious groups in Figure 3.7 below, it can be seen that Muslims have the lowest NAR, followed by SCs, STs, and then the others. This trend is observed starting from the upper primary level because government-initiated programmes such as DPEP and SSA, along with the provisioning of midday meals, have helped in getting children into school at the primary levels and keeping them there. While nearly 80–90% of children from marginalized groups attend primary school, the drop in NAR begins from the upper primary levels and continues all the way to senior secondary.

The caste-wise analysis of out-of-school children reveals that 9.97% of Muslim children were out of school in 2006. This percentage was higher than that of the other categories, like SC (8.17%) and ST children (9.54%). Among Muslim out-of-school

Figure 3.7: Net Attendance Ratio (NAR)

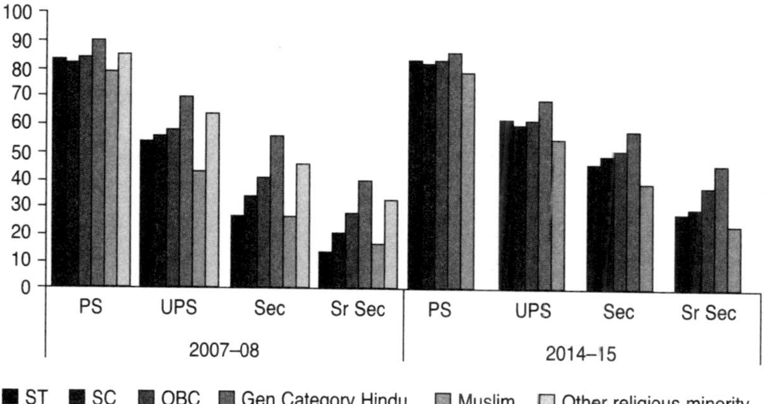

Source: NSSO 2010, 2016.

Table 3.4: Percentage of OOSC by social group (2006 and 2014)

	2006			2014		
	Rural	Urban	Total	Rural	Urban	Total
All Children	7.8	4.34	6.94	**3.1**	**2.5**	**3.0**
SC Children	8.55	6.25	8.17	**3.4**	**2.6**	**2.7**
ST Children	10.11	4.21	9.54	**4.8**	**1.8**	**4.4**
Muslim	12.03	7.17	9.97	n.a	n.a	n.a

Source: SRI-IMRB Survey 2006 and 2014.

children, 46.8% were girls. By 2014, the overall proportion of out-of-school children for all groups had drastically reduced to 3%; even only 4.4% of ST children were not attending school.

If we look at inequity in terms of the choice of schools for available SC and ST children, we see that a majority still opt for government schools. Of these, STs have a higher enrolment in private schools than SCs. SC enrolment in government schools is higher at the upper primary than the primary level; in 2008–09, for example, 80.93% were enrolled in government schools at the primary

Figure 3.8: Enrolment by management

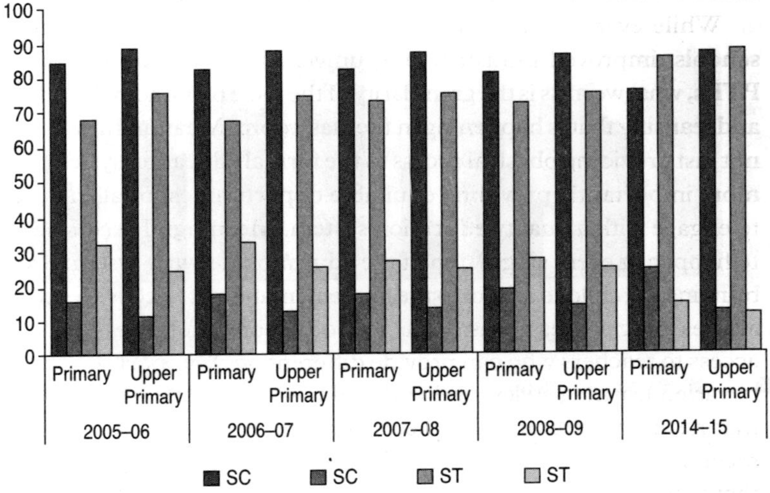

100
90
80
70
60
50
40
30
20
10
0

2005–06		2006–07		2007–08		2008–09		2014–15	
Primary	Upper Primary	Primary	Upper Primary	Primary	Upper Primary	Primary	Upper Primary	Primary	Upper Primary

■ SC ■ SC ■ ST □ ST

Source: NUEPA 2007, 2008, 2009, 2010, 2015a.

level and 86.01% were enrolled at the upper primary level. In terms of SC enrolment in private schools, we can see that the year-on-year enrolment is increasing at both the primary and upper primary levels. It is quite the reverse for STs, whose private school enrolment is decreasing at the primary level and quite stagnant at the upper primary level, and whose government enrolment has increased at the primary level and is also fairly stagnant at the upper primary level.

Meaningful Access

The number of schools being built is on the rise; infrastructure in terms of boundary walls, drinking water facilities, and the number of schools with a girl's toilet has certainly improved; and the Right To Education Act makes elementary education a fundamental right of every child. These are things to be pleased about. While the RTE Act mandates education as a fundamental right, most teachers understand this to have three meanings: 'Do not fail a child, do not hit a child, and do not let children drop out of school,' as succinctly stated by Rukmini Banerji at a conference in 2012.

While everyone is in agreement that there is a need for more schools, improved infrastructure, universal enrolment, and better PTRs, what we miss is the granularity of the issue of the actual teaching and learning that is happening in the classroom. Meaningful access is not just providing physical access to the formal education system, but more importantly providing equitable opportunities for all children to engage with a quality education system. Meaningful access needs to happen at every single step of the education delivery system, from bringing the child to school to the implementation of effective teaching and learning processes inside the school. Meaningful access includes access to teachers who can provide differentiated support that caters to varied learning styles, and who will pay special attention to those that need an extra nudge to keep the pace. And most importantly, meaningful access means providing a safe, gender-sensitive space that will enable girls and boys to express their own identity without fear of mockery or discrimination. Meaningful access can enable reduced dropout rates, smoother transitions between different levels of education, higher learning levels, and most importantly an equal platform for all children to achieve an excellent education.

As a final comment, while the comparability between different data sources is difficult to ascertain, it is important to glean the story that these numbers tell. It would be convenient to dismiss independent surveys as unreliable or unable to measure what is truly happening within the classrooms across our country. But then, the stakes are having our children merely registered or enrolled in schools but coming out no better than when they began. For schools to function as agents of social change and equity, we need to gear the system up to provide enough schools across all levels for all of our children, where they can take part in a high-quality learning experience catering to the differentiated needs of those from varied socio-economic backgrounds. The facilitators of this learning experience (the teachers) need to view themselves as agents of social change and strive to create safe, non-discriminatory classrooms. Involving the parents and community is also important if we are to build a more equitable societal structure.

4 The Cumulative Burden of Exclusion of Tribal Communities

While in percentage terms Scheduled Tribes comprise only 8.2% of the population, there are 104,545,716 (10.45 Crore, 104.5 Million) Tribal people in India. Madhya Pradesh, Maharashtra, Orissa, Gujarat, Rajasthan, Jharkhand, Chhattisgarh, Andhra Pradesh, West Bengal, and Karnataka are the States having larger number of Scheduled Tribes. These states account for 83.2% of the total Scheduled Tribe population of the country. Assam, Meghalaya, Nagaland, Jammu & Kashmir, Tripura, Mizoram, Bihar, Manipur, Arunachal Pradesh, and Tamil Nadu account for another 15.3% of the total Scheduled Tribe population. The share of the remaining states / UTs is negligible. (Chandramouli 2013)

There is a tendency to club all disadvantaged social groups into one; equally significant is the tendency to look at the tribal communities of India as one undifferentiated group. This chapter explores the differences between three tribal-majority regions of India and other areas with significant ST populations. Understanding and appreciating these differences is the first step towards acknowledging the need for different strategies to enhance the educational status of the community as a whole.

(An expanded version of this chapter was prepared for the Ministry of Tribal Affairs of the Government of India and UNDP in 2015 as an input for a proposed Tribal HDR of the Government of India, along with Professor Anugula N. Reddy of NUEPA – who provided all of the data used in this chapter. Unfortunately, the proposed Tribal HDR was not published by the government.)

Since the 1960s, several commissions and committees have been constituted by the Government of India to inquire into the social and economic situation of ST communities. The Tribal Commission of 1960–61 (Chaired by Mr U. N. Dhebar) argued that the historical legacy of different regions influenced access to education, participation in schooling, and most importantly the approach of the education system to the specific challenges faced by each group. At that time, the dominant perception was that developmental and educational interventions should respect the cultural and socio-economic context of the tribal community. Notwithstanding the recommendations of the Dhebar Committee – to pay special attention to the pedagogical and linguistic aspects of tribal education; create a programme to develop and train teachers in tribal areas; provide midday meals, clothing, and books; and most importantly calibrate the school calendar to suit the social rhythm of tribal communities – the unfortunate reality is that not much was done in this direction.

Many issues discussed by the Dhebar Commission were further elaborated by the Indian Education Commission of 1964–65 (known as the Kothari Commission), the Bhuria Committee of 1991, and the Bhuria Commission of 2002–04. In particular, each of these committees/commissions argued that the school calendar and timings have to be modified to suit the agricultural cycle and festivals of tribal communities. The Indian Education commission of 1964–65 also reiterated the need to have focused and time-bound programmes to enhance the pool of teachers in tribal areas. Notwithstanding the recommendations of these various committees, through the 1960s and until 1990 planners and administrators did not pay attention to the specific challenges faced in tribal areas of the country. While there is no comprehensive analysis of why the centre and state governments choose to ignore important recommendations, researchers (Sujatha 1999 and 2002; Jha and Jhingran 2002; Veerbhadranaika et al. 2012) have pointed out that there was no political resolve to address the specific education-related issues of tribal populations in India. Apart from strategies like Ashram Schools and multilingual textbooks (introduced in a few areas on a pilot basis), there has been no holistic approach meant to seriously address the access-, retention-, and most importantly learning-related issues of tribal communities.

This has led to continued neglect of the literacy and education of tribal communities; as a result, there has been only a modest increase in the literacy rates of the ST population, and the gap between ST and other communities remained high (Table 4.1). The momentum, however, picked up through the decades of the 1990s and 2000s, following which some progress in the literacy status of ST communities as a whole can be seen.

Table 4.1: Literacy rate of General, SC, and ST population (1961–2011)

Year	All Social Groups			SC			ST		
	M	F	Total	M	F	Total	M	F	Total
1961	40.4	15.4	28.3	17.0	3.3	10.3	13.8	3.2	8.5
1971	46.0	22.0	34.5	22.4	6.4	14.7	17.6	4.9	11.3
1981	56.4	29.8	43.6	31.1	10.9	21.4	24.5	8.0	16.4
1991	64.1	39.3	52.2	49.9	23.8	37.4	40.7	18.2	29.6
2001	75.2	53.7	64.8	66.6	41.9	54.7	59.2	34.8	47.1
2011	80.9	64.6	73.0	75.2	56.5	66.1	68.5	49.4	59.0

Source: Office of the Registrar General & Census Commissioner, GOI 1961, 1971, 1981, 1991, 2001, 2011.

The acceleration of literacy rates in the 1990s is often attributed to three important developments. First, after the adoption of the 1986 National Policy on Education and India's signing of the 1990 Education for All Declaration in Jomtien, Thailand, the Government of India stepped up efforts to not only open more schools and ensure that all children had a school within walking distance, but also to provide financial and planning support to the state governments through a number of centrally sponsored schemes: the National Literacy Mission (1989), Operation Blackboard (1987), Ashram Schools from primary to Class XII (under the Tribal Sub-Plan 1989), post-matric scholarships and targeted incentives (1992), the District Primary Education Programme (implemented in 1994 for educationally backward districts), and the National Programme for Nutrition Support (1995). Second, the 1990s witnessed economic growth and the rapid expansion of the media (television, radio), leading to changes in the aspirations of the poor – especially tribal communities. The demand for school education became palpable as the government started more schools in rural areas. As a result, many more ST children enrolled. Finally, the 2000s saw the introduction of Sarva Shiksha Abhiyan (2001) and Rashtriya Madhyamik Shiksha Abhiyan (2009), both of which further improved access to schools. In 2009, the Government of India also introduced the Right to Education (RTE) Act, which placed the onus on the government to provide all children with access to schools.

The outcomes of these interventions were fairly positive. However, notwithstanding this positive trend over the last two or three decades, we still have a long way to go before the educational attainments of the ST community are on par with other social groups. This chapter presents an analytical overview of the situation at different levels of education and attempts to analyse the reasons for these persisting challenges.

Literacy Situation in Different Tribal Regions

Statements of the overall literacy of STs hide significant regional variations. This chapter presents the educational situation in four categories of states:[1] Schedule V, Schedule VI, Small/Island states, and other states with significant ST populations. This chapter also uses gender-disaggregated data to underscore the point that the educational situation of ST women is perhaps the best litmus test for the persistence of multiple disadvantages.

The literacy status in Schedule V states is perhaps the most worrisome. As evident from Table 2, rural ST female literacy is less than 40% in Rajasthan, Orissa, and Andhra Pradesh and between 40% and 50% in Chhattisgarh, Jharkhand, and Madhya Pradesh. Equally significant is that several states in the non-Scheduled areas also demonstrate rural female ST literacy rates that are lower than

[1] 'The term "Scheduled Areas" has been defined in the Indian Constitution as "such areas as the President may by order declare to be Scheduled Areas".' Schedule V includes Tribal areas in nine states of India, namely Andhra Pradesh, Jharkhand, Gujarat, Himachal Pradesh, Maharashtra, Madhya Pradesh, Chhattisgarh, Orissa, and Rajasthan. Schedule VI includes tribal dominated areas in four states, viz. Assam, Meghalaya, Tripura, and Mizoram. (Source: Ministry ofTribal Affairs, GOI website, http://tribal.nic.in/Content/DefinitionofScheduledAreasProfiles.aspx)

Table 4.2: Literacy Rates of Scheduled Tribes by State/UT (2001 and 2011)

STATES AND UTs	Rural						Urban						Total					
	2001			2011			2001			2011			2001			2011		
	M	F	T	M	F	T	M	F	T	M	F	T	M	F	T	M	F	T
SCHEDULE V STATES																		
Andhra Pradesh	46.09	24.48	35.43	56.15	37.66	46.88	66.16	45.99	56.39	74.72	58.72	66.82	47.66	26.11	37.04	58.35	40.09	49.21
Chhattisgarh	63.96	38.21	50.95	68.36	47.06	57.57	82.87	59.77	71.71	84.92	68.97	76.94	65.04	39.35	52.09	69.67	48.76	59.09
Gujarat	58.06	34.6	46.45	70.7	51.79	61.29	71.01	51.78	61.76	79.96	65.13	72.71	59.18	36.02	47.74	71.68	53.16	62.48
Himachal Pradesh	77.18	52.5	64.78	82.69	63.36	72.96	92.03	81.15	87.19	92.48	81.94	87.39	77.71	53.32	65.5	83.17	64.2	73.64
Jharkhand	51.67	24.38	38.08	66.6	43.94	55.22	77.83	57.38	67.8	83.22	67.78	75.44	53.98	27.21	40.67	68.17	46.2	57.13
Madhya Pradesh	52.51	27.24	40.01	58.43	40.15	49.32	67.47	45.89	57.23	73.99	59.16	66.73	53.55	28.44	41.16	59.55	41.47	50.55
Maharashtra	64.52	39.88	52.31	72.18	54.13	63.21	82.98	64.7	74.18	86.1	74.13	80.28	67.02	43.08	55.21	74.27	57.02	65.73
Orissa	50.35	22.07	36.13	62.72	39.91	51.08	69.8	45.77	58.12	77.67	60.46	69.08	51.48	23.37	37.37	63.7	41.2	52.24
Rajasthan	61.23	25.22	43.7	66.7	36.1	51.73	75.74	42.97	60.79	81.03	55.57	68.99	62.1	26.16	44.66	67.62	37.27	52.8
SCHEDULE VI STATES																		
Assam	71.29	51.04	61.29	78.04	63.77	70.95	92.43	80.62	86.75	93.75	86.35	90.04	72.34	52.44	62.52	78.96	65.1	72.06
Mizoram	58.72	53.97	56.36	88.57	80.25	84.48	88.95	84.58	86.67	98.37	97.5	97.92	63.49	59.2	61.34	93.59	89.47	91.51
Meghalaya	67.19	43.35	55.46	72.19	69.77	70.98	94.45	89.26	91.97	92.31	90.33	91.26	67.97	44.6	56.48	75.54	73.55	74.53
Tripura	86.11	77.71	82	86.05	70.73	78.44	87.55	96.01	96.77	94.8	89.7	92.22	91.71	86.95	89.34	86.43	71.59	79.05
NORTHEASTERN, NON-SCHEDULE STATES																		
Arunachal Pradesh	54.33	35.83	45.04	67.65	53.3	60.37	85.92	69.05	77.39	90.34	79.38	84.59	58.77	40.56	49.62	71.48	57.96	64.57

Manipur	72.44	57.58	65.09	75.85	66.02	70.97	87.94	74.28	80.94	91.64	83.94	87.67	73.16	58.42	65.85	77.33	67.81	72.58
Nagaland	67.09	57.72	62.55	79.54	72.43	76.04	91.63	85.6	88.7	94.57	90.64	92.59	70.26	61.35	65.95	83.11	76.91	80.04
OTHER STATES																		
Bihar	37.57	13.3	25.91	60.7	39.47	50.32	74.18	55.28	65.67	72.5	57.62	65.33	39.76	15.54	28.17	61.31	40.38	51.08
Jammu & Kashmir	46.44	23.88	35.74	59.27	38.27	49.14	79.01	59.34	70.37	79.57	62.57	71.68	48.16	25.51	37.46	60.58	39.73	50.56
Karnataka	56.92	33.32	45.26	68.54	49.39	58.99	74.39	54.34	64.57	81.88	67.75	74.82	59.66	36.57	48.27	71.14	52.98	62.08
Kerala	70.2	57.28	63.65	79.35	68.96	74.05	84.96	77.7	81.21	92.67	88.21	90.35	70.78	58.11	64.35	80.76	71.08	75.81
Manipur	72.44	57.58	65.09	75.85	66.02	70.97	87.94	74.28	80.94	91.64	83.94	87.67	73.16	58.42	65.85	77.33	67.81	72.58
Nagaland	67.09	57.72	62.55	79.54	72.43	76.04	91.63	85.6	88.7	94.57	90.64	92.59	70.26	61.35	65.95	83.11	76.91	80.04
Tamil Nadu	47.19	29.48	38.41	59.02	43.55	51.32	66.56	50.68	58.6	75.4	62.7	69.1	50.15	32.78	41.53	61.81	46.8	54.34
Uttarakhand	75.29	47.36	61.65	82.49	61.97	72.39	91.55	79.48	85.91	93.73	82.63	88.33	76.39	49.37	63.23	83.56	63.89	73.88
West Bengal	56.6	27.88	42.35	67.21	46.19	56.68	68.57	48.2	58.67	78.4	64.01	71.22	57.38	29.15	43.4	68.17	47.71	57.93
SMALL AND ISLAND STATES																		
A & N Islands	72.68	58.62	65.82	79.75	68.23	74.21	97.01	89.49	93.71	97.96	92.61	95.21	73.61	59.58	66.79	80.87	69.92	65.4
D & N Haveli	53.82	24.6	38.94	70.84	46.3	58.46	81.54	56.73	69.18	88.03	71.1	79.53	55.97	26.99	41.24	73.62	50.27	61.85
Daman & Diu	73.95	51.05	62.83	87.47	71.74	79.65	75.34	55.4	65.72	85.03	70.73	77.96	74.23	51.93	63.42	86.23	71.23	78.79
Lakshadweep	91.26	78.18	84.71	94.69	88.33	91.5	93.29	82.64	87.9	95.97	87.61	91.75	92.16	80.18	86.14	95.69	87.76	91.7
India	57.39	32.44	45.02	66.8	46.94	56.89	77.77	59.87	69.09	83.16	70.32	76.78	59.17	34.76	47.1	68.51	49.36	58.95

Source: Office of the Registrar General & Census Commissioner, GOI 2001, 2011.

50%, namely Bihar, Jammu and Kashmir, Karnataka, Tamil Nadu, Uttar Pradesh, West Bengal, and Dadar and Nagar Haveli. It is significant that in states where the ST population is less than 5% (most notably Tamil Nadu), the situation of small tribal populations remains grim. On the other hand, it is heartening to note that the literacy rates among tribal populations in Schedule VI states is well above national average for both rural and urban areas. This underscores the point that there are significant differences between Schedule V and VI states on the one hand, and between Scheduled areas and non-Scheduled areas with small tribal populations on the other.

Scanning the Census data from 1991, 2001, and 2011, the silver lining is that the gap between the literacy rate of the general and ST populations has been steadily decreasing in all Schedule V and VI states. However, this reduction in the gap of literacy rates between general and ST populations is not so sharp in states with low ST populations like Tamil Nadu (1.10%) and Uttarakhand (2.89%), and island areas like the Andaman and Nicobar Islands (7.50%) (Table 4.3).

Table 4.3: Gap in literacy rate of STs compared to that of the total population

State/UT	1991			2001			2011		
	Total	ST	Gap	Total	ST	Gap	Total	ST	Gap
Schedule V States									
Andhra Pradesh	44.1	17.2	26.9	60.5	37.0	23.4	59.77	42.8	16.98
Chhattisgarh	42.9	26.7	16.2	64.7	52.1	12.6	60.21	50.03	10.18
Gujarat	61.3	36.4	24.9	69.1	47.7	21.4	67.99	52.58	15.41
Himachal Pradesh	63.9	47.1	16.8	76.5	65.5	11.0	73.42	64.64	8.78
Jharkhand	41.4	27.5	13.9	53.6	40.7	12.9	55.56	47.44	8.12
Madhya Pradesh	44.7	18.4	26.3	63.7	41.2	22.6	59.0	41.22	17.78
Maharashtra	64.9	36.8	28.1	76.9	55.2	21.7	72.57	56.01	16.56
Orissa	49.1	22.3	26.8	63.1	37.4	25.7	63.71	43.96	19.75
Rajasthan	38.6	19.4	19.2	60.4	44.7	15.8	55.84	43.09	12.75

(contd...)

(Table 4.3 continued)

State/UT	1991			2001			2011		
	Total	ST	Gap	Total	ST	Gap	Total	ST	Gap
Schedule VI States									
Assam	52.9	49.2	3.7	63.3	62.5	0.7	61.46	61.89	-0.43
Meghalaya	49.1	46.7	2.4	62.6	61.3	1.2	60.16	59.73	0.43
Mizoram	82.3	82.7	0.4	88.8	89.3	0.5	77.3	77.333	-0.03
Tripura	60.4	40.4	20.0	73.2	56.5	16.7	76.34	67.17	9.17
Other States									
Arunachal Pradesh	41.6	34.4	7.2	54.3	49.6	4.7	55.36	54.34	1.02
Bihar	37.5	18.9	18.6	47.0	28.2	18.8	50.44	41.53	8.91
Goa	75.5	42.9	32.6	82.0	55.9	26.1	79.91	71.23	8.68
Jammu & Kashmir	NA	NA	NA	55.5	37.5	18.1	56.35	40.29	16.06
Karnataka	56.0	36.0	20.0	66.6	48.3	18.4	66.53	53.89	12.64
Kerala	89.8	57.2	32.6	90.9	64.4	26.5	84.22	67.29	16.93
Manipur	59.9	53.6	6.3	70.5	65.9	4.7	66.83	62.99	3.84
Nagaland	61.6	60.6	1.0	66.6	65.9	0.6	67.85	68.15	-0.3
Sikkim	56.9	59.0	2.1	68.8	67.1	1.7	72.87	71.28	1.59
Tamil Nadu	62.7	27.9	34.8	73.5	41.5	31.9	71.85	47.23	24.62
Uttarakhand	57.8	41.2	16.6	71.6	63.2	8.4	68.22	45.54	22.68
Uttar Pradesh	40.7	20.0	20.7	56.3	35.1	21.1	57.25	65.14	-7.89
West Bengal	57.7	27.8	29.9	68.6	43.4	25.2	67.42	50.3	17.12
Islands									
A & N Islands	73.0	56.6	16.4	81.3	66.8	14.5	77.32	65.4	11.92
D & N Haveli	40.7	28.2	12.5	57.6	41.2	16.4	64.95	51.95	13.0
Daman & Diu	71.2	52.9	18.3	78.2	63.4	14.8	77.45	69.7	7.75
Lakshadweep	81.8	80.6	1.2	86.7	86.1	0.5	81.51	81.11	0.4
INDIA	52.2	29.6	22.6	64.8	47.1	17.7	63.07	49.51	13.56

Source: Office of the Registrar General & Census Commissioner, GOI 1991, 2001, 2011.

Literacy rates do not always capture recent improvements in school participation or the impact of higher enrolment and completion rates; conversely, they do capture the historical baggage of low participation in schooling. While it is important to start the narrative of education with literacy, more worrisome and challenging issues come to the fore when we discuss participation in school and higher education.

Participation in Schooling

Gross Enrolment Ratio (GER) is a misleading indicator. Over the last ten years, the primary-level GER for all communities has tended to be more than 100. This means that the number of children enrolled in primary classes is higher than the total population of the concerned age group – telling us that many more over-age and under-age children are enrolled in each class, thereby pushing the GER above 100. As educational participation in a community improves and almost all children are enrolled in the age-appropriate grades, GER will come down and be closer to 100. Very high GER is, therefore, not a positive sign; it tells us that our children are not enrolled in age-appropriate grades as mandated by the Right to Education Act of 2009. If we take the last ten years, the GER of STs at the primary level (6–10 years, Classes I to V) has steadily increased from 116.90 in 2000–01 to 136.18 in 2012–13 (MHRD, GOI n.d. [2001]; NUEPA 2013). This increase is comparable to that of the SC community, indicating that formal enrolment in primary school has sharply increased and a large proportion of both over-age and under-age children are enrolled in primary schools.

The censuses from 2001 and 2011 are valuable resources for understanding the educational participation of the ST community in comparison with SC and other communities. In 2001, the percentage of STs who reported having studied up to Matric (Secondary Education) was a mere 3.43% (2.35% male and 1.08% female); their rural-urban gap was significant, with barely 0.73% of STs in rural areas having completed Secondary Education (Office of the Registrar General & Census Commissioner 2001). The situation gets far worse when we move to the Higher Secondary, Non-technical Diploma, Technical Diploma, or Graduate levels and above.

The Census 2011 data on level of education reveal that the ST community has a long way to go before coming close to the age-specific levels of education demonstrated by the forward castes and communities, as evident from Table 4.4 below. In 2011, only 2% of ST adults were graduates – with a significant gender gap (Office of the Registrar General & Census Commissioner 2011). The majority of educated STs have studied only until middle school. While there are rural-urban differences (and notwithstanding the increase in the absolute numbers of STs participating in education), the percentage that actually makes it through the school system and

Table 4.4: Percentage of the population aged 7 years and above by educational level (2011)

Educational Level	General			SC			ST		
	Male	Female	Total	Male	Female	Total	Males	Female	Total
Below Primary	18.2	16.6	17.4	19.1	16.8	18	22.3	18.1	20.2
Primary	18.5	16.7	17.6	20.1	16.5	18.3	18.9	14.2	16.6
Middle	14.4	11.1	12.8	15.0	10.4	12.8	12.5	8.4	10.5
Matric/ Secondary	11.9	8.3	10.1	9.4	6.0	7.7	6.6	4.1	5.4
Higher Secondary/ Intermediate Pre-University/ Senior secondary	8.6	6.2	7.4	6.3	4.0	5.2	4.7	2.7	3.7
Non-technical diploma or certificate not equal to degree	0.14	0.07	0.1	0.06	0.02	0.04	0.03	0.01	0.02
Technical Diploma or certificate not equal to degree	1.0	0.4	0.7	0.7	0.3	0.5	0.4	0.2	0.3
Graduate & above	7.8	5.1	6.5	4.2	2.1	3.2	2.7	1.3	2.0
Unclassified	0.3	0.3	0.3	0.3	0.3	0.3	0.3	0.3	0.3

Source: Office of the Registrar General & Census Commissioner 2011.

moves on to technical or non-technical post-secondary education remains woefully low.

The National Sample Survey (NSS) 55th Round (1999–2000), 61st Round (2004–05), and 66th Round (2009–10) gathered a large variety of information disaggregated by social group. Therefore, these three rounds give us valuable information on the comparative status of ST and other communities. The picture that emerges from a perusal of NSS data confirms the trends shown in the data provided by the Ministry of Human Resource Development, GOI. The current attendance in any educational institution for the 5–14 years, 15–19 years, and 20–24 years age groups increased between the years 1999–2000 and 2009–10 – and this is the case for all ST communities, both rural and urban, and for men and women. Yet there is still an appreciable gap between different social groups, with the status of STs being the most worrisome, followed by that of SCs. As we move from the 5–14 years to the 15–19 years age groups, there is an appreciable drop in the percentage of ST youth attending any educational institution. In the latest round of data, 81.4% (81.4% male and 81.4% female) of ST persons in the 5–14 years age group reported attending schools in rural areas, but this drops to 43% (47.8% male and 37.3% female) when we look at the 15–19 years age group (NSSO 2013). There is a further drop to 9.1% (13.2% male and 5.8% female) when we look at the 20–24 years age group (Table 4.5). This confirms the trends in both GER and dropout rates discussed above.

Table 4.5: Current attendance rates by social group (NSS)

Social Category	Age group								
	5–14			15–19			20–24		
	1999–2000	2004–05	2009–10	1999–2000	2004–05	2009–10	1999–2000	2004–05	2009–10
	55th Round	61st Round	66th Round	55th Round	61st Round	66th Round	55th Round	61st Round	66th Round
Rural Male									
ST	63.2	76.7	81.4	35.5	39.0	47.8	9.4	8.6	13.2
SC	70.1	80.9	86.1	37.9	39.6	52.2	8.6	9.7	12.9
OBC	74.3	84.3	87.8	41.7	48.0	61.9	8.6	10.8	18.8

(contd...)

(Table 4.5 continued.)

Social Category	Age group								
	5–14			15–19			20–24		
	1999–2000	2004–05	2009–10	1999–2000	2004–05	2009–10	1999–2000	2004–05	2009–10
	55th Round	61st Round	66th Round	55th Round	61st Round	66th Round	55th Round	61st Round	66th Round
Others	80.2	87.6	90.2	52.5	54.8	64.1	15.6	14.9	24.9
All (incl. n.r.)	73.9	83.5	87.2	43.7	47.1	58.7	10.9	11.4	18.6
Rural Female									
ST	52.7	67.8	81.4	20.0	24.9	37.3	4.8	5.2	5.8
SC	58.6	73.7	82.1	21.1	28.8	43.5	2.6	3.5	5.7
OBC	62.1	76.3	84.5	23.4	31.5	47.9	2.4	4.1	8.0
Others	73.8	84.2	87.4	36.8	42.7	54.5	6.2	5.9	12.1
All (incl. n.r.)	63.8	76.7	84.2	26.8	33.3	47.3	3.9	4.5	8.2
Rural All									
ST	58.2	72.6	81.4	27.9	32.2	43.0	7.0	6.8	9.1
SC	64.7	77.5	84.2	30.4	34.8	48.3	5.5	6.5	9.2
OBC	68.7	80.5	86.3	33.4	40.3	55.7	5.4	7.3	13.3
Others	77.1	86.0	88.9	45.2	49.1	60.0	10.8	10.5	18.6
All (incl. n.r.)	69.2	80.3	85.9	36.0	40.7	53.7	7.3	7.9	13.3
Urban Male									
ST	80.9	87.1	84.9	58.1	53.3	70.8	24.9	32.1	40.6
SC	79.2	85.0	88.5	48.0	49.5	59.8	17.7	17.0	23.2
OBC	82.4	88.1	90.7	51.2	52.9	64.7	17.5	19.7	30.6
Others	88.0	91.5	93.1	66.0	69.4	76.3	28.3	27.8	36.7
All (incl. n.r.)	84.5	89.0	91.2	58.1	59.3	68.8	23.2	23.2	32.5
Urban Female									
ST	74.0	85.9	84.9	45.2	65.6	68.1	15.4	21.2	21.1
SC	73.9	79.3	90.3	38.5	41.2	60.8	11.2	10.2	15.4
OBC	78.6	87.8	90.2	42.0	50.4	62.2	9.7	11.6	17.7
Others	85.4	91.4	92.6	63.2	67.6	73.4	22.5	21.6	33.4
All (incl. n.r.)	80.8	87.9	90.9	51.9	57.1	66.9	16.6	16.4	24.0
Urban All									
ST	77.6	86.5	84.9	52.3	58.8	69.6	20.1	26.3	31.1
SC	76.6	82.2	89.4	43.8	45.8	60.2	14.6	13.9	19.4
OBC	80.6	88.0	90.5	46.9	51.8	62.6	13.8	15.9	24.3
Others	86.8	91.4	92.9	64.7	68.6	75.0	25.6	24.9	35.2
All (incl. n.r.)	82.7	88.5	91.1	55.2	58.3	68.0	20.1	20.0	28.4

Source: Calculated by A. N. Reddy from NSSO Survey Data, various years.

National studies (EdCil [India] Ltd. and SRI 2010; SRI and EdCil 2014) on out-of-school children (OOSC) done under the aegis of MHRD, GOI and UNICEF India Country Office (New Delhi) reveal that, while STs constitute 8.6% of the population (Office of the Registrar General & Census Commissioner 2011), they accounted for 11% of OOSC in 2009 and 4.2% in 2014. Further, the report clearly demonstrates that there are more OOSC ST children in the older, 11–13 age group. This confirms the findings of other data sources like the Statistics of School Education of the MHRD, GOI and the NSS Survey Reports, which state that a disproportionately large number of ST children are not attending school. For example, Orissa and Rajasthan recorded very high percentages of ST OOSC in the 11–13 age group: 23% male and 26% female, and 24% male and 38.8% female, respectively (EdCil [India] Ltd. and SRI 2010; EdCil 2014).

The most formidable national challenge is to not only enrol ST children in primary school, but to work towards retaining them through all twelve years of schooling and ensure they receive a good quality education and, most importantly, have opportunities to pursue technical or higher education. While the government can certainly claim to have universalized access to primary education, the data reveals that this is not the case at the higher levels.

A recent study on the working conditions of teachers conducted by NUEPA revealed that the proportion of vacant teacher positions reserved for STs is among the highest (Ramachandran et al. 2015). For example, in Madhya Pradesh alone the positions reserved for STs account for close to 21,000 vacant posts. Another recent study on women teachers in Rajasthan revealed the almost negligible participation of ST women in teacher-training courses and the non-availability of higher secondary-level science and commerce streams for girls in backward/tribal areas (Jandhyala and Ramachandran 2015) (Table 4.6).

Analysis of UDISE 2014–15 reveals differential access to arts, science, and commerce streams in tribal and non-tribal blocks (see Table 4.7), with ST girls' access to science and commerce among the worst (NUEPA 2015b). This may be just the tip of the proverbial iceberg; more in-depth study of ST participation in higher education,

Table 4.6: Proportion of ST Teachers, Elementary and Secondary

State	2009–10	2010–11	2011–12	2012–13	Percent ST in Population
Schedule V States					
Andhra Pradesh	6.6	6.8	7.3	6.0	7.0
Chhattisgarh	29.8	31.1	31.5	29.9	30.6
Gujarat	14.9	15.2	14.8	14.1	14.8
Himachal Pradesh	6.1	6.3	6.1	6.1	5.7
Jharkhand	26.1	24.8	26.6	23.8	26.2
Madhya Pradesh	14.6	14.6	14.3	14.5	21.1
Maharashtra	7.4	7.6	7.5	7.1	9.4
Orissa	14.8	12.1	12.8	12.5	22.8
Rajasthan	9.1	9.4	9.4	9.6	13.5
Schedule VI States					
Assam	16.7	16.8	19.4	15.5	12.4
Meghalaya	90.5	89.2	78.9	89.7	86.1
Mizoram	96.8	96.9	96.4	96.9	94.4
Tripura	37.5	37.1	30.1	32.9	31.8
North Eastern States (Non-Schedule VI States)					
Arunachal Pradesh	67.5	67.0	18.6	66.2	68.8
Manipur	37.2	43.1	43.7	41.7	35.1
Nagaland	82.6	82.8	83.4	84.8	86.5
Sikkim	39.4	38.1	38.0	39.4	33.8
Small and Island States					
A&N Islands	7.1	7.4	7.4	7.8	7.5
D&N Haveli	52.5	50.2	13.9	46.5	52.0
Daman & Diu	4.7	4.5	0.7	4.7	6.3
Lakshadweep	96.5	96.4	3.9	95.4	94.8
Other States					
Bihar	3.8	3.9	3.6	3.4	1.3
Jammu & Kashmir	8.8	8.7	8.8	8.8	11.9
Karnataka	4.3	4.6	4.6	4.8	7.0
Kerala	2.5	2.2	1.6	1.5	1.5
Tamil Nadu	0.9	1.0	1.0	0.8	1.1
Uttarakhand	2.7	2.6	2.7	2.7	2.9
West Bengal	4.1	4.4	4.6	4.6	5.8
INDIA	9.4	9.3	9.2	8.7	8.6

Source: GOI, MHRD 2014b, 2012, 2014c.; raw data from MHRD.

teacher education degrees, technical (including para-professional) degrees, and industrial training institutions would perhaps reveal a more worrying picture.

Table 4.7: Enrolment of ST students by stream
(Higher Secondary Grades)

	All Children			ST		
	Boys	Girls	All	Boys	Girls	All
Tribal Blocks						
Arts	48.7	58.5	53.2	62.7	70.0	66.2
Science	34.7	28.3	31.8	24.5	20.7	22.7
Commerce	11.7	10.0	10.9	8.3	6.0	7.2
Vocational	2.6	1.9	2.3	1.6	1.2	1.4
Other	2.2	1.3	1.8	3.0	2.1	2.5
Total	100	100	100	100	100	100
Non-tribal Blocks						
Arts	40.7	52.7	46.3	55.5	61.8	58.4
Science	38.2	29.0	33.9	28.4	24.1	26.4
Commerce	16.5	15.0	15.8	11.6	10.4	11.0
Vocational	3.5	2.6	3.1	3.2	2.6	2.9
Other	1.1	0.8	0.9	1.4	1.1	1.2
Total	100	100	100	100	100	100

Source: NUEPA 2015b.

Again, these all-India figures hide significant regional disparities. It is believed that many higher secondary schools in Schedule V areas do not offer science or commerce streams, and that many girls-only higher secondary schools also do not offer science or commerce. District-wise analyses of the existing data could help us pinpoint the tribal areas that face this problem. Interestingly, while there has not been a policy-level decision about whether to offer science or commerce at certain kinds of schools, administrators at the state and district levels often make these decisions themselves.

A Majority of Tribal Children Study in Government Schools

A perusal of recent data reveals that nearly 64% of male tribal children at the primary level were enrolled in government schools and another

14.4% were enrolled in schools run by tribal and social welfare departments in 2014–15 (NUEPA 2015a). Only a little over 14% were enrolled in private unaided schools. On the other hand, only about 41% of children from the general category were enrolled in government schools, with a negligible amount enrolled in schools run by social and tribal welfare departments.[2] At the same time, a huge proportion of children from the general category – over 47% – were enrolled in private unaided schools.

Similar trends can be observed for other levels of school education, although a somewhat higher proportion of tribal children attend private aided schools at the secondary and higher secondary levels. It may be noted that, as one goes up the educational ladder, the proportion of ST children in school declines. It may be that those tribal children who are able to reach the secondary and higher secondary levels are those who are also able to afford unaided secondary schools. However, the reality is that a majority of tribal children go to government schools and only the small percentage that can afford private unaided schools are able to make it to secondary school.

Table 4.8: Enrolment (GER) of ST students by school management, various levels

Management of School	General		OBC		SC		ST	
	Boys	Girls	Boys	Girls	Boys	Girls	Boys	Girls
Primary								
Government	40.80	46.80	56.80	62.90	67.30	71.50	63.90	66.70
Tribal/social welfare	0.20	0.20	0.80	0.80	1.10	1.20	14.40	15.10
Private Aided	5.30	5.70	5.50	5.70	4.50	4.70	4.90	4.80
Private Unaided	47.20	41.10	33.20	27.20	25.10	21.00	14.30	11.20
Central Government.	1.00	1.00	0.50	0.40	0.40	0.40	0.20	0.20
Others	5.40	5.20	3.30	3.00	1.50	1.20	2.20	2.00
Upper Primary								
Government	37.60	45.60	53.20	60.00	63.90	67.80	56.60	59.00

(contd...)

[2] While tribal welfare schools are located in predominantly tribal areas, children from other communities are not officially excluded.

(Table 4.8 continued.)

Management of School	General		OBC		SC		ST	
	Boys	Girls	Boys	Girls	Boys	Girls	Boys	Girls
Tribal/social welfare	0.30	0.30	1.10	1.20	1.50	2.10	16.20	17.90
Private Aided	13.80	14.00	14.10	13.90	12.30	12.00	11.20	10.60
Private Unaided	43.30	35.20	29.20	22.70	20.70	16.90	14.10	10.80
Central Government.	1.60	1.30	0.50	0.40	0.70	0.50	0.70	0.60
Others	3.50	3.60	1.80	1.80	0.80	0.70	1.20	1.20
Secondary								
Government	30.3	30.3	38.0	38.0	48.4	48.4	41.2	41.2
Tribal/social welfare	0.3	0.3	1.2	1.2	1.6	1.6	15.8	15.8
Private Aided	20.4	20.4	24.4	24.4	20.7	20.7	22.2	22.2
Private Unaided	45.2	45.2	35.0	35.0	27.9	27.9	18.7	18.7
Central Government.	1.7	1.7	0.5	0.5	0.8	0.8	0.9	0.9
Others	2.1	2.1	1.0	1.0	0.5	0.5	1.1	1.1
Higher Secondary								
Government	29.0	29.0	27.9	27.9	39.7	39.7	35.3	35.3
Tribal/social welfare	0.2	0.2	1.0	1.0	1.4	1.4	13.2	13.2
Private Aided	22.8	22.8	27.1	27.1	23.3	23.3	21.9	21.9
Private Unaided	44.5	44.5	42.5	42.5	34.1	34.1	26.5	26.5
Central Government.	1.6	1.6	0.6	0.6	0.9	0.9	1.4	1.4
Others	1.9	1.9	0.9	0.9	0.5	0.5	1.7	1.7

Note: 'Government' includes schools run by state education departments and local bodies. 'Others' includes unrecognized private schools, madrasas, and 'no response'.
Source: NUEPA 2015c.

Why Do So Many ST Children Drop Out?

The available research tells us that a complex interplay of factors pushes ST children out of school. For several decades now, educationists have pointed out that the educational system expects teachers to teach to the curriculum and finish the syllabus within a certain time frame, regardless of whether the children in the class are learning or not. This is a very serious issue in tribal areas, where there is already a language gap between students and teachers. When teachers are not able to address the learning needs of every child, they throw up their hands and teach those who are able to keep pace. Given the specific characteristics of tribal areas, there is a larger prevalence of multi-grade classrooms and more frequent absences of teachers and students. As a result, children fall behind and become passive spectators in the classroom. As time goes by, the cumulative burden of non-learning increases until the children reach a point where they are unable to comprehend what is going on in class.

When habitations are scattered and schools are located in remote areas, there is little school-level monitoring of teaching-learning processes and actual teaching time. Monitoring, when it happens, is confined to inputs: enrolment, midday meals, distribution of incentives, and so on. Institutions created to provide on-site, school-level academic support have instead become data-gathering instruments. These institutions are also often staffed with people who may not have the skills or the aptitude for on-site teacher capacity building. Post-RTE mechanisms like Continuous and Comprehensive Evaluation (CCE) have been reduced to a series of forms that teachers are expected to fill out. Administrators admit that they follow instructions from above, and that they themselves are not educators who could develop systems to monitor children's learning. They need help – hands-on help – to develop an effective monitoring system.

In tribal areas, there is a huge social distance between teachers and students in government schools. A recent inclusion/exclusion study commissioned by MHRD, GOI (Ramachandran and Naorem 2012) clearly brought out the innate prejudices and stereotypes that teachers carry with them to school. Many of them actually believe that tribal children are not motivated to, or cannot, learn. They blame the family and the community. Most importantly, the teachers complain that parents are not able to help their children with their studies. Despite several recommendations about enhancing the pool of local tribal teachers by providing them with opportunities for pre-service teacher education, we have not been able to enhance this pool. It's a chicken-and-egg situation: unless there is a considerable flow of ST students into higher secondary and higher education, it may not be possible to find suitable candidates for pre-service training. And it is this very shortage of teachers that inhibits the educational flow in tribal areas.

In the wake of the implementation of Teacher Eligibility Tests (introduced after RTE) and the high proportion of candidates that failed to clear the examination, there are people who argue that subject knowledge is poor among our teachers. They point out that it might be the quality of the teacher – her/his mastery over subjects, pedagogic skills, and aptitude to teach – that is responsible for poor learning. For example, in Madhya Pradesh over 19,000 reserved posts for teachers are currently lying vacant because of a lack of suitable candidates (Ramachandran et al. 2015). In states like Jharkhand, Rajasthan, Madhya Pradesh, and Chhattisgarh, there is a huge shortage of math and science teachers: and tragically it is in these very states that higher secondary schools in tribal areas rarely offer science or mathematics. Girls' schools across the board primarily offer language and arts subjects. Thus there are many systemic issues that come in the way of building a cadre of teachers from tribal communities.

Given the clashes of scheduling between national and tribal festivals and variations in agro-climatic periods, the actual number of teaching days in tribal areas is low. In most tribal areas, schools function less than 150 days per year. Despite a clear policy of facilitating sub-region-specific school calendars and timings since 1965, teachers' unions have stalled any move to introduce localized time planning. Since the 1965 Education Commission (popularly known as the Kothari Commission), all policy documents have recommended the creation of flexible school calendars in tribal areas. Teachers' unions have argued that holidays and vacations should be standardized because a significant proportion of teachers working in tribal areas hail from non-tribal and urban areas. Following a flexible calendar would mean that they would be unable to align their vacation/holiday time with that of their families who live in urban or non-tribal areas.

The higher education scenario is also quite worrisome. The GER for higher education is 19.4 for all (20.8 male and 17.9 female), and the recent 2014–15 All India Survey on Higher Education (AISHE) points out 'Scheduled Caste students constitute 12.5% and Scheduled Tribes students 4.2% of the total enrolment. 31.6% students belong to Other Backward Classes. 4.5% students belong to Muslim Minority and 2.1% from other Minority Community'

(MHRD, GOI 2016b: iv). The government has not calculated the GER for SCs and STs, as yet.[3] Equally, we do not have a definitive picture of the ST share of enrolment in science-based/technology courses, or in vocational courses that require the student to have studied science and mathematics at the higher secondary level (Industrial Training Institutes, Auxiliary Nurse and Midwife (ANM), Lab technicians, etc.). This remains an important gap in the data; as a result, Ministry of Tribal Affairs, GOI (2014) could not explore the status of ST students in higher and technical education.

As discussed at the beginning of this chapter, these national-level statistics hide significant regional differences. In the following sub-sections, I explore the situation in Schedule V, Schedule VI, and other areas in greater depth.

Schedule V Areas

Schedule V Areas include specifically notified areas (villages, *talukas*, blocks, etc.) in Andhra Pradesh, Jharkhand, Chhattisgarh, Himachal Pradesh, Madhya Pradesh, Gujarat, Orissa, and Rajasthan. These districts have significant ST populations (although the percentage of STs in the state as a whole may not be high) and have certain specified constitutional safeguards. Almost all Schedule V Areas are educationally backward and, given the scattered nature of the habitations in these areas, access to schools has historically been poor. An overwhelming majority of Educational Backward Blocks are tribal. While access to primary education in Schedule V Areas has improved significantly in the 2000s, access to the secondary and higher secondary levels continues to be worrisome. The overall educational status of Schedule V Areas is far poorer than non-Scheduled or Schedule VI areas, as evident from the data on literacy rates.

[3] The Gross Enrolment Ratio (GER) for higher education in India is calculated for the age group of 18–23 years. The higher education GER at the All-India level is 21.6 for males and 18.9 for females. The GER for Scheduled Castes and Scheduled Tribes has not been calculated, as the Office of the Registrar General & Census Commissioner has not yet released the age-wise population of these categories from the 2011 Census.

Before we go into depth on school participation in Schedule V Areas, it is important to reiterate the specific issues that frame educational access in these areas, namely: (a) these areas are among the poorest in the country; (b) they are primary rural; (c) they are characterized by scattered habitations; (d) they primarily rely on agriculture, the collection of minor forest produce, wage labour, and seasonal migration; (e) the health and nutritional status of children is precarious; and (f) they have a higher proportion of small and single/two-teacher schools. This larger context of local livelihoods and health–nutritional status frames the educational participation in these areas.

Table 4.9: List of Schedule V Areas per state

State	Schedule V Areas
Andhra Pradesh	Visakhapatnam, East Godavari, West Godavari, Adilabad, Srikakulam, Vizianagaram, Mahboobnagar, and Prakasam (only some *mandals*)
Jharkhand	Dumka, Godda, Devgarh, Sahabgunj, Pakur, Ranchi, Singhbhum (East & West), Gumla, Simdega, Lohardaga, Palamu, and Garwa (some districts are only partly tribal blocks)
Chhattisgarh	Sarbhuja, Bastar, Raigad, Raipur, Rajnandgaon, Durg, Bilaspur, Sehdol, Chindwada, and Kanker
Himachal Pradesh	Lahaul and Spiti districts, Kinnaur; Pangi Tehsil and Bharmour sub-Tehsil in Chamba district
Madhya Pradesh	Jhabua, Mandla, Dhar, Khargone, East Nimar (Khandwa), Sailana Tehsil in Ratlam district, Betul, Seoni, Balaghat, and Morena
Gujarat	Surat, Bharauch, Dangs, Valsad, Panchmahl, Sadodara, and Sabarkanta (parts of these districts only)
Maharashtra	Thane, Nasik, Dhule, Ahmednagar, Pune, Nanded, Amravati, Yavatmal, Gadchiroli, and Chandrapur (parts of these districts only)
Orissa	All of Mayurbhanj, Sundargarh, and Koraput; parts of Raigada, Keonjhar, Sambalpur, Boudh, Kondmals, Ganjam, Kalahandi, Bolangir, and Balasor
Rajasthan	All of Banswara and Dungarpur; parts of Udaipur, Chittaurgarh, and Siroi

Source: Ministry of Tribal Affairs website.

It is also noteworthy that ST populations speak several different languages and are not aligned with the dominant language of the state. This adds another critical dimension to the education of these populations. Since the 1960s, almost all of the committees and commissions constituted by the government to inquire into the status of education have argued that the medium of instruction in the early grades has to be the students' mother tongue and, more importantly, that teachers should be familiar with tribal languages. To this end, almost all of the commissions recommended the implementation of a time-bound and focused programme to enhance the pool of qualified teachers from ST communities. This remains a major challenge, and contributes in no small measure to low comprehension, poor learning, and high dropout rates.

The gap between the literacy rate of the whole population and that of STs, especially female STs, remains fairly significant for most of the tribal populations of Schedule V states. Even in Himachal Pradesh, where the literacy rates of STs are relatively high compared with the other Schedule V states, the gap between the female literacy of the general population and that of STs is quite high. Equally, rural-urban differences are quite sharp.

Recent discussions on the number of out-of-school children and of working children/child labour also reveal that Schedule V states are particularly vulnerable to these issues. A recent study on the multiple deprivations of the ST population that conducted a detailed analysis of the availability of schooling facilities in Jharkhand, Orissa, Chhattisgarh, and Madhya Pradesh revealed that

> the difference in provisioning of education in tribal dominated areas to villages where they are negligible is remarkable in Odisha and Chhattisgarh. The gap in availability of schooling facility between villages that are tribal dominated and others increased as one goes up from primary to middle to secondary school levels, with the tribal concentrated villages having very low proportion of secondary schools per village. (Rustagi, Mishra and Mehta 2011)

A recent GOI-commissioned survey of OOSC (SRI and EdCil 2014) reveals that the situation in Schedule V states is perhaps the

Table 4.10: Percentage of out-of-school children in Schedule V States in 2014

State	SC	ST	OBC	Others	All
Andhra Pradesh	0.98	0.30	0.44	0.00	0.91
Chhattisgarh	3.88	0.00	0.00	-	3.75
Gujarat	1.82	3.32	0.00	0. 00	1.94
Himachal Pradesh	0.21	-	-	0.00	.21
Jharkhand	1.33	4.92	1.91	-	2.02
Madhya Pradesh	3.08	10.65	0.00	-	3.78
Maharashtra	0.81	0.76	0.00	0.97	0.81
Orissa	6.12	6.34	0.00	-	6.1
Rajasthan	4.84	6.85	0.00	2.88	5.02
India	4.84	6.85	0.00	2.88	5.02

Source: SRI and EdCil 2014.

most worrying, because a larger percentage of ST children are out of school (Table 4.10).

This insight is confirmed when we look at the GER and dropout rates of ST students at various levels of schooling. As evident in Table 8, which provides data for the enrolment of ST students at different stages of education, with the exception of Himachal Pradesh the GER drops as we move from primary to upper primary to secondary and higher secondary. The situation in Jharkhand and Orissa is perhaps the worst, with the ST GER dropping to a low of 10.11 in Jharkhand (11.12 boys and 9.12 girls) and 17.22 in Orissa (boys 20.13 and girls 14.19) at the higher secondary level (NUEPA 2015c).

It is therefore important for planners and administrators to understand the reasons for STs' poor educational participation beyond the elementary level. This is where there are significant data gaps; while we have overall estimates of the ratio of secondary/upper secondary schools to primary/upper primary schools, it would be valuable to calculate this specifically for the tribal-dominated educational blocks. This may give us a clearer picture of STs' educational access. Second, we also need to estimate the percentage of single- and two-teacher schools, schools with less than 50 students, and the actual, physical presence of teachers in rural habitations.

A ground-level analysis of schools, their facilities, the availability of teachers, how frequently the schools are inspected/visited, and the percentage of women teachers could perhaps give us greater insights into the actual functioning of the school systems of Schedule V Areas.

Schedule VI Areas and other North-eastern States

The Sixth Schedule covers parts of the North-eastern states, such as Assam, Meghalaya, Tripura, and Mizoram. These are entirely tribal areas, where the entire state or parts of the state have special provisions for educational administration. Tribal communities are not a minority here; these areas do not face some of the problems discussed for Schedule V Areas. If we look at the data (from the Census, DISE, and other sample surveys), it becomes quite evident that the tribal literacy rate among women is significantly higher in the North-eastern states such as Mizoram, Meghalaya, and Nagaland where tribal groups constitute a large percentage of the general population (quite unlike the Schedule V states of Rajasthan, Orissa, and Madhya Pradesh). In the North-eastern states, the male-female gap in literacy levels and educational participation is also lower than that of Schedule V Areas. This may be because tribal societies in general, and some Schedule VI Areas in particular, are characterized by women's participation in all spheres of social and economic life. This is the case for both matrilineal and patrilineal tribal societies.

There are, of course, noticeable differences between the Schedule VI Areas. Mizoram tops the literacy list with a ST literacy rate of 91.5% (93.59% male and 89.47% female), followed by Meghalaya, Tripura, and Assam (Office of the Registrar General & Census Commissioner 2011). All of these states have literacy rates that are far higher than the national average. The gender gap is also low, except in Tripura where it is 9.17%. With the exception of these two indicators of educational progress, Schedule VI Areas are perhaps as deprived as the Schedule V ones in all other aspects of education. Gross Enrolment Ratios at the primary level are high, but at higher levels the picture is not so rosy. The dropout rates remains high, with over 70% of students in Assam, Meghalaya, and Tripura dropping out before they complete Class X. In Meghalaya, just over half of the schools are accessible by all-weather roads; most schools throughout the region do not have access to electricity. An important feature of Schedule VI Areas is that a majority of the schools have less than 50 students, and many of them are single- or two-teacher schools. This poses a huge challenge to teachers and administrators alike.

Teacher absence remains an area of concern. The Meghalaya Human Development Report (Government of Meghalaya 2008) points out that teacher absence continues to be a serious issue. In Mizoram, substitute teachers (*Aikal*[4]) are common (Education Reforms Commission, Mizoram 2010). Similarly, as the Arunachal Pradesh Human Development Report points out,

> Absenteeism among teachers is a serious concern. This was observed by the SHDR survey team, which found that teachers tend to remain absent on a more or less "rotational basis". This is especially true in remote areas. In the multi-teacher schools, one or two teachers are always absent in a cyclical manner through the year. The main reason for this absenteeism, according to the villagers, is the absence of proper residential accommodation and the non-availability of food that the teachers (who are usually from outside the State) are used to. Many teachers are not used to living in the temporary thatched houses provided to them. The recruitment of teachers from the villages or surrounding areas, especially in the more remote areas, is an option but, this will only be possible once education reaches these remote areas and local people are available for employment. (Government of Arunachal Pradesh 2006: 64)

One of the unique features of Schedule VI Areas is the presence of autonomous district councils, which are mandated to be local

[4] This is a unique Mizo term for a proxy teacher. Literally, *Ai* means 'instead of' or 'in lieu of' and *Kal* means 'to go'. So *Aikal* refers to a person who goes in place of another to carry out a task. Nothing is given in writing to the person as an evidence of being a substitute. Hence, a teacher who uses a proxy teacher continues to be a teacher in absentia unofficially. The salary is shared between the absentee teacher and his/her proxy. The arrangement works out well for both, at the cost of the system. *Aikal* is widely prevalent, particularly in rural and semi-urban areas. Urban schools too have their share of the *Aikal* system (Chaudhuri and Mathur 2014).

self-government institutions. For example, in Mizoram, according to the provision of the Sixth Schedule of the Constitution of India, three groups of minorities – Mara, Lai, and Chakma – were given autonomous district councils in the southern part of the state to look after their own interests and welfare. These district councils came into existence in 1972, when the Mizo District Council was dissolved and the Pawi-Lakher Regional Council was split into three District Councils, namely the Mara, Lai, and Chakma District Councils. These Councils are provided with certain powers by the Sixth Schedule of the Indian Constitution and have full authority within their jurisdictions over the management of primary education, with effect from 1980, and middle school education, from 1995. However, in practice, the Education Department of the state continues to play significant administrative and academic roles in the primary and middle school education of the district councils. In fact, the department makes all of the educational rules and regulations; the powers of the district councils are restricted, since the State Education Department exercises control over the final allocation of funds (Education Reforms Commission, Mizoram 2010). Similar experiences are reported from other Schedule VI Areas. The district councils do not have much effective authority over school education, as the education department provides the budget according to the strict norms set by the state government or SSA. Nagaland attempted to reform this administration by introducing what is now known as communitization in 2001, whereby the management and operational control of various social sector programmes of the government were transferred to the local tribal council or community. In education, the day-to-day management and monitoring of schools is done by the village-level tribal council.[5] However, similar initiatives meant to decentralize the administration of development projects are not evident in the other Schedule VI states.

Across the region, administrators and researchers point out that the quality of education is a major challenge. For example, in Arunachal Pradesh, while the gross enrolment in and attendance

[5] For a detailed description of communitization, see Jandhyala and Ramachandran 2007.

at school give some idea about participation in the school system, the final performance of students in examinations indicates only limited success of the education process. The performance of children in the Class X and XII board examinations shows that, of the children that are able to go to school, only a handful actually pass. The pass percentages in 2003 were 43.58% in Class XII and only 27.43% in Class X. The low success rate, especially at the Class X level, points to the inefficient use of resources and poor teaching in the schools. A pass percentage of 27% means that roughly one in four students passed the examination (Government of Arunachal Pradesh 2006: 60).

The learning crisis is further revealed when we look at the transition from primary to higher education. Up to the elementary level, enrolment in Schedule VI Areas is better than the All-India average, and there is no gap between the ST population and the rest of the population. For example, the literacy gap is low or negative in most states: Assam: -0.43, Meghalaya: 0.43, Mizoram: 0.03, and Tripura: 9.17 (Office of the Registrar General and Census Commissioner, GOI 2011). What is noticeable is that the GER remains high and dropout rates are equally high – and in these there is not much difference between Schedule VI Areas and the rest of the country. As we move to higher secondary, diploma, graduate, and higher levels, the advantage evident in elementary education starts slipping and the situation becomes similar to the rest of India.

Another serious issue framing school education in Schedule VI Areas is that textbooks and teaching-learning materials rarely reflect the environment in which STs live, their lives, their heritage, their stories, their culture, or their livelihoods. For this reason, the children cannot connect to the material that they are given, leading to incomprehension and alienation. There is a direct impact on the self-esteem and self-confidence of children who do not see their culture reflected in any nationally produced educational material. Children do not see any connection between the textbooks and their own lives.

The overwhelming consensus across all of the surveyed literature is that the transition from upper primary to secondary and then to higher education remains the biggest challenge. The poor availability of vocational and technical education opportunities further compounds this problem because there are very few colleges and universities. This has led to high outmigration from the region in pursuit of higher education, a trend that is visible in most educational centres like Delhi, Bangalore, Chennai, Pune, and Mumbai. The aspirations of the North-eastern youth for higher education and formal employment remain unfulfilled in their respective states.

Other Areas

There are small, yet significant tribal communities in Bihar, Goa, Jammu and Kashmir, Uttarakhand, Uttar Pradesh, West Bengal, Kerala, Karnataka, and Tamil Nadu. The situation of the ST community in these states is different, with literacy rates ranging from a low of 40% in Jammu and Kashmir to a high of 72% in Goa. Even in forward states like Tamil Nadu and Karnataka, the literacy gap between STs and the rest of the population is more than 22%. In Kerala, the gap is close to 17% (Office of the Registrar General & Census Commissioner, GOI 2011). This tells us that, wherever there are tribal communities, their literacy and educational situation is far worse than that of the general population. In states like Tamil Nadu where they form only 1.10% of the population, the gap in literacy clearly reveals neglect of the ST community. In terms of STs' access to schooling and enrolment and dropout rates, there are noticeable differences from the general population – underscoring the point that the situation of the ST community (wherever it is a very small minority) merits special attention and context-specific planning and implementation.

The situation of island states is also unique. At one end, there is Lakshadweep, where 94.8% of the population is ST; at the other end, there is Daman and Diu, where STs make up only 6.32% of the population. While the percentage of STs is small in the Andaman and Nicobar Islands (7.50%), the communities there are not only poor, but also notified as Particularly Vulnerable Tribal Groups (PVTG).[6]

[6] Particularly Vulnerable Tribal Groups were identified by the GOI in 1961, and include tribal communities that were living in a pre-agricultural system of existence at that time (i.e., hunting and gathering, zero or negative population growth, and extremely low levels of literacy in comparison with other tribal groups). 75 tribal

The literacy rate also varies, from a high of 81% in Lakshadweep to a low of 52% in Dadra and Nagar Haveli.

There is considerable evidence that the literacy and educational status of the Andaman and Nicobar Islands' six PVTGs vary considerably. The Great Andamenese, Onge, Jarawa, and Sentinalese live in the Andaman group of islands, while the Nicobarese and Shompen live in the Nicobar group of islands. If the island state is taken as a whole, the literacy level of 86.63% (90.27% male and 82.43% female) is much higher than All-India levels and significantly higher than tribal communities in other parts of the country. However, given that STs comprise only 7.5% of the population (Office of the Registrar General and Census Commissioner, GOI 2011), this data hides the inequality between the tribal and non-tribal populations of the islands. The non-tribal population is mostly from West Bengal, Tamil Nadu, and Kerala.

Conclusion: Cross-cutting Educational Issues

Access to school beyond the primary level

One of the formidable challenges across all tribal areas concerns access to all levels of schooling. The last National Social Survey to address these issues (NSSO 2010) reveals that, as we move to higher levels of schooling, the relative access of STs goes down, especially in rural areas (see Table 4.11). Close to 30% of ST students have to go more than 5 km to reach a secondary school. Far fewer SCs have to travel that far. The implications of this are fairly serious for children from scattered habitations, especially when we take into consideration hilly terrain and the absence of all-weather roads.

Language and content

Starting from the 1950s, most commissions have pointed out the importance of using students' mother language in the early grades. 'The Constitution of India allows the use of tribal dialect (mother

groups have been categorized by Ministry of Home Affairs as PVTGs, residing in 18 States and the Andaman and Nicobar Islands. (Source: http://www.tribal.nic.in/pvtg.aspx)

Table 4.11: Access to schooling by social category

Distance to School	Percentage of households having access to school by distance											
	Rural				Urban				Total			
	ST	SC	OBC	All	ST	SC	OBC	All	ST	SC	OBC	All
Primary Level	90.1	92.6	92.2	91.7								
D<1km	90.1	92.6	92.2	91.7	90.8	91.6	92.3	92.3	90.1	92.4	92.2	91.9
1km≤D<2km	6.5	6.1	6.3	6.5	7.0	7.5	6.8	6.7	6.6	6.4	6.4	6.6
2 km ≤D<3km	2.0	1.0	1.1	1.2	1.8	0.8	0.8	0.9	2.0	0.9	1.0	1.1
3km≤D<5km	0.3	0.2	0.2	0.2	0.1	0.0	0.1	0.1	0.3	0.2	0.2	0.2
D≥ 5km	0.8	0.0	0.0	0.1	0.0	0.0	0.0	0.0	0.7	0.0	0.0	0.1
Upper Primary Level												
D<1km	51.8	61.4	64.2	61.6	78.9	80.0	81.5	82.5	54.5	65.4	68.6	67.5
1km≤D<2km	15.2	18.1	16.3	17.1	15.7	16.5	14.4	14.1	15.3	17.8	15.8	16.3
2km≤D<3km	15.1	12.5	11.2	12.2	3.5	2.8	3.1	2.5	14.0	10.4	9.1	9.4
3km≤D<5km	9.5	5.6	5.5	5.8	1.6	0.6	0.8	0.7	8.7	4.5	4.3	4.4
D≥ 5km	8.1	2.2	2.6	3.1	0.1	0.1	0.2	0.2	7.3	1.8	2.0	2.3
Secondary Level												
D<1km	22.1	32.1	29.7	30.7	63.5	63.9	65.8	68.6	26.2	38.9	38.7	41.5
1km≤D<2km	11.2	17.2	16.2	16.6	22.3	26.7	23.4	22.1	12.3	19.3	18.0	18.2
2km≤D<3km	19.1	20.4	19.6	19.7	7.6	6.3	6.2	5.6	17.9	17.3	16.2	15.7
3km≤D<5km	17.5	15.3	17.0	15.8	4.8	1.8	3.6	2.5	16.3	12.4	13.6	12.0
D≥ 5kms	29.8	14.9	17.5	17.1	1.7	1.3	1.0	1.0	27.0	11.9	13.3	12.5

Source: NSSO 2010.

tongue) as the medium of instruction in case the population of the said tribe is more than one lakh. But this has not been adopted on the grounds of feasibility and viability of introducing and sustaining such a change' (Sujatha 1999: 47). This remains a huge challenge despite the repeated observations of government-appointed committees and commissions. Most recently, the National Curriculum Framework (NCF) (NCERT 2005) underscored the importance of appropriate pedagogic practices and mother-tongue education in the early grades, stressing the need to 'enter the world of formal learning through the language of [the] home and environment' (Veerbhadranaika et al. 2012: 32). While there have been small-scale initiatives in Orissa and

Andhra Pradesh, multilingual textbooks and education continue to elude us. Even where there is material, the ability of teachers to use it remains a huge problem. Again, since the Dhebar commission there has been a tabled proposal for implementing time-bound initiatives to increase the pool of schoolteachers from tribal communities. However, this remains an unimplemented recommendation.

A related issue concerns the depiction of tribal communities in educational material. 'The representation of Adivasi in stereotypical negative terms as "backward" people, outside the pale of progress, has meant that elementary education has only reinforced stereotypes and further marked the Adivasi in discriminatory terms' (Veerbhadranaika et al. 2012: 32). Another site of alienation is the content of our textbooks, which give primary importance to urban-centred subjects and do not weave in the histories and cultures of the tribal people of India.

A third important issue relates to the availability of science education in higher secondary schools located in tribal areas. As Leela Visaria points out with respect to Gujarat,

> In none of the villages in half of the predominantly tribal blocks is there any school that offers science education for students at the higher secondary level and yet in many of these blocks ITIs have been established [...] It is reported that the non-tribal students from non-tribal blocks are enrolling themselves in such ITIs in great numbers' (Visaria 2014: 106).

Small schools, multi-grade classes, and one or two teachers

There is enough quantitative evidence from the DISE and UDISE (NUEPA 2015c) to show that the percentage of small schools (less than 50 students) continues to be fairly significant. Equally, 11.79% of primary schools and 8.65% of elementary schools have just one teacher. Some states have a very high proportion of single-teacher primary schools: 30.19% in Rajasthan, 23.68% in Andhra Pradesh, 58.15% in Arunachal Pradesh, and 37.25% in Goa. Given their settlement patterns, many tribal villages have small schools with one or two teachers. As a result, they are invariably multi-grade. This issue has been highlighted in almost all evaluations and reviews of the Sarva Shiksha Abhiyan and in research studies on small schools or the education of tribal communities (Veerbhadranaika et al. 2012; Nambissan 1994; Azim Premji Foundation 2004; Rustagi, Mishra, and Mehta 2011). These reports and evaluations also mention that teachers are not trained to handle multi-grade classrooms and that the teaching-learning materials that are produced do not lend themselves to use in multi-grade situations.

Impact of poor nutrition and malnourishment on education

Another under-researched and ignored area is the impact of poor health and nutrition on the cognitive development of children. This remains a serious issue in many predominantly poor tribal regions of the country. Notwithstanding the midday meal and the Integrated Child Development Services (ICDS) programmes, the health and nutritional status of very poor tribal communities remains an area of concern. Mitigating the vicious cycle of malnutrition, poor health, and high dropout rates demands greater coordination between the education department, the ICDS programme, and the health department.

Challenges faced in areas affected by conflict

There are very few in-depth studies available about the impact of conflict on children's participation in education in India. Many tribal areas of India are affected by Maoist insurgencies (popularly known as Naxal-affected areas). A set of six studies was commissioned by an NGO known as NEGFIRE,[7] but unfortunately the draft reports were never published. There are also areas that are affected by communal violence, but the tribal areas affected by this kind of conflict are fewer. In recent memory, the situation in Kandhamal, Odissa was reported in the media (Young India, Odisha 2010). Research reveals that sustained conflict – even low-intensity conflict – can have a deep impact on education. A Save the Children Alliance study done in 2006 documented the following

[7] New Education Group – Foundation for Innovation and Research in Education (NEGFIRE) commissioned a series of studies (which were never published, but circulated as draft reports) on the education of children in conflict zones. Due to the sensitive nature of these studies they were not uploaded on their website.

impact of armed conflict on schools and children (International Save the Children Alliance 2006):

> Schools are destroyed or commandeered by security forces. This essentially implies that the schools do not function; even if they do function in an alternative site, they are irregular and subject to frequent closure due to rain or extreme weather.
> Teachers are irregular; in many districts of Chhattisgarh and Jharkhand, contract teachers are locally recruited to manage the schools. There is little monitoring or oversight because officials and supervisors are not able to travel to these regions.
> In areas where frequent armed attacks on government properties are common, parents do not send their children, especially small children and girls, to school.
> When livelihoods are threatened, children either migrate with their parents or join them in survival duties like the collection of forest produce.
> Children are traumatized, and those who directly witnessed violence or were attacked continue to be terrified. There is no mechanism to provide counselling and support to such children.

An example of the trauma experienced by children has been documented in a 2011 research study conducted by the G. B. Pant Institute in Allahabad:

> Tribal communities make up about 32 percent of Chhattisgarh's total population, and about 79 percent of the population in Dantewada and Bijapur districts in southern Chhattisgarh. Maria Gonds and Dorla tribes are the two main tribal communities in this region. Naxalites commenced their activities in the Bastar region of Madhya Pradesh in the1980s. A combination of political, economic, and social factors in this region, including economic exploitation of tribal communities, poor relations with the police, and absence of government facilities and state institutions, contributed to the popular support and growth of Naxalism. Some people also report that Naxalite methods have gradually become increasingly authoritarian, undemocratic, and marked by human rights abuses including extra-judicial killings,

> beatings, and extortion. Over time, this has created resentment among some villagers. Over a period of approximately two-and-a-half years, between June 2005 and the monsoon season of 2007 (June to September), government security forces joined Salwa Judum members on village raids, which were designed to identify suspected Naxalite sympathizers and evacuate residents from villages believed to be providing support to Naxalites. The government relocated thousands of villagers to government-run makeshift Salwa Judum camps near police stations or paramilitary police camps along the highways. (G. B. Pant Social Science Institute 2011)

This research involved visiting schools and talking to the communities. In Bastar alone, 31 schools were occupied by security forces and the grounds of seven Ashram schools were being used by paramilitary forces. In Kanker, 264 schools were shifted to resettlement colonies (Salwa Jhudum camps). The number of schools that were destroyed by the Naxals is even higher: 87 in Dantewada, 42 in Bijapur, 30 in Narayanpur, 48 in Kanker, and 22 in Bastar.

Similar situations have been reported in several states in the North-east. One study on Manipur (Meetei 2011) lists the number of rebel or armed groups and the impact this violence has had on the people of the state. However, these studies were not rigorous or done over a substantial period of time – nor were they published. Therefore, I would like to simply flag this issue as one that needs further research.

Displacement and its impact on education

The High Level Committee Report (Ministry of Tribal Affairs, GOI 2014) has rightly highlighted displacement as a very important phenomenon that has affected the lives and livelihood of tribal communities: 'Given that the Scheduled Tribes constitute about eight per cent of the country's population, they are clearly disproportionately represented in the number of displaced persons' (Ministry of Tribal Affairs, GOI 2014: 50). It is important to flag the impact of displacement on the education of children in tribal areas, especially for Particularly Vulnerable Tribal Groups (PVTG).

One example is the situation of groups who have been displaced from reserve forests/wildlife parks. To begin with, the areas inhabited by PVTG are underserved by schools; when they are displaced, they are at a loss for both livelihood and education. Even in relatively better-developed regions of the country, the situation of tribal populations in such enclaves remains precarious. For example, the

> Baiga families were relocated to an area where their traditional forest livelihood of collecting Sal and Tendu leaves as well as bamboo was no longer feasible. In Tamil Nadu, the Kattunayakans, a tribe of honey-collectors, who live on the fringes of the Mudumalai Wildlife Sanctuary, a 321-square-kilometre park, are being prevented from collecting honey as well as other produce from the forest area due to immense harassment from forest guards. (Ministry of Tribal Affairs, GOI 2014: 62)

The first and most glaring impact of displacement is that the children are pushed out of school and into labour. As Sahoo Madhulika observes in her research study on the impact of displacement on tribal children of Orissa,

> The impact of displacement on children is all the more precarious. It was revealed that displacement in many cases has led to child labour in Odisha especially, when they are not resettled and rehabilitated properly. Research indicated that, in most of the time, displaced children find themselves separated from their close relatives and community people; they are vulnerable to exploitation and abuse. They are generally weakened by the long-distance resettlement that they are forced to travel and they have to thus endure the difficulties. They are more prone to the easy targets by the smugglers, traffickers, and/or armed groups. Furthermore, such displaced children can also become victims of discrimination and many of their fundamental rights may be violated due to their displacement. This article has tried to examine the impact of displacement on tribal children in Odisha, which has further tried to study the case of child labour and child abuse in post-displacement situation. (Madhulika 2013: 61–65)

Similarly, a study by A. B. Ota noted that the children of displaced families suffered from a loss of education. While the distance to primary schools may have decreased marginally, the distance to high schools and colleges increased substantially in the study area (Ota 2006). While most research work on displacement in other parts of the world mentions its impact on education, this remains an under-researched area in the Indian context.

Agency, Voice and Governance

The most challenging governance issue is the existence of a dual system of administration for education, which is split between the tribal welfare department and the education department. One of the issues that have been highlighted since the 1990s is that the tribal welfare department lacks expertise in education, and for that reason tends to manage the schools under its jurisdiction like it manages health, law and order, or forest rights. There is no cadre of educational administrators. Similarly, teachers' access to professional development opportunities, new pedagogies, and ongoing academic support has also been reported to be problematic.

A second issue, one that has been with us since the 1950s, has to do with the school calendar. Local tribal festivals/fairs and the heavy monsoon months do not coincide with national holidays or the annual vacation calendar. As a result, schools officially close during national or state holidays/vacations and are also effectively closed during tribal festivals and fairs – resulting in a drastic fall in the number of working schooldays. This not only affects learning, but breaks the continuity of teaching-learning processes. The problem cited by educational administrators is that teachers' unions are not willing to localize the school calendars, even though that has been a stated policy of the government since 1968. However, the real issue is that the tribal community and its leaders lack voice in the political system. While tribal communities are considered vote banks, given the peculiar electoral politics in India tribal leaders seem to be more interested in obtaining reservations for jobs and quotas in educational institutions. Be it because of the leadership in tribal communities or that of the mainstream political parties, the festering neglect of education in tribal areas is still to receive the attention it merits.

The issue of small schools with one or two teachers also remains a governance challenge. The time has come to review this policy; in view of improvements to roads and infrastructure, state governments could explore whether nodal schools could be strengthened and children picked up and dropped by bus. This is already being done by private schools, even in remote areas. Larger schools with one teacher for each class could also be monitored more effectively, and the problem of multi-grade teaching could be addressed.

Nomadic tribes move a lot and migrate from one part of the state to another, or even to other states. As pointed out in one report (Ministry of Tribal Affairs, GOI 2014), communities like the Bakkarwals and Gujjars of Jammu and Kashmir move to higher altitudes in summers and lower altitudes in winter. 'While residential schools have been set up in lower altitudes, the younger children still travel with their parents and do not stay in the school hostels' (Ministry of Tribal Affairs, GOI 2014: 186). Similar problems have also been documented in Gujarat and Andhra Pradesh. This remains a major challenge that can only be addressed through negotiations between community leaders, the education department, and the Ministry of Tribal Affairs.

Finally, and most importantly, the issue of the language of instruction at the primary level is also caught up in political rhetoric and administrative apathy. The importance of early education in a child's own language has been demonstrated across the world, and yet in India this issue is not taken seriously. Even states that have a majority tribal population have not paid attention to the need to start education in the child's own language and gradually build bridges to the local language of the state, Hindi, or English.

5 The Intermeshing of Gender and Equity in Education

Improvements in gender and social equality in education are influenced by three interlocking sets of issues: (i) systemic issues; (ii) the content and process of education; and (iii) the economy, society, and culture. There is almost unanimous acceptance of the fact that gender, as a category, needs to be seen within its larger social, regional, and local context. India is both a land of rich diversity and one of sharp disparities. The interplay of socio-economic inequalities and gender relations creates a complex web that either promotes or impedes the ability of girls to go through schooling, depending on the location.[1] While economic disparities and social inequalities are certainly important, a number of researchers argue that regional characteristics and cultural beliefs and practices also play an important role (Colclough, Rose, and Tembon 2000). As Shireen J. Jejeebhoy and Zeba A. Sathar point out,

> the cultures of South Asia are largely gender stratified, characterised by patrilineal descent, patrilocal residence, inheritance and succession

[1] 'There are well-known substantial differences in well-being across social groups in India [...] Average per capita income of SC/ST at all-India level is about one-third lower than among other groups. Headcount poverty among other non-deprived groups (in 1999/2000) was 16%, 30% for minorities (Muslims), 36% for SC and 44% for ST [...] Deprived groups also have lower literacy than other groups [...] Ranking on neonatal, post-natal, infant, child and under-5 mortality indicators for socially excluded groups are similar to those of other indicators (e.g. IMR for SC and ST are about 84 and 62 for non-deprived groups)'. (de Haan 2004: 9)

> practices that exclude women, and hierarchical relations in which the patriarch or his relatives have authority over family members. Levels and patterns of female autonomy vary considerably within the region [...] region plays the major conditioning role. (2001: 687)

For example, the social and educational status of Muslim girls in Kerala and Tamil Nadu is appreciably better than that of Hindu girls in Rajasthan or eastern Uttar Pradesh. Similarly, the social relationships and status of Dalits vary by region or state, and sometimes even within districts. The position of women in different communities and their access to education also differs as we move from region to region, social group to social group, and between urban and rural areas.

In India, it is therefore important to understand the intermeshing of poverty, social inequality, and gender relations. The three intersect in different ways in different regions of the country – with reinforcing each other in some areas and offsetting each other in others. Understanding and unravelling this complexity is the biggest challenge today.[2]

In the early years after the World Education for All conference in Jomtien, Thailand (1990), most of the analysis of gender issues in education started with mapping the provision (supply) of and the 'demand' for schooling, i.e., the readiness of families to send their children, especially girls, to school. However, over the years there was a realization that in education one cannot separate supply and demand: wherever there were accessible schools that were regular and functioning well, demand increased (Wazir 2000). Conversely, where schools were dysfunctional or indifferent, where not much teaching was happening, and where communities did not have

[2] Naila Kabeer and Ramya Subrahmanian highlighted the specific forms of disadvantage that are faced by girls and not boys. These disadvantages are related to roles deriving from the gendered division of labour, girls' reproductive cycles, and perceptions about the risk of and their vulnerability to sexual violence. These gender specific forms of disadvantage (Kabeer and Subrahmaniam 1996: 3) need to be identified and distinguished from those that are gender intensified to facilitate the identification of suitable social policies (Subrahmanian 2002: 12).

confidence that the school was a safe place for their daughters, demand fell sharply. There was recognition that the availability of well-functioning schools is an important variable influencing household decisions (Ramachandran et al. 2004; Subrahmaniam 2002; PROBE Team 1999). Research also pointed out that push factors and what happens inside the school are equally important. Studies highlighted how the attitudes and practices of teachers and the behaviour of children from relatively forward social groups affected the ability of children from socially disadvantaged groups to remain in school (Jenkins and Barr 2006; Nambissan 1996, 2000, and 2001). Further, NGOs like the MV Foundation systematically gathered information on push factors, recognizing the need to interrogate classroom practices and teacher/administrator attitudes. Equally, the work done in the run-up to the formulation of the National Curriculum Framework (NCF) 2005 revealed that the curriculum can reinforce social biases/discriminatory practices and gender stereotypes (NCERT 2006; Ramachandran 2004). Three decades after Jomtien, it is now recognized that the prevailing gender gap in educational achievement is more than just the combined effect of supply and demand factors. Greater social awareness in the community, proactive campaigns to promote girls' education beyond the elementary level, improved schooling facilities, a better-functioning school system, appropriate interventions to address the educational needs of specific groups of girls, and most importantly more gender-sensitive interventions inside the classroom and specific incentives make a big difference in the educational outcomes of girls.

Recent government reports based on NSS 71st Round (2014) highlighted the nutritional status of children in the country as a whole (NSSO 2016). The National Family Health Survey Round 3 (NFHS-3) data reveal that 38% of Indian children under 3 are stunted (too short for their age); 19% are wasted (too thin for their height); and 46% are underweight (too thin for their age) (IIPS 2006). This recent evidence highlighted the alarming nutritional status of Indian children despite the ICDS programme and the midday meal scheme. Given the prevailing gender relations and the attitudes and practices towards girl children, chances are that girls account for a larger proportion of these malnourished children.[3]

When there are natural disasters like drought, children experience both short-term and persistent hunger for a period of time (Ramachandran et al. 2004). Hunger, poor health, and frequent illness impede regular attendance in school. However, the link between health and education is yet to be addressed, especially for children from the poorest quartile of the population. The workload of children from very poor households before and after school also comes in the way of studying/reading after school hours. This problem becomes particularly severe for first generation school-goers, and among girls – especially if they are the oldest children in the family.

Despite attempts to address issues of accessibility, social and physical access continues to be a problem for children living in relatively inaccessible locations, especially at the upper primary and secondary levels. Notwithstanding the initiatives undertaken by the government to enhance access, general household characteristics like income, caste, occupation, and the educational level of parents continue to determine children's access, attendance, completion, and learning achievements. Children from rural families with substantial land, non-agricultural occupations, and higher educational levels have greater access to education than do children from landless, agricultural wage-earning families and migratory groups. In fact, the invisibility of these landless and migratory groups in the macro data has contributed to their alienation from the educational processes. Those most affected are:

Severely disadvantaged communities and those residing in tribal, hilly, desert, and remote habitations

[3] Malnutrition continues to be a significant problem for children and adults in India. While there have been some improvements in the nutritional status of young children in several states, nutritional deficiencies are still widespread. Most striking has been the increase in wasting, or low weight for one's height, among children. NFHS-3 found 19% of children were wasted, an increase from 16% seven years earlier. Over the same period there was virtually no change in the percentage of children who were underweight (47% in NFHS-2 and 46% in NFHS-3). NFHS-3 also found a remarkably high prevalence of anaemia – 79% – in children aged 6–35 months. The prevalence of anaemia is even worse than it was in NFHS-2 (74%) (IIPS 1999, 2006). Anaemia in India is primarily linked to poor nutrition.

New/seasonal migrants into cities
Children living in places where there is social strife and conflict
Children of families displaced due to natural or man-made disasters
Children of sex workers, people affected with HIV/AIDS, and single women (widows, deserted/separated women, unwed mothers)
Among all of the above, the situation of girl children is more worrisome than that of boys

Even where formal access has been provided, providing quality education remains a big challenge, especially in more deprived rural and urban areas – thereby leading to poor learning achievements. While overall issues of access, infrastructure, functionality, quality, and attitude affect all children, given the prevailing social inequalities and hierarchies these factors affect poor children, especially girls, much more than they affect the more privileged sections of society.

The Snakes and Ladders that Frame Gender and Education

'We are prisoners of myths and ideologies of our own creation – sometimes we are so vehement about that we think is right that we refuse to listen!' This is what we discovered about ourselves and about the education community in general during an in-depth qualitative study that I led in 2003, which explored the factors that facilitated or impeded successful primary school completion among the poorest households in Uttar Pradesh, Karnataka, and Andhra Pradesh (Ramachandran et al. 2004). Among the research tools used were a series of structured activities with children – both in and out of school – to understand their point of view. What emerged was a fascinating matrix of snakes and ladders.

As we entered the house of one of our 'sample households', we saw school bags hanging neatly in a row on a *kutcha* (makeshift) wall. This is not a classroom, but the one-room house of a Muslim family in an urban slum in Sitapur District of Uttar Pradesh. The primary earner, the father, is a rickshaw puller who believes in the value of education and ensures that all six of his children (two boys and four girls) go to school – a local private school – regardless of his economic situation. Their mother is bedridden with a serious illness. The eldest daughter does all of the housework, puts her mother's medicines into a small plastic bag, pins the bag to her blouse, places a pot of water and some food next to her, and runs to school! This bright child of 12 years is overworked and shoulders a significant amount of responsibility on her slender shoulders. Yet at the same time she is determined to complete high school!

Across the three states, we realized that being the eldest girl is indeed a big issue: the dice is loaded against them. Similarly, not

Table 5.1: Factors that facilitate (ladders) or impede (snakes) successful primary school completion

LADDERS (Facilitating factors)	
Mild 🕊, *Strong* 🕊🕊, *Very Strong* 🕊🕊🕊, *and Extremely Strong* 🕊🕊🕊🕊	
A bright and welcoming school	🕊🕊🕊🕊
A school where children can learn	🕊🕊🕊
School within reachable distance *(boys)*	🕊🕊
School within reachable distance *(girls)*	🕊🕊🕊
Having teachers who are affectionate, kind, and empathetic	🕊🕊🕊🕊
Having an adult in the family who values education	🕊🕊🕊
A father who is caring and concerned about the welfare of the child	🕊🕊
Adults in the family who share work (i.e., care of siblings, fetching water)	🕊🕊🕊🕊
Having fewer siblings, but not being the eldest	🕊🕊🕊
Being in good health	🕊🕊
Being the youngest child	🕊🕊🕊
Role models/success stories where education has resulted in tangible benefits in terms of social status and/or livelihood/upward mobility *(girls)*	🕊🕊🕊🕊
Role models/success stories where education has resulted in tangible benefits in terms of social status and/or livelihood/upward mobility *(boys)*	🕊🕊🕊
SNAKES (Impeding factors)	
Mild ↓, *Strong* ↓↓, *Very Strong* ↓↓↓, *and Extremely Strong* ↓↓↓↓	
GIRLS AND BOYS	
Having a parent with a disability	↓↓

(contd...)

(Table 5.1 continued.)

Having a mother who goes to work for long hours	↓↓↓
Having a large family and many siblings to care for	↓↓↓
Alcoholism in the family (father/mother)	↓↓↓
Being from a lower caste/disadvantaged community	↓↓
Being sick or disabled	↓↓↓
Having to care for cattle or other livestock	↓↓
Death, disability, and illness in the family	↓↓↓
Regularly working during peak agricultural cycles	↓↓↓
Drought/migration/natural disasters/riots/violence	↓↓↓
Hunger (every morning)	↓↓
Hunger (persisting)	↓↓↓
Teachers that have children do personal chores	↓↓↓
Teachers that beat children and give out other harsh punishments	↓↓↓
Where education *has not* led to tangible benefits in terms of social status/livelihood/upward mobility of local youth (i.e., negative role models)	↓↓↓
GIRLS	
Being a girl	↓↓
Being the oldest child	↓↓↓↓
Having a sibling with a disability	↓↓↓↓
School is far away	↓↓
Burden of work (at home/outside)	↓↓↓
Early marriage	↓↓↓
Teacher drunkenness (safety of girls threatened)	↓↓↓↓
BOYS	
Having no sisters	↓↓
Having a sibling with a disability	↓
School is far away	↓
Burden of work (at home/outside)	↓↓↓
Indebtedness leading to short-term bondage	↓↓↓↓
Gambling in the neighbourhood	↓↓↓↓
Teacher addiction (learning bad habits)	↓↓↓

Source: Ramachandran et al. 2004.

having sisters was described as a big snake that eats up a lot of time! Bringing up younger siblings was also a snake. We worked with the children from this and other families to map out the snakes and ladders framing their ability to continue in school. What came out of this exercise was a fascinating matrix (Table 5.1).

Many children above the age of 6 that attend government schools reported working both before and after school. The burden of work was most severe for the first-born – especially the eldest daughter. Apart from caring for siblings, children support their families by taking care of cows/goats; fetching fuel wood, fodder, or water; running errands; and looking after sick family members. As a result, children either become late for school or miss it altogether. The gendered division of work, and the added responsibility of household work for the older girls in the family, were marked in all three states. We also met boys who were in temporary bondage to pay off a loan taken by their parents. Engaging in full-time work during weekends and holidays is a fairly routine activity for these children.

Heterogeneous Gendered Realities and Multiple Disadvantages:[4]

The quantitative data used to measure gender gaps or the progress towards gender equality in education reveal little about the texture of the inequalities that are inherent in the society and reinforced by the prevailing educational system. Coming to grips with gender and social equity issues in education requires a framework that can capture children's heterogeneous gendered realities and multiple disadvantages. Gender is part of a complex social and institutional structure in India; it is therefore necessary to look at gender inequalities in education within the broader framework of social, economic, and location-specific inequalities, on the one hand, and the prevailing school system on the other. A global study on girls' education summarized the issues thus:

> Already excluded because of their gender, many girls face multiple barriers, making it more difficult for them to enrol in and complete

[4] This phrase has been taken from NCERT 2006.

> primary school and continue on to secondary school [...] reaching excluded girls generally means higher costs and alternative policies and strategies because their needs differ from those of majority population. (Lewis and Lockheed 2007: 19–20)

Feminist scholars and others working on social exclusion have tried to tease out the texture of these forces of exclusion by identifying their levels, sources, and the forms they take. While the government has provided schools in most areas, discrimination or non-inclusion in those schools takes different forms for people from different social groups and in different locations. Table 5.2 below tries to capture this interplay between socio-economic factors and the prevailing schooling system.

Using this framework, it is possible to explore the texture of gender and equity issues by seeking answers to the following questions:

> How are equity and gender understood?
> Is the government's focus primarily on input variables, process indicators, or output indicators?
> To what extent do the teaching-learning process and curriculum affect the realization of equity goals in school education?
> How important is the deployment of the required number of teachers?
> Do teacher attitudes, prejudices, and mindsets make a difference?
> How can we assess the experience of schooling, particularly regarding issues of caste-/community-based inclusion and exclusion; child abuse (physical, emotional, sexual); and corporal punishment?

How the Government Monitors Gender- and Social Equity-related Goals

Starting from the 1990s, the government and its donor partners have used various input variables to track educational progress. However, there was no systematic effort to monitor processes or track outcomes and outputs. Even when outputs were measured, they were rarely followed up with analyses of how children learn and the factors that contribute towards effective learning in classrooms.

Table 5.2: Heterogeneous gendered realities and multiple disadvantages that frame girls' access to meaningful education

Prevailing school system ➤ **Socio-economic Factors ➤**		Physical access	Teachers and teaching	Learning	Monitoring and supervision	Accountability	Transition to next level
Poverty	In abject poverty Above poverty line	No access to schools or school not within reach (walking or bus)	Teacher attitude/ prejudice	No support at home, no money for tuitions	Does not capture the workload of girls or child labour/trafficking	No voice in the community or school	Poor learning, cannot afford tuitions
Location	Rural Urban Tribal	Teachers absent, irregular, take turns; parents fearful of security	High teacher-pupil ratios lead to low teaching time	School not supervised; little teaching-learning	No oversight; school-level committees ineffective	Schools not visited by officials; no oversight	Accessing next level difficult
Community	Tribal SC Muslim Others	Formal access, but social distance may prevent children from enrolling	Teachers unfamiliar with tribal languages Culture/caste/ community/ class biases	Social/cultural/ language barriers	Not a priority	No voice in village or school	No girls-only high schools
Occupation	Home-based Outside, waged labour	Work burden on children	Attitude/ prejudice of teachers	Poor learning, irregular in school	Not factored in as an area to track, monitoring	Parents cannot participate in meetings	Children drop out and join

(*contd...*)

(Table 5.2 continued.)

Prevailing school system ➤ **Socio-economic Factors ➤**		Physical access	Teachers and teaching	Learning	Monitoring and supervision	Accountability	Transition to next level
				Seasonal absenteeism while working	system insensitive		parents in work
Migration	Seasonal	Long absences	Teachers indifferent, not equipped to handle children joining mid-stream	Irregular in school	Not factored in as an area to track	No voice in community	No hostel facilities
Displacement	Permanent Periodic	Move to new location	Teachers indifferent	Irregular in school and language can also be a barrier	Not factored in as an area to track	No voice	No hostel facilities
Conflict	Communal Insurgency	Poor access due to insecurity, fear	Teacher attitudes/ prejudice; teachers absent	Low learning due to fear, insecurity, and schools shut	No monitoring	No voice in the administration	Fear of movement

(contd...)

(Table 5.2 continued.)

Prevailing school system ➤ **Socio-economic Factors ➤**		Physical access	Teachers and teaching	Learning	Monitoring and supervision	Accountability	Transition to next level
Violence & abuse	Home School Society	Poor access due to fear, low self-image	Corporal punishment, verbal abuse	Low learning due to fear, insecurity	Not part of monitoring protocol	Not accountable to children or parents	Leads to dropping out
Gender relations/ attitudes	Eldest/older children Girl child	Difficult to access due to burden of work Reaching of menarche leads to being pulled out of school or missing class during period	Teacher attitudes, sexual abuse, harassment; gender and social stereotypes in textbooks and the classroom	No time to study at home, burdened by work	No specific monitoring done through data/ information that is disaggregated by both gender and social group	Gender issues not taken into account	Leads to dropping out
Disability	Mental Physical	Identification and assessment of requirements Identification and assessment of requirements;	Ignored, not equipped to handle special requirements	No support	No specific monitoring	IED programme not gender sensitive	Drop out

(contd...)

(Table 5.2 continued.)

Prevailing school system ➤ **Socio-economic Factors ➤**		Physical access	Teachers and teaching	Learning	Monitoring and supervision	Accountability	Transition to next level
		physical access depends on supports/aids					
Health	HIV, Scabies, visible symptoms	Prejudice in school/ community	Teacher attitudes	No support	Not specifically monitored	Not part of monitoring protocol	Drop out

Source: Author.

Since the mid-2000s, a number of national and international organizations have been arguing for the regular testing of students. The ASER Reports, for example, are excellent examples of using nation-wide, sample-based testing to highlight the issue of poor learning. The response from the government (and a significant section of the education community) has been predictable, however: they either denied the findings or rejected them on methodological grounds. Notwithstanding these reservations, people across the board agree that poor learning outcomes are indeed a serious issue that needs urgent attention. While this is a step forward, the factors that positively contribute to or impede learning have still not been adequately explored. The hard reality is that the majority of children who are left behind in the learning race are from socially marginalized groups.

In this context it is important to explore how governments and major international donor agencies track the progress that is made toward achieving gender and social equality in school education. In Table 5.3, I draw upon a rigorous review of the SSA monitoring reports, known as Joint Review Missions (JRM reports) from 2003 to 2014.

Unpacking Access

The first and most important critique of how gender and equity issues are monitored concerns the SSA's very limited understanding of access. As seen in Table 5.3, 'access' is limited to physical access to a school and enrolment of students. Most research on education and equity underscores the importance of the following key supplementary variables. First, having physical access to a school is not adequate; the school must be functional and have proper facilities, adequate teachers, and the necessary teaching and learning materials. Second, the notion of physical access is meaningless in a highly unequal society, where the question is whether all children have equal access. How is the school dealing with communities that live on the margins of society, or those who are apprehensive about sending their children to school? For example, in a recent study in Rajasthan (Jandhyala and Ramachandran 2015), we would that Muslim families living in Ajmer were wary of sending their

Table 5.3: Indicators commonly used to monitor educational progress

Access	• Enrolment, class-wise and level-wise (primary, upper primary), disaggregated by social group and gender • Sample survey findings on out-of-school children (OOSC) • Information on average dropout, retention, attendance, and transition rates • Opening of new schools and school infrastructure
Learning	• Periodic national-level student assessments carried out by NCERT • Purchase and use of teaching and learning materials (TLM) • Pedagogy and classroom processes that are normally used
Teachers	• Numbers of teachers hired, disaggregated by regular and contract teachers • Number of female teachers • Pupil-teacher ratios (PTR) • Teacher attendance • Teacher training (numbers as a proxy indicator of academic support)
Community involvement	• Whether School Management Committees/Village Education Committees (SMC/VECs) have been constituted • Structure of SMC/VECs • Meetings held and issues discussed
Management	• Convergence between different departments and NGOs for community mobilization, enhancing school quality, and providing school facilities • Civil works and infrastructure
Finance	• Allocations of the GOI • Allocations of the state governments • Expenditures

Source: Aide Memoire of Joint Review Missions 1–20, available at http://ssa.nic.in/page_portletlinks?foldername=monitoring.

daughters to a secondary school that was just across the road, but technically outside of their locality. Similarly, tribal Sahariya[5] families in Rajasthan and Madhya Pradesh hesitate to send their children to government schools because they fear discrimination.

[5] A tribe that lives in both Rajasthan and Madhya Pradesh that is an extremely poor and socially disadvantaged community. They typically live outside of villages in primitive stone structures, with little access to sanitation. They are shunned by the population and until the 1960s did not interact with outsiders. They collect forest produce to survive.

Table 5.4: Unpacking access

Indicator	What is covered
Enrolment	• Enrolment rates across gender and social groups • Gendered gaps in enrolment • Transition from primary to upper primary schools (UPS) • Trends in enrolments in government and private schools • Enrolment data on children with special needs (CWSN), the urban poor, and children belonging to migrating families
OOSC	• Number of out-of-school children (OOSC), and among them those with disabilities and those from different social groups such as SC, ST, and Muslims – all disaggregated by gender • Strategies adopted to reach out to OOSC • Definition of OOSC
Drop out	• Dropout rates disaggregated by gender and social group • Dropout rates at critical points like Classes V to VI, upper primary to secondary, and after enrolling in secondary • Agreed definition of drop out • Reasons for dropping out: poverty, domestic work, sibling care, migration, inability to learn, corporal punishment, discrimination, harassment
Retention	• Overall retention rates • Retention of girls, SC/ST, and Muslim children • Retention of students, especially in the upper primary and secondary levels
Attendance	• Overall attendance rate • Enrolled but not attending • Sporadic and persistent absenteeism • Reasons for absenteeism: migration, ill health, distance, discrimination, irrelevant teaching methods
Transition	• Transition rate from primary to upper primary • Main reasons that inhibit transition from primary to upper primary, including the non-availability of upper primary schools and socio-economic reasons, especially for girls
School infrastructure	• 'Access' has been understood mainly as physical access • Emphasis on overall growth in school infrastructure • Functioning toilet for girls and CWSN continues to be a challenge • Ramps for CWSN

Source: Ramachandran and Chatterjee 2014.

In the course of a recent review of the SSA programme, we realized that the Joint Review Mission reports were invariably silent on several important issues. For example, location is an important determinant of equity because it determines the quality of the schools that are available to children. According to NUEPA (2013b), more than 57% of primary schools and 56% of upper primary schools in rural areas are located more than 10 km away from the Block Resource Centre (BRC). We also know that remote locations have adverse Pupil-Teacher Ratios (PTR), and that almost 71% of single teacher schools are located 10 km away from the BRC. It would, then, be a fair assumption that these schools are most likely catering to disadvantaged groups, are likely to have higher dropout rates, and may have serious implications for girls' participation. Unfortunately, when the government gathers information for monitoring or evaluation purposes, the physical location of schools (rural, remote, tribal, urban) is rarely correlated with indicators like the physical quality of the school, availability of trained teachers, availability of adequate number of teachers, and the practice of having more than one class of children sit together in a multi-grade setting.

Similarly, while government reports routinely mention that the dropout rates for SC/ST and Muslim children are high, and that among them the dropout rates of girls are higher than those of boys, the reasons for this are rarely discussed. The few monitoring reports that do attempt to analyse this phenomenon lack analytical depth. For example, one group of researchers observed in 2011 that the dropout rates among girls have gone down, while dropout rates among boys went up slightly in some areas (De et al. 2011). However, there is almost no reference to or discussion about the reasons for this trend. In the years following the December 2004 tsunami in Tamil Nadu, government officials noticed that many more boys were dropping out after Class VIII. It was only during the course of an in-depth study on the impact of the tsunami (that I did on the request of the Government of Tamil Nadu in 2006) that I understood the reasons for this. Apparently, before the tsunami one motorized fishing boat had been shared by a group of families and all of the men would go deep-sea fishing together. Interestingly, in the post-tsunami relief and rehabilitation process many boats were distributed and in some places each family ended up with their own fishing boat. As a result, young boys – most over the age of 12 or 13 – joined their fathers in fishing, thereby dropping out of school. As this example shows, the reasons for the dropping out boys or girls could be different in different situations and communities.

We know that private school enrolments have been increasing, but we do not know who has more access to private schools. It is a prevalent assumption that private schools provide better education compared to government schools and that boys have more access to private schools (De et al. 2011; Goyal and Pandey 2009). If this were the case, it could create a gender imbalance in both government and private schools. Again, there is no mention in the JRM reports of either a strategy to counter this phenomenon or a debate on this issue. Similarly, when children with special needs (CWSN) are discussed, or when the urban poor are identified as the most challenging group to target for education, there is almost no discussion about their details or profiles. Are the neglected CWSN invariably girls? Do most of the urban poor come from some specific social group(s)? It is indeed sad that, despite having so much information on gender inequalities and caste-poverty linkages, the analysis of data on the most deprived is done in a generic manner, with little reference to gender, caste, or place of origin, and most of the discussion is limited to data on enrolment. Admittedly, the monitoring and evaluation reports of SSA *have* flagged some important issues, but they do not delve into inter- and intra-state variations or equity issues related to gender, location, socio-cultural profile, and economic status.

Learning and Teaching

Along with improved access, equity, and retention, the quality of the schooling experience is closely linked with improved learning outcomes. As the 11th JRM succinctly states,

> Universal enrolment, attendance, retention and inclusive education are necessary components to ensure equity in education, however, it is ultimately the quality of the schooling experience for the children, the classroom processes and activities and improvement of learning levels that are of essence in achieving education of equitable quality and moving towards the goals of Education for All. (SSA 2010)

While this statement holds true, ironically the approach to learning and teaching has been fragmented (Table 5.5).

Table 5.5: Learning and Teaching

Indicator	Issues and concerns flagged in JRM Reports
Student assessment and learning outcomes	• Large-scale evidence of rote learning • Increasing use of Continuous and Comprehensive Evaluation (CCE) as a continuous assessment tool, but confusion about CCE itself • Learning levels are low, especially among disadvantaged groups • Listing of learning interventions that have been implemented
TLM, curriculum, textbooks	• Curriculum and textbook reforms • Increasing use of bilingual languages
Classroom processes	• Listing of innovations in different states, like activity-based learning, computer-aided learning, pictorial dictionaries, and mobile libraries • Traditional methods of teaching largely employed by teachers
Recruitment of teachers	• Trends in number of single-teacher schools over the years • Numbers of teachers recruited
Female teachers	• Steady increase in the number of female teachers over the years • Proportion of female teachers in educationally backward areas is not known
PTR	• Low PTR ratios, especially in remote & tribal areas • High PTR in the most populous states, where PTR can go up to over 100 students per teacher
Teacher training	• Data on number of teachers who have been trained and are yet to be trained • Extensive use of traditional methods in teacher-training programmes • Poor content of training programmes • Lack of focus on equity and gender issues in training programmes
Attendance & accountability	• Insufficient time being spent on child-centric activities • Insufficient data on teacher accountability

Source: Ramachandran and Chatterjee 2014.

On several occasions, monitoring reports have highlighted that learning levels are low, particularly among children from vulnerable groups. However, most monitoring reports are largely disappointing when it comes to tracking the learning levels of children. One possible reason could be that, apart from NCERT's National Achievement Survey (NAS) data, there has been very little documentation of learning levels in government schools. In addition, there is an unwillingness to refer to data from independent studies. Consequently, SSA monitoring reports lack a cohesive discussion about the reasons for low learning levels among students, even though other studies have highlighted factors such as the socio-economic status of the family, education level of the parents, and the child's gender, that have a bearing on the educational outcome for a child (De et al. 2011; Reardon 2011).

Again, government reports are quite robust when it comes to reporting on the various programmes and interventions that have been introduced under SSA to improve learning achievement levels. But we don't know which children are being targeted for such interventions, how many children have benefitted from them, or what their impact on learning levels has been. Similarly, all states are required to incorporate gender and equity issues into their school curriculum and textbooks, but monitoring reports are weak on demonstrating the ways this has been done or how gender and SC/ST stereotypes have been removed from textbooks.

A major part of student learning and achievement depends on the effectiveness of their teachers; one of the first steps to map this effectiveness is the recruitment of teachers. According to norms set by the Right to Education Act (RTE), there should be a minimum of two teachers in a school that has 1–60 students. However, according to NUEPA (2015c), 11.47% of primary schools in rural India are single-teacher schools, of which, and, as stated above, about 71% are located in remote locations. This has quite a few implications, including: a) these schools probably have poor facilities; and b) the number of female teachers will be low. In fact, it has been acknowledged that low female teacher ratios are a problem for many schools, especially those that are situated away from main roads. Schools in these areas also have high PTR. Almost 40%

of primary schools and 30.58% of upper primary schools in rural areas are struggling with adverse PTR (NUEPA 2013b). It is quite possible that schools with high PTR also have students who come from disadvantaged and vulnerable groups, are first-generation learners, and are facing language barriers. However, PTR has not been examined from a gender and equity lens, and there is only limited discussion about the condition of schools that are located in remote areas. This clearly demonstrates that, while there has been an improvement in overall enrolment across India, the enrolment of girls still needs to be the focus to drive the universalization of elementary education in its truest sense.

Teachers are one of the main determinants of the quality of education. Encouraging teachers to perform well is a complex process that not only involves teacher recruitment and distribution, but also their working environment, training, professional development, and remuneration (SSA 17th JRM 2013). A major issue is the didactic nature of the teacher-training process: as pointed out in the 6th JRM (SSA 2007), there continues to be persistent use of lecture-based training methods, with little focus on participation, reflection, or skill development. This style of teaching eventually gets transferred into the classrooms and, as a result, very little time is spent on child-centric activities (Research, Evaluation and Studies Unit, Technical Support Group 2010). In addition, most training programmes lack a focus on equity and gender issues or the challenges of multi-grade classrooms. Even though JRM reports have repeatedly emphasized that the content of teacher-training programmes needs to be revised, there is hardly any evidence of change.

Finally, there is no information about teachers' attitude and behaviour towards students. If we only look at the JRM reports, we cannot know what is happening inside the classrooms and how teachers' own prejudices and biases interplay with teaching in the classroom – although there have been numerous research studies highlighting the impact of teachers' prejudices, especially towards SC/ST children (De et al. 2011; Nambissan 2009; Ramachandran and Naorem 2012). In conclusion, we know how many teachers are getting recruited and trained, but we don't know what role they are playing in creating an inclusive classroom.

Other indicators used to monitor gender and social equity goals

As evident from Table 5.3 above, indicators such as the role of School Management Committees (SMC) and communities, programme management, and finance are regularly included in reports, but not often linked to equity issues. Even though these indicators have not been discussed at length in this chapter, I wish to make two observations that are important for this discussion of gender and equity. Community participation and School Management Committee (SMC) involvement has been positioned as an important strategy to achieve the equity goals of SSA. The belief is that involving people, especially women and parents from disadvantaged groups, will help bridge the gap between schools and communities. While it has been mentioned in JRM reports that women and parents from minority communities do not participate actively in SMC meetings, there is a lack of seriousness concerning the gender and equity issues that influence how SMCs are constituted, how meetings are conducted, and whether members are oriented and trained to function effectively. Instead, the focus is mostly on the formal constitution of SMCs. Similarly, discussions of finances are dedicated to the allocations made by GOI and state governments and schools' overall annual expenditures. There is not much information about the status of allocations to districts that are grappling with high PTR, high rates of OOSC, inadequate infrastructure facilities, etc., or the impact of the money spent on the various innovations and strategies that have been implemented for children from disadvantaged and vulnerable groups.

A major problem with the monitoring mechanisms used by the government is that they look at gender and equity issues superficially. This is partly due to the limited understanding and articulation of equity in most government programmes, especially SSA, which understands equity as creating 'equal opportunity' (MHRD, GOI 2010). This definition itself is inconclusive because it does not specify what 'equal opportunity' means. Does it mean that all children will have equal access to a physical school, that all children will be treated fairly and equally in the classroom, that all children will have equal access to all resources and basic minimum facilities, or, most importantly, that all children will have an equal opportunity to learn? It remains unclear.

Input indicators like enrolment, number of schools, status of infrastructure, and number of teachers, remain important for measuring the progress made towards achieving the equity goals of SSA. Even when SSA does look at other indicators such as retention, teacher recruitment, or single-teacher schools, they are mostly discussed individually and not as a part of the quality-equity continuum. Additionally, issues such as diversity in the classroom, the social gap between teachers and children, and issues of exclusion and discrimination are not integrated with an analysis of equity and quality.

It is long established that gender, social, and economic status exert strong influences over the education of a child (Table 5.6). A child is at a greater disadvantage if he/she is living in a rural area and comes from a poor family. This situation becomes worse if the child is a girl, especially an older girl.

Table 5.6: Different children, different chances: A summary of findings

Rural/Urban	Less likelihood of rural children enrolling in preschool and completing primary school
Income	Poor children have lower chances across location, gender, and caste, including children from urban slums
SC	Lower chances than non-SC/ST children for all measures, including preschool and regular school enrolment
ST	Even lower chances than SCs
Gender	Disparities increase as girls grow older and affect completion and repetition rates
State	Children from northern and eastern states are less likely to enrol in pre-school, or to complete primary school

Source: World Bank 2004: 34.

Other factors such as caste, family income, and parents' occupation and educational level also significantly contribute to educational inequalities (Saihjee and Ramachandran 2002). Similarly, first generation learners are at a greater disadvantage because of the limited support for schoolwork available at home.

All of these factors exert a significant influence on children's access, attendance, completion, and learning achievements. Yet the JRM reports are weak when it comes to making connections between learning achievements and various socio-economic factors.

Likewise, discussions of CWSN and children from urban poor or migrant families remain superficial in most JRM reports. We know very little about these groups or who constitute them. With respect to CWSN, monitoring reports are largely restricted to the number of children identified as such that are enrolled in schools, and the various incentives that have been provided to them. Monitoring reports have also acknowledged that the identification of such children and the training of special educators continue to be major challenges. However, from the reports we cannot know the gender or social group composition of CWSN, which children among the CWSN have more access to schools, what inclusive practices are in place, and what attitudes teachers and other children have towards CWSN.

Similarly, we know that children from migrant and urban poor families are the hardest to reach. Beyond some isolated initiatives that have been introduced to address the educational needs of this group, not much is known. There is absolutely no discussion of street children, children with HIV/AIDS, child labourers (full-time, seasonal, or part-time), children living in areas of conflict, children affected by war or natural disasters, or children who have been victims of physical, mental, and sexual abuse. A possible reason for the limited discussion of this group is the lack of visible policies or guidelines concerning them. Equally, there is often a hesitation to talk about the workload of children in diverse poverty situations, including those who are enrolled in schools but frequently absent due to seasonal or after-school work. This hesitation stems from the refusal of the government to acknowledge the wide prevalence of child labour across the country. Children who are formally enrolled are known to support their parents by working before and after school, during vacations, and when the family is under economic distress. Notwithstanding this information, it is incorrectly assumed that children who are enrolled in school do not work.

Summing Up

An important element of schooling is contributing to the overall development of a child, building her self-esteem and confidence, and enabling her to stand with her head held high among both her family and society. However, the nature of our schools, the attitudes of administrators and teachers, and the practices of exclusion and humiliation – especially of girls and boys from socially disadvantaged groups and the children of extremely poor people – do not always lead to the healthy overall development of a child. A recent qualitative study across six states of India exploring the inclusion or exclusion of children in schools, classrooms, and among peers (Ramachandran and Naorem 2012) threw up disturbing findings. Blatant and subtle forms of discrimination were found in almost all dimensions of school: (i) how children are seated in class; (ii) to whom the teacher gives attention; (iii) who the teacher ignores; (iv) who cleans the toilets and does menial tasks; (v) who gets to drink water first and who has to wait in line; (vi) who is asked to lead the assembly; (vii) who can perform special duties during school functions; (viii) who eats the midday meals; etc. The list is endless. Issues such as gender, caste, poverty, appearance, skin colour, overt symptoms of illness or disease, cleanliness, and the language spoken at home emerged as important markers of (in)equality. A girl child from a poor Dalit or ST family and who is not 'properly' dressed is at the bottom of the pecking order in schools. The attitudes and practices of teachers in the classroom and in the school reflect prevalent social practices. Adolescent boys and girls that we interviewed and worked with said that the situation is much worse today than their parents had described in stories about their own time in school.

Another deeply disturbing trend is the incidence of sexual harassment and sexual abuse in our schools. Even though newspapers have reported many cases where schoolteachers were accused of raping or sexually molesting their students, there is a lot of resistance from the government about doing any systematic research on the prevalence of sexual abuse in schools. During the course of the inclusion/exclusion study, we asked girls and boys (separately) in

FGDs about sexual abuse and violence, both verbal and physical. After a hushed silence, some of them talked about what they had heard or seen. Women teachers also talked about their own experiences. Dalit and *Adivasi* women teachers had to face the dual burden of caste and gender, which makes them far more vulnerable. These issues are rarely talked about and there is little research-based evidence that we can go by. One recent study revealed,

> children were physically punished, often quite brutally, mentally tortured, humiliated and even sexually abused in schools. Corporal punishment is rampant in both low cost private, private elite and government schools, both English and vernacular medium. The difference being that the number of private schools where such cases are reported far outnumber the government schools and the similarity being that majority of these schools, irrespective of type appear to cater to disadvantaged sections of the society. (Nawani 2013: 23–26)

In the last few years, studies by my colleagues and me have observed that subject choice at the higher secondary level remains a huge problem in many states – especially in rural and tribal areas (Jandhyala and Ramachandran 2015). Science, mathematics, and commerce are not offered in girls' higher secondary schools in states like Rajasthan, Gujarat, Uttar Pradesh, and even in Delhi. If and when girls express interest in science or commerce, they are advised to enrol in a 'boys' school' (in reality, a co-educational one). However, given the prevalence of sexual harassment (euphemistically called 'eve-teasing'), parents are wary of sending their daughters to such schools. This has major implications for the girls' career choices. The girls we interacted with in secondary schools wanted to become nurses or lab technicians, but given their lack of access to science subjects in higher secondary school, these aspirations remain unfulfilled.

The relationship between gender and education is not easy to unravel. Given the layers of inequalities and the growing prevalence of caste- and community-based mobilization, there is a need to go beyond the standard indicators of enrolment and transition and deeper into the schooling experience of children. Maybe one way to do this is to develop a framework that can enable us to factor in a child's physical location, identity (gender, caste, religion, and occupation of parents), and economic status. This could enable policy makers and administrators to develop a deep and nuanced understanding of the factors that facilitate or impede successful completion of schooling and become ready to venture into higher, vocational, and technical education. This is what I have attempted to do in this chapter.

PART II

BEHIND THE SMOKESCREEN OF DATA

6 Segregated Schools in an Unequal Society

Part I of this book attempted to capture India's historical baggage of social and economic inequalities that has reinforced educational disparities for over 50 years. This chapter attempts to capture the texture of this inequality and how it manifests on the ground. Based on six qualitative studies done in 2001 and 2002,[1] this chapter explores how social inequalities are reinforced and sometimes exacerbated by our segregated school system.

District Primary Education Programme

The District Primary Education Programme (DPEP) was initiated in 1994 as part of the larger Social Safety Net Credit Adjustment Loan provided to India in 1991 under the Structural Adjustment Programme of the World Bank. Starting from the policy guidelines of the 1986 National Policy on Education (NPE) and drawing upon the experience of a range of primary education programmes,

(I would like to acknowledge the input of Aarti Saihjee for a different version of this chapter published in *Economic and Political Weekly* in 2002)

[1] All from Ramachandran 2004. Saihjee, 'Long live the alphabets! Reflections from Betul District, Madhya Pradesh'; Das, 'More unequal than others: Evidence from Surguja District, Chattisgarh'; Mahajan, 'The hidden picture: A case study from Hisar District, Haryana'; Periodi, 'Second generation issues in equity and education: Learning from Kolar District, Karnataka'; Jandhyala, 'So close yet so far: Primary schooling in Warangal District, Andhra Pradesh'; and Ratnam, 'The weft and the warp of public education: A tale of two primary schools in Cuddalore District, Tamil Nadu'.

the DPEP Guidelines of 1994 stated that holistic planning and management is necessary to achieve universal primary education, and that it 'should incorporate a gender perspective in all aspects of the planning and implementation process' (MHRD, GOI 1994). The guidelines recognized the importance of mainstreaming gender issues, including emphasizing the education of girls to promote gender equality, and making them an integral part of DPEP; the need for a gender focus when tackling the problems of access, retention, and achievement levels; and the importance of reaching out to children from the most disadvantaged groups/communities. For this reason, educationally backward districts with female literacy below the national average were considered a priority. Equally, the project stressed the need for educating socially disadvantaged groups. DPEP adopted a two-pronged strategy for meeting these gender and social equity goals, namely:

> Make the education system more responsive to the needs and constraints of girls and children from disadvantaged communities
> Create a community demand for girls' education and the enabling conditions for greater participation of the community

Insights from Micro-studies

Six micro-studies done in Andhra Pradesh, Chhattisgarh, Haryana, Karnataka, Madhya Pradesh, and Tamil Nadu generated rich data on the local context (political, social, and economic) within which the DPEP initiatives were implemented, creating a textured and nuanced understanding of 'education' in rural India. This chapter briefly describes the methodology used in these studies and tries to capture the similarities and differences in the DPEP experience across the six states, highlighting the chasm – at times widening and sometimes narrowing – between rhetoric and the reality of policy initiatives.

The methodology adopted for these micro-studies was primarily qualitative and interpretive in nature, and attempts to capture the local dynamics that frame social and gender equity issues in primary schooling in DPEP districts across each state. The selected districts were DPEP Phase 1 districts (with the exception of those in Andhra

Pradesh) and had sizeable SC/ST populations, so that the research would reflect the social complexity of local contexts. Similar criteria were used to select each research block and *Panchayat*. The selection was done in tandem with state and district authorities; however, with the exception of Madhya Pradesh, the choice of *Panchayat* was left to the discretion of the principal researcher to preclude any obvious bias from the government.

Since DPEP was initiated in phases and not all districts were at the same level, it was decided to identify five *Panchayats* from DPEP 1 districts and one where the Andhra Pradesh Primary Education Programme (APPEP) had been operating for more than ten years. This strategy was successfully used in all states with the exception of Madhya Pradesh, where the state government controlled the selection of the *Panchayat* and the researcher could not influence the selection. While Betul did meet the criterion outlined for the selection of the district and the block, the same cannot be said for the *Panchayat* and, as we shall see, this had important implications for the actual micro-study.

Each micro-study was designed to yield an in-depth qualitative analysis of one *Panchayat*, focusing primarily on the educational facilities and infrastructure available at the primary level, including:

Age-specific demographic data of children in the 6–11 and 11–14 age groups

Listing of schools: government primary schools (GPS), Alternative Schools (AS), Education Guarantee Scheme schools (EGS), and private schools

Detailed enrolment data for various classes from one GPS, one AS/EGS, and one private school (recognized and/or unrecognized)

Focus group discussions with the Village Education Committee (VEC), Mother-Teacher Association (MTA), a women's group, and the most deprived/poorest community within the village (a Dalit community, a migrant group, or any other group that is perceived by the community as the poorest in the village)

Cohort study of one batch of students currently studying in Class IV or V in government and EGS schools

In-depth interviews with two teachers from the GPS, one from the EGS, and one from the private school

Classroom observation for two to three days and informal interactions with children and their parents

The methodology sought to combine direct observation, informal conversations, open-ended interviews, and focus group discussions to capture the local dynamics of access, enrolment, and retention in primary schooling at the micro level, especially over the last five years. The primary foci of these studies were the children, teachers, parents, and community (especially in its manifestation as school committees), and their experiences of schooling. School registers were examined and were essential for documenting current enrolment patterns according to gender, caste, and community, as well as for the cohort analysis of the primary classes (Classes I to VI) of academic year 2001–02 to ascertain trends in retention and dropouts. Open-ended interviewing was actively encouraged to explore and understand the attitudes, opinions, feelings, and behaviour of individuals, especially teachers. Direct observation was most helpful for documenting classroom processes, to understand the teaching and learning processes, and to flesh out the impact of the social identities of teachers and students in framing these interactions.

Focus group discussions (FGDs) with women, SC/ST social groups, and VECs were also used to capture the intangible social and gender dimensions of the DPEP that were partly framed by people's perceptions and subjectivity. It was observed during these FGDs that shared experiences provided a fertile ground for discussions, thus providing a higher volume and quality of information compared to groups with widely varying experiences. All of this information was further supplemented by informal conversations with key people, primarily in their roles as parents and community elders. Given the time constraints (five days in each village), the researchers relied heavily on key informants to provide an overview of both the *Panchayat* and the DPEP interventions implemented over the previous five years and their impact on primary schooling. The main strength of this methodology was its flexibility to adapt to specific local contexts by giving researchers the space to redefine the issues in the process of doing research.

Insights from Qualitative Micro-studies

All six micro-studies clearly indicate that the local demand for children's education is growing by leaps and bounds. This is not to say that the demand did not exist before, but rather that DPEP, especially through its household survey (known as *Lok Sampark Abhiyan* or LSA) in Madhya Pradesh and Chhattisgarh and similar surveys in Andhra Pradesh and Karnataka, has been able to isolate and highlight the out-of-school children – and in the process has tried to convince the larger community of its collective responsibility to educate children, as opposed to letting the burden fall on individual families. Thus it is not surprising to find that, while peer pressure seems to work as an effective mechanism to ensure school enrolment and attendance in Madhya Pradesh, the Chhattisgarh study observes that 'the increasing emphasis laid on education and on sending children to school by different government functionaries has made the community "silent" on this aspect (non-attendance), so it even becomes difficult to assess the reasons behind children not attending school' (Das 2004: 186).

Who goes to which kind of school?

Physical access to a functioning school, at least on the surface, has emerged as a non-issue in all of the studies except for the one in Chhattisgarh, where it is still an issue because of dysfunctional – as opposed to non-existent – schools. In large part, physical access was eased by the introduction of new government and private schools in the south (Tamil Nadu, Andhra Pradesh, and Karnataka), as opposed to the predominance of private schools in Haryana and government EGS schools in Chhattisgarh and Madhya Pradesh.

However, an urgent need to unpack the term 'access' was observed, since it seemed to gloss over several important issues. One such issue was that, within the larger context of universal access, *there was an emerging trend of children from different social backgrounds attending different schools*. In Andhra Pradesh, there was a divide between the GPS near the Dalit habitations, which was only attended by SC students, and the GPS of the main hamlet, which had very few SC students. The youth in the SC colony stated categorically that, even if SC children tried to seek admission in the other GPS, they were discouraged and told to attend the school in their own colony

(Jandhlaya 2004). A similar divide was observed in Tamil Nadu, and was made all the more significant by the glaring disparities between the infrastructures of the two schools (Ratnam 2004). In both of these studies, local private schools were vying for students, and ended up receiving students who were mostly from relatively well off Other Backwards Caste (OBC), and forward caste (FC) families. In Andhra Pradesh, there was no clear information about 155 children between the ages of 6 and 14; according to the resource person, they were probably attending private schools outside the village. This seemed to be a valid assumption, given that six private school buses visited the research village (Jandhlaya 2004).

In all of these studies, there was a strong belief that private schools instil discipline and are better organized since each class has its own teacher. Above all else, the main advantage is that the child learns English in private schools. Despite evidence to the contrary and the high fees charged, the belief and perception that private schools offer 'quality' education seems to be strengthening. This is best illustrated by the situation in Haryana. Enrolment in private schools in the research *Panchayat* revealed their exclusivity in terms of social status as well as a large gender divide. While the village has an almost equal ratio of SC to OBC and forward caste populations, more than 90% of the children in the government school were from the SC community and more than 90% of those in the private schools were from OBC and forward castes. Despite this high number, the proportion of forward caste girls was extremely low in all schools, partly because of a low sex ratio in the age-specific population of the village. This was particularly noteworthy among the forward castes – the Bishnoi, Jat Sikhs (Jat), and Mehta (Bania) communities. We know that sex ratios among the forward communities in Haryana are particularly skewed against girls. This phenomenon of FC girls missing from village schools can also be attributed to their enrolment in the 'convent' private schools in a nearby town, which have a transport pick-up facility (Mahajan 2004). The Karnataka study also highlights the growing lure of private schools, but captures the struggle between them to establish credibility in the community – especially when the local GPS can boast of similar or greater facilities than the private school and is supported by the village elite (Periodi 2004).

A different manifestation of the new dynamics of school segregation is evident in Madhya Pradesh, where children from the ST community primarily attend EGS schools while the locally dominant Kurmi community (OBC) send their children to the GPS. The issue of segregated schools was further complicated by the presence of a separate girls' primary school (a consequence of political negotiations by the Kurmi community), which not only reinforced the parallel school systems in the village but also reflected the different socio-cultural attitudes of the social groups: the gender-segregated GPS mainly catered to the OBC community and allowed for the education of Kurmi girls, while the co-educational classes of the EGS reflected the relaxed gender norms evident in the tribal communities. Despite the lack of tangible evidence regarding the 'performance' of these different schools, local opinion clearly considers the GPS as imparting 'quality' education in comparison to the EGS and to be more 'suited' to the needs of the OBC children (Saihjee 2004). In an interesting turn of events, the Chhattisgarh study turned the dichotomy evident in Madhya Pradesh on its head: the EGS schools were monopolized by the OBCs and the tribal elites, whereas the dysfunctional local GPS was transformed into the preserve of the poor tribal population (Das 2004).

Given this empirical evidence, it became apparent that the democratization of access to schools is ironically accompanied by the reification of a child's caste, community, and gender, which comes to define which school he or she attends. This has considerable equity implications, since it seems to further reinforce the stratification already existent in the Indian educational system. Of particular concern was that, despite the absence of objective criteria by which to actually rank these schools (with the probable exception of Chhattisgarh (Das 2004)), the emerging school ranking seems to reflect larger social divisions in society and is primarily contingent on 'who goes to school where'.

Reaching the hardest to reach

Despite efforts to universalize access, several groups of children remain hard to reach and constitute the majority of out-of-school children (OOSC). These groups are child workers, children with special needs, and adolescent girls. Child workers were targeted by the government primarily through alternative schools, with the intention of mainstreaming them into existing formal structures. These groups were prominent in Andhra Pradesh, Haryana, and Chhattisgarh, and many of them fall into the category of 'indentured labour'. In Andhra Pradesh it was observed that most of the out-of-school boys were bonded child labour, called *jeetagallu*, where boys are bonded to well-off families for agricultural work, watching the fields, or grazing livestock: 'The use of the term of out-of-school children has tended to gloss over the issue of the existence of bonded children in the village. It is only in extended conversation with the teachers and community that the problem is named' (Jandhyala 2004: 222). Although the Andhra Pradesh government had identified strategies for mainstreaming out-of-school children through bridge classes at non-residential centres at the habitation level, residential bridge camps, community campaigns and drives, and closer monitoring of out-of-school children at the village level, it was observed that the *jeetagallu* consistently fall through the cracks. For instance, the cohort analysis revealed that there were five male dropouts (or long absentees). Of these dropouts, two had gone back to being bonded labour. In 2002, the researcher was informed that there were still about 32 children in the SC colony who were out of school. They included eight girls and 26 boys, 21 of whom were bonded labour. What makes the issue more complicated is the fact that this practice is not exclusively observed in inter-community relations; it is also prevalent between SC families. One of the reasons that the issue of bonded labour has not been able to draw special attention is that it is seen as the responsibility of the mobiliser attached to the Non-residential Bridge Course (NRBC), not the primary schoolteacher (Jandhyala 2004).

Similar experiences are reported from Haryana. Vandana Mahajan (2004) observes that there is a clearly visible practice of child labour that had not been addressed at all by the school, VEC, Cluster Resource Centre (CRC), or Block Resource Centre (BRC). During discussions with Class V boys at the GPS, the issue of child labour bonded/contracted to local landlords emerged as an important area of concern. Several young boys (in the 11–14 age group) had been given away as *Palis* (short-term bonded labourers) to the landlords on a yearly contract of Rs 3000 to 5000, plus food, some grain, and clothes. They do a range of household and farm work, and some are enrolled (on paper) in the AS school. Although the teachers in the GPS were at pains to explain that child labour was no longer a problem in their village, there was plenty of evidence during our interactions with and visits to the poorer and SC communities in the village. According to the Head Teacher's own rough estimate, at the time of research there were 100–150 *Palis* in the village. These children are from the SC community and work as *Palis* for the landowning general class families of the same village (Mahajan 2004). The Chhattisgarh study also mentions the presence of such a practice, though it does not document it in detail (Das 2004). There is no doubt that the issue of equity takes on a new dimension in the presence of child labour, whether in inter- or intra-community interactions.

A glaring absence of strategies to cater to the special needs of children with disabilities was observed in all of the research *Panchayats*, with the exception of Andhra Pradesh. This is primarily because Andhra Pradesh's Warangal District is one of those selected by DPEP for a pilot Integrated Education (IED) initiative. The IED initiative in the research *Panchayat* started in 1999, when a survey was undertaken to identify the 'special needs' children: five handicapped children in the SC colony, of which two were hearing impaired, two were mentally challenged (of whom one was also physically handicapped), and one had behavioural problems. As of this research, only one hearing impaired girl in Class IV and one mentally challenged boy in Class V were in school, while the physically handicapped slow learning boy was enrolled in the Non-residential Bridge Centre. Records showed that the remaining children were enrolled but not attending school. While the identification had been done and aids had been provided, especially to the hearing impaired and physically disabled children, teachers were not equipped to deal with their special needs – especially those of the slow learners. Despite this, these children seemed comfortable and well-integrated into the regular classrooms, and it was observed that the other children comfortably interacted with them. There was little evidence of any support provided by the *mandal* resource group that was formed to help and assist the teachers to deal with these special needs; instead, the effort was primarily an individual choice facilitated by the fact that the headmistress was sensitive to the issue because her own daughter had a speech impediment (Jandhyala 2004). This is a far cry from Madhya Pradesh, where – despite initiatives at the district level – the concept of disability itself is unclear (Saihjee 2004), or Chhattisgarh, where the discourse on special needs is absent (Das 2004).

Infrastructure

DPEP's gender and equity strategies included improvements to the school infrastructure and the provision of toilets for teachers and students. It was argued that improving the infrastructure would have a positive impact on the enrolment and retention of children in school. The six micro-studies confirmed that an overall improvement in school infrastructure and facilities does make a difference. Conversely, a lack of proper infrastructure – like the broken-down building of the GPS in Chhattisgarh – exerted a negative influence on teachers, students, and the community. Low attendance, lack of interest among teachers, and the indifference of the community were linked to this crumbling school building (Das 2004). In Tamil Nadu, the poor facilities (a one-room thatched building) of the school in the Dalit hamlet was cited as proof of discrimination. A dirty swamp surrounded even the new building constructed recently by the Adi Dravida Welfare Board, and the hand pump and toilet were right in the middle of a cesspool (Ratnam 2004). Differences between schools in the same village raise doubts in the minds of the poor about the intention of the government to provide equal opportunities to all, irrespective of caste, community, or economic status.

Another interesting thread running through these six micro-studies is the issue of toilets. In almost all of the schools, the availability of toilets was considered unimportant, at least at the primary stage, for children (both girls and boys). The teachers mostly used the toilets. Discussions with children revealed that they do not see toilet facilities as either an encouraging or an inhibiting factor for school attendance. On the other hand, the availability of drinking water was an important issue, whether from a hand pump, well, or bore-well. In Chhattisgarh, the EGS school children fetched water from a well that was considered 'dirty', and a few children were sent to the government primary school (about a kilometre away from the EGS school) to fetch water from a hand pump for the teachers (Das 2004). In all six *Panchayats*, children in government primary schools did not bring water bottles with them, but almost all of the children attending private schools (observed in Haryana, Tamil Nadu, and Karnataka) were seen carrying water bottles and lunchboxes to school.

Inside the classroom

Classroom processes, child-centred learning, and pedagogic renewal received a considerable amount of attention in the DPEP. While a considerable amount of work has been done in this area, a coordinated and multidimensional approach (simultaneous work on curriculum, textbooks, teacher attitudes, pedagogic renewal, and teacher-pupil ratios) has not been followed evenly in all DPEP 1 districts. This unevenness is clearly reflected in the six micro-studies.

While almost all of the teachers interviewed in the course of the micro-studies had participated in DPEP training workshops, not all of them were convinced about the value of child-centred pedagogy. Many of them had not internalized the spirit of this new approach and seemed to apply the 'techniques' mechanically. For example, in Tamil Nadu, the teachers were familiar with the new approach but were not convinced. One of them even used his cane to order children to sing (Ratnam 2004)! In Karnataka, the teachers' familiarity with the vocabulary and techniques did not always lead to changes in classroom processes – but corporal punishment had definitely decreased and there was little fear associated with attending school (Periodi 2004). This was also the case in Haryana and Madhya Pradesh. There was no evidence of any impact in Chhattisgarh.

In most of the *Panchayats*, there was not much change in teachers' attitudes towards girls or children from very poor/socially disadvantaged groups. It is not that teachers discriminated against all girls: social class, cleanliness, and performance also influenced their behaviour. There was a clear bias in favour of the better performers and relatively better-off children (out of the mainly poor families that attended government schools). The double impact of gender and social status was evident in the attention given to the children, the allocation of work in the school, and classroom seating and interactions.

Discussions with teachers in all six *Panchayats* revealed that there was not much understanding of the family circumstances or special needs of first generation learners. Similarly, teachers have little appreciation of the work burdens of the children and their inability to devote time to studies when at home. In particular, teachers were not sensitive to the predicament of girls who put in several hours of work in the house every day, or of boys who graze cattle before coming to school. They seem to dismiss such children as low performers or bad students.

Discussions with the teachers did not reveal periodic or ongoing academic support given by the Block, Cluster, or *Mandal* Resource Centres. For example, in Andhra Pradesh the *mandal* resource person was engaged in checking enrolment at the school and non-residential centre, thereby playing an inspectorial role rather than an academic one (Jandhyala 2004). In Chhattisgarh, the BRC and CRC were guardians of data and records, and again did not seem to provide sustained or regular academic inputs (Das 2004).

Another interesting revelation was the minimal (if any) use of teaching-learning materials (TLM) in the classroom. In some areas, funds for TLM had been discontinued for the academic year 1999–2000.

VEC and community participation

Community participation was an integral part of the DPEP vocabulary. However, all six micro-studies reveal that it has become synonymous with the VEC or SMC (or any other formal committee instituted for alternative schools). These committees are meant to be the primary interface between the school and community, but evidence from the field indicates that they either do not function at all, or seem to function with a mechanical intensity rather than a genuine sense of participation and commitment. The VEC was dysfunctional in Chhattisgarh: the last meeting of the village education committee had been held more than a year before the study was done and the records did not include members' signatures. The resolution that had supposedly been approved in previous meetings – repairing the school building – had not been carried out (Das 2004). In Haryana, the functioning and responsibilities of the VEC were restricted to occasional get-togethers and the celebration of national days like Independence Day and Republic Day. School-related issues were apparently not discussed in any of the village forums, whether a VEC, Gram Panchayat and Gram Sabha, or Village Development Committee. Seven of 12 VEC members participated in discussions with the field study team, along with three special invitees – but only two of the members had enrolled their children in the government primary school. The children of the remaining members were either going to private schools or were below the age of 6. They admitted that the VEC does not meet regularly; this is not surprising, given that the children of most of the members attend private schools. In this year, it had met only once – to give approval for the auctioning of trees inside the school campus, which had fallen during a storm (Mahajan 2004).

Further, despite the mandated composition of VECs as an attempt to ensure gender and social equity, in reality they seem to uphold the interests of the dominant group. As per DPEP Haryana norms, the VEC of the village should have representation from the SC community, women, and two parents of children studying in the GPS, with the Head Teacher of the GPS as its Member-Secretary. In reality, the official VEC had been 'replaced' by another VEC that had been constituted by the Block Office of the education department, resulting in a local power tussle. Membership in the Village Education Construction Committee (VECC), which was formed out of the VEC, was sought after because of the substantial funds transferred to VECC for the proposed school building:

> It was quite shocking to see that members of the VEC were openly discussing the possibility of misappropriation of school building funds in front of the research team. In the view of the VEC members, its functioning and role is limited by the local level politics of using these bodies to exert influence, extract benefits and distribute patronage. Participation of the women members on the VEC was also not evident. Most of its members had no knowledge of their role and importance of the VEC to address issues related to children's education. (Mahajan 2004: 210)

Such politics tend to create a hierarchy between the various school committees based on the political clout they wield. In Andhra Pradesh, the teachers of the GPS in the Dalit *basti* (slum) were aware of the play between caste and class factors in determining the facilities for the school. They were unequivocal in stating that the school in the main village stands to gain from its proximity to the high school, as well as the fact that a majority of its students come from the dominant backward caste (BC) social group of the village. Consequently, their school committee is stronger and able to lobby for more benefits, better infrastructure, space, etc. This view was expressed during discussions with the community in the SC colony as well (Jandhyala 2004). Similarly, in Madhya Pradesh caste identities were reflected in the school committees, with the VEC representing the OBC and the SMC representing the ST – and with the latter lacking the resources to demand facilities for its school (Saihjee 2004).

In Karnataka, the existing VEC was considered to be very active and efficient and claimed a high level of community participation. In reality, however, the community was reduced to a few elites who, according to Vani Periodi, 'participate, take decisions and implement according to their wish and others follow' (2004: 228). She further adds that, though there were young people who were active and vocal in the SC community, their voices were not heard by the VEC. When they spoke out in the Gram Sabha or in front of visiting officials, their elders were called and warned by the village elite. At the time of the study, however, the state government had withdrawn the VEC and instead introduced a School Development and Monitoring Committee (SDMC) comprised of parents of children attending the school. This new committee had not yet been constituted, but was already a source of much concern and friction in the community. The VEC was rather critical of this development, as it challenged their authority. They were blunt in expressing their displeasure about ordinary people occupying the same position as them (Periodi 2004).

In the case of the Village Level Committee (VLC, same as VEC) in Tamil Nadu, the situation was even more polarized. The committee hardly met, as some members lived and worked in Chennai. The headmaster was keen to have an operational VLC and to keep the members informed about the school's needs, concerns, and events.

However, he admitted that convening a meeting was difficult. Instead, he met with each member of the VLC individually, wrote a report in the format of a meeting, and circulated it among the members for their signature. He then met the members again to get their views on the draft and transcribed the minutes in a book. Each member was expected to sign the minutes and pass it on to the others. However, the interesting aspect of the Tamil Nadu study is not the in-absentia functioning of the VLC. It is the only case study that actually documents genuine community participation by women's self-help groups (SHG) and local Dalit activists (high or middle caste women were not members of the SHGs). Their participation was neither mechanical nor bureaucratized; in fact, it was often a highly political activism, especially in relation to the school in the Dalit area. According to Ratnam, this is due to the presence of a village unit of the Dalit Panthers of India. Young Dalit men spend time in the local school and teach a few classes so that the teacher can concentrate on other ones. Their participation has helped the teacher cope with a multi-grade classroom. They also pitch in with extra-curricular activities and provide continuous moral support to the teacher (Ratnam 2004).

The teachers of both schools reported that the women's groups should be credited for the high rate of enrolment and retention at the primary level. It is important to underscore the point that Dalit, Adi Dravida, and other backward caste children from poor families attended government schools. For this reason, women (also from backward castes) had been enthusiastic participants in enrolment campaigns at the beginning of every academic year for the previous four years. Women's self-help groups (comprising mostly backward caste and Dalit women) were not only active, but very clear about their role: they saw themselves as the support structure for the school. They kept in touch with the headmaster throughout the year, and their engagement with the school ranged from leading a procession of students for enrolment, supplying sweets during school functions, and persuading parents to make extra efforts to enhance their wards' academic performance, to demanding improved facilities for the school in *panchayat* meetings. Those who lived close by allowed the girl students to use the toilets in their houses. They provided drinking water to the students and persuaded reluctant neighbours to also extend these facilities to the children. The Tamil Nadu study also clearly marks the emergence of women's voices and the growing recognition that they are major players in the enterprise of getting children to school (Ratnam 2004). Apart from Tamil Nadu, we did not see any concrete evidence of DPEP efforts to encourage women to be more active at the local level through creating MTAs, holding mother-daughter fairs, or keeping a one-third reservation in the VEC (also reinforced by the *Panchayati Raj* Institutions (PRIs)). The situation on the ground, at least in the six *Panchayats* covered, does not suggest active community participation.

The mechanical interpretation of community participation has not fostered the genuine participation of the families of children who attend government schools. Karnataka has made a beginning by dissolving the VEC and replacing it with a School Betterment and Management Committee composed of parents of the children attending the school (Periodi 2004). It remains to be seen whether this new committee will have the teeth to monitor the regular functioning of the school.

Competing inequalities

The undeniable fact is that people – be they parents, teachers, children, or community members – do not have one homogeneous identity and hence cannot be subsumed into the generic category of 'community'. While there are distinct social and economic groups like SCs, STs, or OBCs, these categories also need to be further broken down for each specific context. Similarly, though gender is no doubt a source of stratification, it is also stratified along the lines of caste, class, and community. Thus, the experiences of a poor ST girl will be qualitatively different from those of an OBC girl, and so on. In fact, in some areas the situation of girls in forward communities could be worse than that of SC girls. Similarly, in areas where boys migrate for work (like in the hotel belt of Karnataka), girls may have greater access to schools by virtue of the fact that they remain in the village, while boys are often compelled to drop out of school and leave for the city. It was observed that DPEP's emphasis on broad categories of social stratification actually tends to render invisible the competing inequalities that define children's lives and experiences of primary schooling. One overlooked primary source of inequality is the social and economic geographies of marginal settlements within the context of the larger *Panchayat* and their impact on access to 'functional' schools. All of the studies presented here point to the consequences of this spatial organization.

Another overlooked issue is that all members of a social group do not necessarily share the same economic standing – there may be differences within the community. For instance, in Andhra Pradesh the practice of bonded labour occurs between some families of the SC community, with children working in the homes of richer SC families in the same village. There is also a local perception that the relatively better off SCs who live on the periphery of the main settlement and whose children attend the main GPS are of a different sub-caste and hence different from the rest of the SCs in the village (Jandhyala 2004). Similarly, the Majhwars in Chhattisgarh are classified as ST, as are the Kunwars. However, the latter exhibits a very high rate of literacy, while the former are landless labourers and barely 5% of their children make it to school (Das 2004). Again, the situation of children from Valmiki households is different from other SC groups in Haryana. Among Sikhs, the social and economic situation of Jat Sikhs is far better than that of the Mazhabi Sikhs (an SC community) living in the same village (Mahajan 2004).

Some Hard Questions

It needs to be reiterated that, with the exception of Chhattisgarh, almost universal enrolment in primary schools has become a reality. The number of never-enrolled children – both girls and boys – is almost negligible. The presence of functional upper primary and secondary schools exerts a significant influence on children's and parents' motivation to continue their education. The cohort analyses also reveal that the dropout rate at the primary level is very small and that almost all children study up to Class IV (in Andhra Pradesh) and Class V (in other states); the problems begin at the last stage of the primary and in the post-primary level. Hence the availability and accessibility of middle and high schools could dramatically change the educational scenario in rural India. The micro-studies reveal that, while the better off have the resources to send their children to schools outside the village, it is the poor who are left behind. Quality remains a challenge in almost all areas: learning achievements, pedagogic renewal, gender- and social class-sensitive classrooms, and teacher attitudes. All of these need a lot more work if we are to ensure a good quality education.

Decentralized micro-planning under DPEP was expected to address context-specific issues and tailor the programme for the hardest to reach. However, the six micro-studies revealed that this was not happening, even where village-level surveys had been done, like in Madhya Pradesh, Karnataka, and Andhra Pradesh. The overwhelming finding of the six micro-studies is the need for *Panchayat*-level micro-planning to meet the educational needs of the most deprived in the area. The Village Education Committees and/or School Betterment and Management Committees have not been mandated to actively participate in planning, nor do they have any decision-making power. Further, the children of most VEC members do not go to government primary schools. The members thus have little interest in the school or in the marginalized groups in the village. The composition and role of VECs need to be reviewed (as has been done in Karnataka), and these committees need to be given more teeth.

DPEP made a good beginning and definitely made primary education a priority. Yet inequalities have not been addressed and the most deprived children still do not have meaningful access to schooling. Educational projects may not be able to directly provide solutions to the gender and social inequalities that define every aspect of the Indian social, economic, and political fabric. Yet, their success is inextricably entwined with their ability to critically engage and dialogue with the larger weave of Indian society.

Since the completion of these micro-studies in 2003, I have had the opportunity to travel to the Udipi, North Kanara, Bellary, and Davangere districts of Karnataka. As of 2013, there were a negligible number of out-of-school children in the 6–11 age groups in these areas. Primary schools were functioning and teachers were regular. However, what disturbed me was meeting a large number of working children in the 11–14 age group, and an even larger number from the 15–18 age group. Discussions revealed that many of them had dropped out after Class IV, V, or VI – with poverty and the need to work cited as reasons for dropping out, even in the more educationally forward districts of Udupi and Uttara Kannada. The lack of relevance of education (especially beyond the primary stage) and of employable or self-employment related skills, insecurity about the future, and economic pressures continue to push children out of school and into work. This observation is certainly quite different from the situation in Bijapur, Raichur, and Koppal, where a recent household survey revealed a large proportion of out-of-school children in the 6–11 age group (EdCIL and SRI 2010). However, it is interesting to note that even when a district (like Udupi and Uttara Kannada) achieves universal enrolment and participation at the primary stage, ensuring eight years of basic education continues to pose a major challenge.

In 2003, a team of researchers was involved in documenting the backward and forward linkages that strengthen primary education. We covered two districts in Karnataka, Uttar Pradesh, and Rajasthan, one in Orissa, and the cities of Hyderabad, Mumbai, and Kolkata (Ramachandran 2003). Again, the overwhelming impression we gathered was that parents recognize the value of primary education and are eager to send their children to primary schools regardless of their economic situation. In the more educationally backward areas of the country, the availability of a functional primary school remains a problem. In urban areas, the accessibility of schooling for children in difficult circumstances and from migrant families continues to pose a major challenge. In almost all of these areas, a significant number of children drop out at the primary stage, with an overwhelming number of girls dropping out after Class IV or V. Again, the accessibility and availability of post-primary education remains a problem in many areas.

The learning from a range of research studies done over the last 20 years reiterates that there are no shortcuts or magic formulae to address the fundamental problems of access, equity, and gender disparities, on the one hand, and quality, content, and relevance, on the other. Education has to be perceived as an integral part of people's survival and fight for a life of dignity and self-respect. Different components like physical access, mobilization, quality, teacher management, and pedagogic renewal need to be addressed simultaneously – thereby ensuring confluence and synergy. An integrated approach is necessary for meaningful change and a lasting, sustainable impact. Also, five years of schooling may enable children to learn rudimentary reading and numeracy skills, but are insufficient to gain mastery over literacy and numeracy skills or for cognitive development. This is of particular importance to communities that have historically been denied access to education and do not (as yet) have ready access to reading material. Internationally, eight years of basic education is recognized as essential.

Primary education is not a stand-alone activity and the government has to acknowledge the importance of looking both backward and forward. Preschool is a critical input into primary education and, conversely, upper primary education is also necessary to encourage and motivate children and their families to send children to primary schools. Moving one step forward, the availability of relevant and good quality vocational and life skill educational opportunities is essential to generate the necessary momentum for primary and upper-primary education. The age span we are looking at is 3 to 18, with one level feeding into the other and higher levels creating the necessary suction effect for earlier stages.

There is a consensus that the primary responsibility of basic education has to be shouldered by the government, because it alone has the mandate, the ability, and the wherewithal to respond to the educational needs of the poor. This study – like several others – shows that it is the very poor who go to government schools. Enhancing the capabilities of the poor and providing them with the tools to negotiate this unequal world from a position of strength requires political commitment and societal support. Investing in the development and growth of those who need it most is indeed the need of the hour. DPEP has made a beginning with respect to primary education; the question is whether this momentum will be sustained at higher levels and for another decade. If it is, India may well be able to achieve the goal of Universal Elementary Education by the year 2020 – as hoped for by the government.

What Does Education Actually Mean?

What does education really mean? Having the skills to read with comprehension? Signing one's name? Being able to fill in forms and write letters? Over the years, we have embraced a minimalist notion of basic education and literacy. Time-bound campaigns that teach people to read generate a great deal of enthusiasm, but in the absence of a larger literate environment the adults who learn these skills quickly lose most of them. Similarly, children who notionally complete five years of education have also been known to lose their reading and writing skills. The experience of Cuba, China, and some of the South American countries was different: adult literacy programmes were located within the larger effort to ensure universal elementary education, thereby shutting the tap before mopping the floor. This created a literate environment in the households.

At the heart of the debate on literacy and learning is the timeless question of the goal of education. Is it to enhance the ability of people to negotiate an increasingly unequal, divisive, and polarized world from a position of strength? Is it to enable people to reflect critically on their life situation and understand the world they live in to make informed choices? Good quality education, in essence, involves creating a system that enables both adults and children to learn to know, learn to do, learn to live with others and appreciate interdependence and diversity, and above all learn to 'act with ever greater autonomy, judgment and personal responsibility' (Delors et al. 1998).

The tragedy in India is that our inability to address the fundamental goals of education does not auger well for our democracy.[2] In an

[2] The relationship between education and democracy is like an invisible spiral that helps those in control remain there and those at the bottom continue to be alienated and voiceless. Through its potential of enhancing people's capabilities, education impacts the substantive practice of democracy more than the formal system of elections. As Dreze and Sen argue:

> 'Capability' refers to the alternative combinations of functioning from which a person can choose. Thus the notion of capability is essentially one of freedom – a range of options a person has in deciding what kind of life to lead. Poverty, in this view, lies not merely in the impoverished state in which the person actually lives, but also in the lack of real opportunities – given

increasingly polarized world, religion, race, caste, and language identities are being reinforced. Government schools – as evident from the insights of these six micro-studies – no longer provide a common shared space for children of different backgrounds, races, and communities. Children today grow up without getting an opportunity to mix with children from other religions and social groups. While children from those middle class and affluent families with greater access to world media may potentially be exposed to different viewpoints, the majority of poor children not only attend a school where they mix only with their own kind, but also have little access to print, electronic, or visual media. In the 21st century, we seem to have lost sight of the real power and true value of education and learning.

> by social constraints as well as personal circumstances – to choose other types of living. (Dreze and Sen 2002: 35–36)

Lack of access to education of an acceptable quality actively inhibits the development of citizens' abilities to engage with democratic institutions, thereby denuding the practice of democracy (Ramachandran 2005).

7 What Facilitates Successful Primary School Competition?

This chapter is based on the findings of a qualitative study done in 2002–03 with the funding of the World Bank (New Delhi) Educational Resource Unit. This study explored the factors that contribute to or impede successful primary school completion among children living in diverse poverty conditions. The initial part of this chapter briefly summarizes what we actually observed in the field regarding the health, education, and nutritional status of the children, and the role played by their family, community, and the local infrastructure and service delivery system. The next part identifies the social, economic, and political factors that exert considerable influence – separately and in tandem – over the ability of a child to complete primary education, as they play out along the integrated child development continuum. The conclusion explores the possible implications of these findings for making and implementing policies.

At the outset, it is important to clarify that education and the overall development of children was approached from selected vantage points across five social domains: the individual child; her or his family; the community; institutions (preschool and primary school); and other services (healthcare, sanitation, water, transport, etc.). This social domain approach helped us understand the causes and social

The World Bank has published a much longer and more detailed version of the study on which this essay is based, and for which I was the principle investigator (Ramachandran et al. 2004). I also wrote a summarized version of the main insights with Kameshwari Jandhyala and Aarti Saihjee, which was published in *Economic and Political Weekly* (Saihjee, Jandhyala, and Ramachandran 2003.)

processes that are implicated, wholly or partially, in children's full participation in schooling (e.g., poverty, class, gender, birth order, ethnicity, lack of schools or transportation, poor health, etc.).

We attempted to explore the critical intersection between sociological analyses of these domains and the life-cycle perspective of child development. The research focused on children in discrete age groups along the integrated child development continuum: conception to birth, 0–3 years, 3–6 years, and 6–11 years. The adoption of a life-cycle perspective helped us highlight the cumulative impact of neglect or support of a child along the continuum, and the different needs of children from varied age groups. This dual yet integrated focus on physical development and psychosocial factors enabled us to explore the continuous and cumulative nature of the social and economic exclusion that poor children face from the moment of their conception, and its possible impact on their ability to complete primary education.[1]

This study is based on detailed qualitative research conducted in both rural and urban areas between September and November 2002 in the three states of Andhra Pradesh, Karnataka, and Uttar Pradesh. We relied on detailed observation and formal and informal interviews to collect the data. The study was exploratory and illustrative, focusing on children, their families, the larger community, and the extant local education and health services within the context of the larger socio-economic situation of the selected states and the specific study area. Six villages and six urban settlements were selected for the study to reflect a diversity of location, demographic characteristics, remoteness/accessibility, caste and community composition, and the functioning of Integrated Child Development Services (ICDS) and primary education facilities (a profile of the sample villages and urban settlements is given in Tables 7.1 and 7.2 below). Hence the study is not and does not purport to be an evaluation of existing programmes

[1] Successful primary school completion requires not only the completion of five years of schooling. It also implies the ability to read and write with comprehension and being an active learner, as well as developing a self-learning capacity, the ability to articulate/communicate, and developing positive self-esteem and self-confidence.

on the ground. Further, the sample is skewed in favour of children living in diverse poverty situations, with a primary focus on the bottom quartile of the socio-economic ladder.

Table 7.1: Profile of sample villages

Village	Population	Caste/ Community	Infrastructure	Occupation	Location
Village 1 Uttar Pradesh	2294	OBCs are dominant SCs Few Muslim families	Permanent roads, erratic, electricity open wells for water, few personal toilets	Agricultural wage labour is a mainstay of the poor A few children work to repay loans	30 KM and 7 KM from main roads
Village 2 Uttar Pradesh	1600	OBCs are dominant SCs are poor Few Brahmins	Permanent road 2 km away, erratic electricity, hand pumps, few toilets	Wage labour, road construction, seasonal migration to city	20 KM from Block HQ 2 KM from Road
Village 1 Karnataka	1704	STs are dominant SCs, Muslims, and OBCs	All-weather road, electrified, bore wells/ hand pump	Agricultural labour, construction work	11 KM from Taluk HQ
Village 2 Karnataka	5094	SCs are dominant STs and OBCs	All-weather road, electrified	Agricultural labour, wage labour including construction	14 KM from Taluk HQ
Village 1 Andhra Pradesh	1205	100% Tribal	Proper road up to village, good bus service	Agricultural labour, road work and other wage labour	10 KM from Mandal HQ
Village 2 Andhra Pradesh	1115	SCs, BCs, STs, and OBCs	Two villages Proper road up to village		4 KM from Mandal HQ

Source: Ramachandran et al. 2004.

Table 7.2: Profile of sample urban settlements

Urban settlement	Population	Caste/ Community	Location/ Infrastructure	Occupation
Urban 1 Uttar Pradesh	3510	OBCs, Muslims, and SCs in almost equal proportions	Periphery Brick road, electrified, hand pumps, public taps, and public toilets Open drains	Wage work in PAC, vending, petty business, domestic work. Children work as apprentices to tailors and mechanics
Urban 2 Uttar Pradesh	9136 in the entire ward 250 families selected	OBCs, SCs, Muslims, and a few FCs	Centre of town Paved roads, electrified, hand pumps Mix of open and covered drains	Construction workers, masons, painters, vendors, wage workers
Urban 1 Karnataka	1105 in selected Ward	Adikarnataka[2] SCs, STs, and OBCs	Periphery of town, makeshift approach road, municipal bus twice a day Open drains Municipal lorry comes once every 3 months to pick up garbage	Dependent on wagon loading – near the railway station where coal and iron ore wagons are loaded
Urban 2 Karnataka	698 in selected Ward	SCs, STs, Muslims, and OBCs	Centre of town All facilities – road, piped water supply (below ground level with taps in pits), electricity, public community toilets, etc. Open sewer that runs through the settlement	Manganese and iron ore mines Occupational diversity – construction work, tailors, drivers, domestic labour
Urban 1 Andhra Pradesh	1313	SCs, BCs, and OBCs (including Muslims)	Centre of town with all facilities	Daily wage work Domestic work Petty trade

(contd.)

[2] One of the most backward indigenous tribal communities of Karnataka.

(Table 9.2 continued.)

Urban settlement	Population	Caste/ Community	Location/ Infrastructure	Occupation
Urban 2 Andhra Pradesh	1192	SCs, BCs, OBCs, STs	Periphery – 6 KM from city centre	Daily wage, auto drivers Construction Beedi[3] rolling

Source: Ramachandran et al. 2004.

From Conception to Birth

Children in this stage of development have been the focus of a number of health and family welfare programmes: the ICDS programme since the mid-1970s; the Maternal and Child Health (MCH) and Family Planning (FP) programme of the 1980s; the Child Survival and Safe Motherhood (CSSM) programme in the 1990s; and more recently the Reproductive and Child Health (RCH) programme. Despite these programmes,

- Across all three states, few women took any special nutrition during pregnancy; if anything, *most women reported that they consciously ate less in order to ensure a smooth delivery*. Most believed that it was all right to eat less and have smaller babies, since it would be easier to deliver them. Many women interviewed said that they 'fatten them [the children] up' after they are born.
- Nearly all deliveries across the three states took place at home assisted by a *dai* (traditional birth attendant or TBA) or a relative/ neighbour. Trained midwives were either not within their reach or families did not think it was important to access one. Even though macro data indicates a higher incidence of institutional and trained midwife-assisted deliveries in Karnataka and Andhra Pradesh, our research shows no significant difference across the three states.[4] It is likely that, as we go down the

[3] Local hand-rolled cigarettes.

[4] Interestingly, the official statistics compiled by the Registrar General of India from the records of Auxiliary Nurse Midwives (ANMs) or other sources and those collected by the National Family Health Survey (NFHS) are different from the responses of women. The households consistently report a much higher proportion of deliveries taking place at home.

socio-economic ladder, inter-state variations on key indicators of maternal and child health become minimal. Institutional delivery still remains uncommon.

- Most women continued to breastfeed their children until the next child was on its way, even up to 24 months. Special nutrition during lactation was clearly a novelty for most women. Since most children were breastfed or fed milk until they were able to pick food up on their own, the activity could as well be called eating and not feeding: the caregiver did not consciously make an effort to feed the child, but the child was encouraged to pick food up and eat it.
- Most women in the surveyed households were thin, anaemic, and ate once or at best twice a day. Given the severe drought prevailing during the time of our fieldwork in 2003, especially in Karnataka and Andhra Pradesh, food was limited to rice/broken wheat with salt and some chillies. Children were fed the same food. Many women reported they had been given iron/folic acid (IFA) tablets by the ANM. However, given the very low awareness about its benefits, consumption was irregular. Very few had taken a pre-delivery tetanus toxoid injection, especially in Uttar Pradesh.[5]

Perhaps the most disturbing finding of this study relates to the low level of awareness about health, including antenatal care, immunizations, and nutrition during pregnancy. Despite the presence of the Aanganwadi worker (AWW) and monthly visits by the ANM, nutrition and health education activities are practically non-existent – at least among the poorest women in the surveyed villages and urban slums. While general awareness levels were certainly higher in urban slums, there were few differences in practice. Focus group discussions usually revealed that the monthly visit of the ANM was limited to one specific location in the village, and that she rarely visited the distant settlements inhabited by SC or ST groups. Similarly, in urban areas service providers bypassed recent migrants, especially those at the periphery of the slums.

[5] NFHS Round II revealed that only 52.89% of rural and 76.34% of urban women were given iron/folic acid tablets during pregnancy. In Uttar Pradesh, only 32.91% of women received IFA tablets (IIPS 1999)!

Not surprisingly, the workload of women in the poor households was fairly high and even more so in the case of those engaged in daily wage labour, and there was no evident respite for pregnant women. Many women also feared losing children due to illnesses that could easily be prevented by vaccinations. Regarding birth control, very few women used an IUD or Mala D (contraceptive pill); the dominant measure was female sterilization. The health situation of women and children, by all indicators, was very poor.

The initial phase of child development is without a doubt dependent on the overall well-being of the mother. Given that the mothers in our sample villages and urban slums belonged to poor families, the picture that emerged was dismal – of women having babies too soon, too often, and too many. While the average age of marriage in the different states as a whole were 19 (Uttar Pradesh), 20 (Karnataka), and 18 (Andhra Pradesh), the mean age of marriage in

Table 7.3: Key reproductive health indicators for selected households

	Uttar Pradesh	Karnataka	Andhra Pradesh
Number of mothers profiled	18	18	18
Average age of marriage	13.1	15.1	14.8
Average age of first conception	17.2	16.3	18.8
Average number of pregnancies per woman	6.2	5.4	4.3
Average number of successful deliveries per woman	5.9	5.3	4.1
Successful deliveries	89 out of 114 pregnancies	89 out of 92 pregnancies	65 out of 76 pregnancies
Abortions/miscarriages	3 out of 114 pregnancies	3 miscarriages reported	4 out of 76 pregnancies
% of children who died before reaching 5 years of those born alive	20.8%	10.9%	14.9%
Average number of children per woman	4.7	4.5	3.5
Deliveries at home/assisted by *dai* (TBA) or relatives	109 out of 110	83 out of 89	68 out of 76
Deliveries in hospital	1 out of 110	6 out of 89	8 out of 76

Source: Household Survey Data.

the profiled households was much lower: 13, 15, and 15, respectively. Clearly, very poor households in all three states demonstrated a much lower age of marriage than the state average. The combination of low awareness and the poverty of individual households contributes to both the poor health of mothers and the low birthweight of babies. We could not ascertain the birthweights, as most of the children in the surveyed households were born at home; however, the children appeared visibly small and malnourished. Our investigation confirms the macro-level knowledge that one-third of all babies in India are born with low birthweight (as compared to Bangladesh, where the proportion is one-half, and sub-Sahara Africa, where it is one-sixth) (Ramalingaswami, Jonsson, and Rohde 1996). This has important implications for the physical survival of children as well as their ability to grow and learn.

From Birth to 3 Years Old

Out of the 85 children in the 18 households surveyed in Uttar Pradesh, 13 were below the age of 3. Most of those who were 2 years old or younger were still being breastfed. They hardly ate any food – partaking in what was cooked in the house, mostly roti dipped in

Table 7.4: Nutritional status of children aged 6–60 months

State	Number of children	Level of malnutrition	Wt for age (Under-nutrition)	Ht for age (Stunting)	Wt for Ht (Wasting)
Uttar Pradesh	22	Severe (-3SD)	7	3	1
		Moderate (-2SD)*	10	6	6
Karnataka	18	Severe (-3SD)	4	9	0
		Moderate (-2SD)*	11	13	3
Andhra Pradesh	23	Severe (-3SD)	1	2	0
		Moderate (-2SD)*	3	3	0
Total for 3 states	63	Severe (-3SD)	12	14	1
		Moderate (-2SD)*	24	22	9

Source: Household Survey Data.
* Includes –3SD.
** The above 'Z' Scores (Wt for age, Ht for age, and Wt for Ht) were obtained using NCHS standards through the 'EPINFO' package of the WHO of measured anthropometrical measurements of height and weight and reported approximate age in months in Andhra Pradesh Schools.

tea! As evident from Table 7.4 on the nutritional status of children aged 6 to 60 months, mortality rates (in the surveyed households) for the under-5 age group were high in all three states. Even though the studied sample was very small and statistically insignificant, we suspect that there is higher deprivation and vulnerability among the population living in urban slums in all states. Contrary to our experiences in Karnataka and Uttar Pradesh, the situation in the centrally located slum in Andhra Pradesh was far worse than in that located on the periphery, especially with respect to under-5 mortality. Was this because of greater environmental pollution, abject poverty, and the lack of coping mechanisms vis-à-vis the pressures of modern life, despite better access to health facilities?

Though there was not a severe water shortage in most of the study areas, hygiene was poor and most children were unclothed or unkempt, with uncombed hair, unwashed faces, and un-bathed bodies. Running noses, skin rashes, and boils were a fairly common sight. Close to 64% of the children in Uttar Pradesh and 83% of the children in Karnataka reported that they had suffered from some ailment during the last three months. In Andhra Pradesh, 37% of the children were suffering from some ailment during the field visit. The incidence of waterborne diseases was high, despite our finding that a majority of the houses obtain their drinking water from tube wells and protected water supply schemes. This could be due to a lack of sanitation and poor personal hygiene, as indicated by the lack of good practices for water dispensing and hand washing with soap at appropriate times, at either the household or school level. Exposure to smoke was very high in all observed habitations, as firewood was the main fuel and there was no separate cooking space. These factors may also be contributing to low birthweights.

Table 7.5: Immunization status of observed children

	BCG and Polio	Only Polio	NA
Uttar Pradesh (n = 36)	11 (30.56%)	18 (50%)	7 (19.44%)
Karnataka (n = 25)	12 (48%)	12 (48%)	1 (4%)
Andhra Pradesh (n = 57)	28 (49.12%)	23 (40.35%)	6 (10.53%)

Source: Household Survey Data.

A shocking finding in Uttar Pradesh was that out of the 36 children from all of the surveyed families, 11 (30.56%) were partially immunized, 18 (50%) had received only polio drops, and 7 (19.44%) had received no immunization at all! Regular child health and immunization services were also non-existent. Unlike the common perception, there was no discernible difference in the immunization status of Muslim families profiled in Uttar Pradesh. The situation in Karnataka and Andhra Pradesh was somewhat better. While nearly all children had received polio drops and approximately 40% had received anti-Tuberculosis BCG shots, other vaccine-preventable diseases (DPT, Measles) did not seem to be on the priority list of service providers (Table 7.3). During our discussions and interviews, it emerged that families do not resist immunization services if they are provided within their hamlet, *but they rarely go out of their way to get their children immunized.* Awareness about the importance of vaccines in disease prevention was low, as was knowledge about the links between drinking water, sanitation, and disease.

Ramrati of Village 2 of Uttar Pradesh used powdered 'Septran' (Sulpha drug) tablets bought in the village shop to manage her 1-year-old son's fever. In Karnataka, the medicine shop in Village 2 stocks antibiotics and Sulpha drugs, which the shopkeeper dispenses after ascertaining the symptoms. Small babies are also given powdered drugs dissolved in water. The situation was similar in Andhra Pradesh. This is a very disturbing phenomenon because untrained shopkeepers and pharmacists dispense these drugs without a doctor's prescription.

The impact of malnutrition was evident in children who were underweight and small and had unhealthy reddish-brown hair and cracks at the edges of their mouths, indicating a vitamin deficiency. The situation seemed particularly alarming in the tribal village in Andhra Pradesh, where around 59% of the observed children were suffering from some ailment.

These problems notwithstanding, four out of six rural children and two out of three urban children below 3 years of age in Uttar Pradesh (from both Muslim and Hindu families) attained age-appropriate development milestones: they were seen crawling, walking, running,

Table 7.6: Status of children in surveyed households

	Uttar Pradesh	Karnataka	Andhra Pradesh
Children below 36 months	**12**	**17**	7
Children enrolled in ICDS	1	3	None
Children 3–6 years	13	19	15
Children 3–6 years enrolled in ICDS	5	13	10
Children 6–11 years	36	32	31
Enrolled in school	30	24	30
Dropouts	2	6	1
Never enrolled in ICDS	4	2	-
Children in 11+ age group	**24**	**14**	5
Enrolled in primary school	4	1	2
Dropped out before completing primary	5	NA	NA
Completed primary and then dropped out	4	2	-
Enrolled in upper primary	2	1	-
No information		-	3
Never enrolled in primary school	9	1	-

Source: Household Survey Data.

communicating their needs, laughing, and playing. The sheer joy of being a child and running free was visible in their bright eyes and shrill voices. It was heartening that these children were in constant interaction with their older siblings and neighbouring children and families. These experiences provided them with a rich repertoire of communicative and social stimulation. While their stimulation from parents who are terribly overworked may be limited, this seems to be more than compensated for by siblings and others in the extended family and community.

When poor and weak mothers give birth to children in the absence of family, community, and institutional support, an inter-generational process of poor health, nutrition, and education is set in motion – in which a majority of Indian children are caught willy-nilly. These very handicaps have long-term effects and become more evident further along the child development continuum.

The Preschool Period: Ages 3 to 6

This phase in the life of a child is critical for physical, emotional, and intellectual development. Nutrition, which is vital for children at this stage of development, is unfortunately deficient in poor households. While children are able to feed themselves, the inability of parents to provide a balanced diet, or even basic food at regular intervals, emerges as one of the most important areas of concern for this age group. This is also a period when the child becomes a little more independent, and is able to eat by herself, ask for food, and communicate her needs. At the same time, she becomes more susceptible to infections as she wanders around.

Measurements taken in the course of the study (Table 7.2) showed that the majority of the children observed in poor households seemed to be moderately to severely malnourished. Though appropriate indicators for rural and urban poor have still to be worked out, the children looked underweight for their age or height, their skin was coarse or rough, their hair discoloured, and many of them (especially those between 6 months and 4 years) were pot-bellied with extremely thin limbs. While the children were very active and alert, playing and running around, it was apparent that greater attention needs to be paid to their overall nutritional status.

Most experts argue that a child can be positively prepared for schooling through supportive experiences in the home, neighbourhood, and community. Our household observations, however, suggest that there is little structuring of children's activities by adults or parents, whether they be play-, language-, or health-related. When it comes to participation in household activities, care of younger children, economic participation, food processing, and other productive activities, the parents' messages are unambiguous, assertive, directive, and sometimes even autocratic. Children's socio-economic participation in household and productive activities seemed far ahead of age expectations: even young children were taking responsibility for a range of sibling care-, household-, farm-, and cattle care-related work (Sharma and Chaudhary 2003).

The observed children had too little to eat, too few times a day. A limited number of children accessed ICDS services, with the location of the ICDS centre and the ICDS Worker's caste (and its relationship with that of the children) determining access. Further, the quantity and regularity of the distribution of supplements remained a major source of worry in Uttar Pradesh. While this did not seem to be a major issue in Karnataka and Andhra Pradesh, discussions with women revealed that only a fixed number of children were admitted to each centre; many were turned away because the centre could not accommodate more children. Fortunately, in some areas like the two villages visited in Karnataka, the government has responded by opening an additional ICDS centre in the village (one in the Golla community in Village 1, and one in the SC community in Village 2). This was not the case in Uttar Pradesh.

Even in a relatively better-performing state like Andhra Pradesh, children's access to ICDS centres was far from satisfactory. For example, out of the 22 children in the 0–6 year age group, there were seven children below 36 months and 15 in the 3–6 years age group. None of the children below 3 years of age had utilized the services of the ICDS centres. In the 3–6 years age group, only ten of the 15 children were enrolled in the ICDS. This includes one disabled boy in an urban centre household. Of the remaining five children who were not enrolled, four were in the tribal village. However, *it is worth stressing that enrolment does not automatically imply access to services.* Some children were merely enrolled and not necessarily benefiting from the ICDS programme.

The problem of irregular food supply was most acute in Uttar Pradesh. Our qualitative study confirms the findings reported in larger surveys:

> According to the National Institute of Public Cooperation and Child Development (NIPCCD) report (1992), 27 per cent of AWC [ICDS Centres] did not receive any nutrition supplement for over 90 days, i.e., the average disruption in their supply was for 63.7 days [...] *Similarly, the provision of double ration to severely malnourished children is often ineffective, because such children are often incapable of eating much at one go.* Again the morning hours appear to be the time when pregnant and lactating women are busy with domestic chores and hence attendance at AWCs for accessing these services,

> especially supplementary nutrition, tends to be poor. (Gupta 2003: 134, original emphasis)

Apparently, the situation on the ground is not very different in 2016, especially in Uttar Pradesh. There have been several new reports on dysfunctional ICDS programmes in Uttar Pradesh.

This analysis was confirmed during our investigations. Discussions with ICDS workers revealed that there was no system for identifying the severely malnourished so as to provide them with double rations. While we saw pregnant and lactating mothers coming to ICDS centres in Karnataka and Andhra Pradesh, we did not come across even a single woman who came to the ICDS centre in Uttar Pradesh. Most distressing was that in Uttar Pradesh the fortified wheat porridge *(dalia)* that was meant to be distributed through the ICDS centre was being sold in local shops – with people buying it both for their own consumption and to feed their cattle!

While recording the food habits of children in the three states, we were alarmed to see that in Uttar Pradesh the children usually ate plain *roti* with salt – in a region where a wide variety of green leafy vegetables are grown and locally available. Consumption of meat and fish was at best sporadic, with boys getting a very small piece and girls making do with the gravy. Interviewees in Andhra Pradesh and Karnataka also reported that their main diet was polished rice with some chutney or diluted *saru*.[6] It needs to be stressed that *state averages camouflage the extreme inequalities that prevail between and within communities*. Since this study focused on children in diverse poverty situations, our findings may appear far more alarming than indicated by state or national averages.

The preschool education component of the ICDS programme, although evident on paper, was non-existent in reality. While we did find play material in most ICDS centres, *it was locked up in cupboards*. This seems be the result of a fairly common belief that expensive (relatively speaking) and attractive things are too precious to be handled by children, and need to be brought out for display only during inspections. The same was the case with library books in schools.

[6] *Saru* is a thin broth made of lentils and tamarind juice.

During our visits, children were usually sitting quietly in rows with a slate and chalk in their hands. In some villages we could not interact with the children directly because they dispersed soon after the meal. The attitude of the teacher and the helper also left much to be desired: they were indifferent, if not shouting at and scolding the children. Only in Karnataka did we observe some preschool play activities. Clearly the ICDS programme, as it is operating in the villages, did not seem to play a significant role in enhancing the preparedness of children for school – unless of course sitting quietly in rows with a slate and chalk can be considered preparation!

Mothers' committees were introduced to strengthen Information, Education, and Communication (IEC) activities (education and awareness), and to make the ICDS workers accountable to the community. In all of the villages/slums surveyed, however, these committees were dysfunctional – even though they did exist on paper. We also found that villages with other women's groups (Mahila Samakhya groups or Self Help Groups) were more aware of the ICDS programme and what it was supposed to deliver, although even in these villages awareness rarely translated into active collaboration.

The visibility of the ICDS programme was high, both in the villages and in the urban slums. People were aware of the ICDS worker and helper. Focus Group Discussions invariably turned spirited, in most cases leading to vocal criticism about the helper (who is expected to escort children to the centre), other services (looking after the children), irregular supplies of nutritional supplements (in Uttar Pradesh), and the timings of the centre. The ICDS programme, however, does not seem to enjoy the same importance as other schemes, especially in the eyes of the government. The status of the ICDS worker, as compared to a schoolteacher or ANM, is low. Consequently, even if the worker is motivated and wants to be active, she has little space to take the initiative. She also receives little support from her superiors.

Many ICDS workers (especially in Uttar Pradesh) were not trained and essentially learned on the job. They look upon the ICDS as a nutritional programme and their role as the one who distributes food and maintains records. Interviews with ICDS workers in all areas revealed that recordkeeping was seen as a key function of their job.

The worker does not make any conscious effort to enrol the most deprived strata. Though Karnataka and Andhra Pradesh have sanctioned additional centres in more marginal hamlets/settlements, this was not the case in Uttar Pradesh, where one ICDS centre catered to the entire village regardless of the number of children in the appropriate age group. As a result, only 40 children were enrolled per centre in Uttar Pradesh, and even this was not based on who needed the services. *Caste was a serious issue with the ICDS worker and ICDS helper (who does the cleaning and cooking)*. If both were from a forward caste, then children from Dalit families will had little access. Parents in the sample households felt that they could make no demands on the worker, as she was upper caste. Almost all of the families we interacted with expected the ICDS helper to come and fetch the kids, which did not happen. Therefore, if the centre was at some distance and mothers were overworked, the children did not reach the ICDS centre.

As children reach preschool age, education and access to education becomes a defining variable in framing their life-chances, along with health and nutrition. This is an exploratory phase of the child development continuum, where preschool exposure and adequate nutrition (as an integral part of service delivery) become essential inputs for the holistic development of children. Evidence from the ground indicates a rather contrary picture, marked by the erratic provision of nutrition by the state and a rather weak and ineffective preschool component. Politics of caste identity and status, too, emerge as an important factor for denying children and parents access to nutrition supplements and for the effective functioning of the ICDS centres.

Going to School: Age 6–11

The good news on enrolment

In the 1990s, the issue of access to primary schools received considerable attention in all three states. Most children in the surveyed households were enrolled. Many of them were also attending school, with varying degrees of regularity.

In Uttar Pradesh, the eighteen poorest households selected for the study had a total of 85 children, of whom 60 were in the 6+ age group. Of these, 40 children were in primary school or had completed primary education, thirteen never went to school, and seven had dropped out. An interesting pattern that emerged from the educational history of these families was that, of the 36 children in the 6–11 age group, 30 were in primary schools. *This included all of the children (including girls) in the 6–11 age group from Muslim households*. In the 11–14 age group there were eleven children, of which four were in primary school at the time of the study. Of the six who had completed primary schooling – which included one Muslim girl and two Muslim boys – only two (one boy and one girl, both non-Muslims) had moved to the upper primary level. Of the remaining thirteen children in the 14+ age group, only one had completed primary schooling, another was still in primary school, and eleven had never gone to school or had dropped out. The families reported that the educational situation had significantly improved in the previous five to seven years.

In Karnataka, out of the 30 boys and 30 girls from the eighteen households surveyed, 57 children (29 girls and 28 boys) were either enrolled in the ICDS centre or in primary school. One boy had dropped out in Class IV, and one boy and one girl were not going to the ICDS centre. Discussions with mothers and with adolescent girls and boys revealed that campaigns like *Chinnara Angala* (to get children back to school), *Ba Marali Shalege* (Monday morning processions of children with songs and slogans asking children to come back to school), and *Pratibha Karanji* (encouraging creativity) had made a significant impact. Almost all of the children in both village and urban settlements were in school. The state government also provided free uniforms, textbooks, and school bags to all SC and ST children up to Class V, with the same extended up to Class X for children of *devdasis*.[7]

[7] *Devdasis* are a group of women who are dedicated on attaining menarche to the local goddess. Once they are dedicated to the goddess, then they become sexually available to the men in the village. *Devdasis* have also been the custodian

In Andhra Pradesh, out of the 58 children observed in the eighteen households surveyed, 31 were in the 6–11 years age group, of whom 30 were enrolled and attending schools. Only two out of the five children in the 11+ age group were in school. Enrolment in the ICDS centre was low. Out of the 22 children in the 0–6 age group, none below 3 years old was enrolled in ICDS, while eleven out of the fifteen in the 3–6 age group were enrolled in the ICDS centre.

Enrolment is obviously not a big issue anymore: attendance, transition, completion, and learning outcomes are emerging as bigger issues. It is these that are intimately related to our preferred proxy indicator of child development, namely the completion of primary schooling. The cohort study, based on official school records in all of the schools visited, gives a fairly positive picture. For example, in Karnataka 71% of the children who entered Class I in 1998, in all six sample schools, reached Class V in 2002. However, once we moved beyond the 'official' records, conversations with teachers revealed a different and more complex picture. In all three states, the teachers were reluctant to admit that children had dropped out; hence, many children were shown in the registers and some were marked as long-term absentees. In some schools, they were also marked present, especially when the distribution of dry rations was linked to attendance! Teachers often became defensive when asked about dropouts or long-term absentees.

Attendance and the work burdens of children

Children from poor households do not attend school very regularly. They tend to be absent for a range of reasons. In Karnataka and Uttar Pradesh, parents said that it was difficult for them to 'force' their children to go to school, especially when nothing much happens there. Some parents said that they needed their children at home for small chores, especially during the heavy/peak

---

of some forms of dance and music in India. The government passed a law in the 1950s outlawing this custom, and yet it persists. Till date, the children of *Devdasis* are given a number of government incentives like uniforms, textbooks, bags, etc.

agricultural seasons, when a child was sick, or when they had to migrate for work.

Many children above the age of 6 in the surveyed households who were attending government schools reported that they work both before and after school. *The burden of work was most severe for the first-born, especially for the eldest daughter.* There was no appreciable difference between Muslim/Hindu and Tribal/non-Tribal families. Apart from sibling care, children supported their families by calling for cows or goats; fetching fuel wood, fodder, or water; running errands; and looking after sick family members. As a result, children either became late for school or missed it altogether. *The gendered division of work and added responsibility of household work for older girls was marked in all three states.* In Andhra Pradesh and Uttar Pradesh, we also met boys who were in temporary bondage to pay off a loan taken by the parents. Engaging in full-time work during weekends and holidays was a fairly routine activity among the children.

If this is the situation with respect to children and work, what implications does it have for primary schooling? The community is aware of the need to educate their children, and yet the children are burdened with work before and after school, and during holidays and vacations – both household/domestic work and sibling care-related chores, as well as work outside the home or in home-based occupations. This obviously directly impacts the learning abilities of the children, especially when they put in long hours of work every day. Most children in poor households do not get time to revise/read their schoolbooks, especially girls who are higher in the birth order. Given the nutritional status of most poor children, their energy levels are low, impacting their ability to concentrate in school. The impact of working while also attending school needs to be examined with reference to its impact on learning outcomes.

Learning outcomes

In Uttar Pradesh, we observed that most children in Classes III, IV, and V were unable to read fluently from their textbooks, nor could they solve simple addition or subtraction sums. For instance, in the urban sample from Uttar Pradesh, eleven (out of eighteen) children were attending the government primary school (GPS).

Five children who were attending Classes II, III, and IV showed poor learning levels. Three children from Class II were unable to recognize alphabets or numbers. The two children in Class III were also unable to read, write, or count, although they knew certain lessons by rote. 'Earlier, Class II-pass could read postcards, now they can't even write their names', complained one father during the FGD in urban Sitapur. 'What is the use of sending him to school? I pulled him out after Class IV and he now helps me with my work.'

The situation was not dramatically different in Karnataka or Andhra Pradesh. Yes, a few more could read – but on further investigation, we realized that the children with literate parents (especially the mother), or those who attended private tuition classes were the ones most able to read. Children who were first-generation school-goers barely managed to recognize alphabets and could, at best, read a few words. Group discussions in the community revealed that parents felt the quality of teaching had declined, and that the teachers did not really care if the children of the poor learned to read or not. Also, teachers were not made accountable for the learning outcomes of children, especially in the primary and middle schools where there are no board examinations. Other qualitative and quantitative research studies have made similar observations (PROBE Team 1999; Ramachandran 2004; Rana, Rafique, and Sengupta 2002; Jha and Jhingran 2002).

Is Schooling Really Free for the Poor?

There is widespread agreement that access to education has improved significantly in the last decade, and that even very poor parents want to send their children to school. There is also growing evidence that private schools are mushrooming and that children are being sent to private schools if the parents can afford it. In some cases, parents even decide to send their children to private schools when they really cannot afford it!

In Uttar Pradesh, we found that the teachers collect 'fees' – official and unofficial. Community members and students reported that teachers in government schools often asked children to give Rs 5 to 10 during national festivals, for the issuing of Transfer Certificates, or for the release of scholarship money. If payment for schooling is necessary, parents consciously opt for private schools. However, it is important to note that the 'fees or charges' in government schools were far below the private school fees. The perceived poor quality and poor learning outcomes of GPS often seem to be instrumental in the emergence of these private schools. In one study village in Uttar Pradesh, we came across a private school that *did not expel children even if they did not pay the fees regularly – making it more attractive for parents.*

Private tuition classes profiled in Karnataka were priced at around Rs 30 per month. Parents seemed to believe that children learned faster and were able to cope with their studies if enrolled in tuition classes. One mother said: 'I send my child to a private tutor for two hours because it is better to have a short time of concentrated learning than many hours of not learning at all!' The collection of unofficial fees, compelling children to buy guidebooks, and private tuitions are gradually becoming more common. What is noteworthy is that such practices have become more widespread in the last decade, the period when community demand for perceivably good quality education has also grown.

Hunger in the Classroom

Discussions with children revealed that many of them do not eat anything in the morning, especially girls, who have little time after their morning chores. In Andhra Pradesh, several girls said they only drink water before coming to school and eat only during the lunch break! Given that most poor households have to gather fuel wood, families cook two meals a day, one around mid-morning and one after sundown. Where children do eat before coming to school, the staple diet seemed to be *roti* and salt in Uttar Pradesh, and a piece of *jawar roti* or rice with chilly chutney in Andhra Pradesh. *On most days, almost 10% to 15% of children (most of them girls) came to school without eating.* When we explored further and asked why so few children ate in the morning, some girls said they did not have time to eat! The lack of an adequate meal before attending school, usually referred to as short-term hunger, has an adverse impact on a child's performance in school, and her ability to concentrate and learn new concepts. In this situation, the provision of a hot midday meal becomes all the more necessary.

Incentives: Do They Make a Difference?

Some teachers, especially in Uttar Pradesh, argued that incentives like dry rations had little effect on retention or achievement levels, and that the students who enrolled for incentives were only regular on the days prior to distribution. They were of the opinion that scholarships should be handed out at the end of the academic session to ensure that the children would not drop out midterm or attend class at random intervals. The teachers did not think that rations for poor children ensured food security at home or prevented them from taking on extra work. However, both the community and the teachers were all praise for the midday meal programmes in Karnataka and Andhra Pradesh.

Incentives like textbooks, uniforms, and scholarships, though welcome, were issues over which parents and teachers complained and argued. While we did not come across any case of delayed distribution of textbooks, the distribution of scholarships and uniforms left much to be desired. In Uttar Pradesh, free textbooks meant little because most children purchased 'guide books', right from Class I. Parents insisted that teachers forced them to buy these books, while the teachers averred that parents bought them on their own. The fact remains that most children had them in their school bags and copied the questions and answers into their notebooks. Some of the children could not even read what they had written!

Teachers' Attitudes and Development of Self-esteem and Confidence

What value does education add to the lives of children? There is little disagreement over the role of education in enhancing the self-esteem and self-confidence of children. Yet, discrimination and/or differential treatment in school can and does affect the overall confidence levels and self-esteem of children. Both children and parents stated categorically that teachers treat poor children different, and that they do not appreciate the predicament of children who have to work before and after school. Constant reinforcement of caste- and community-based negative stereotypes also have long-term effects on children, who internalize these perceptions and see themselves as responsible for their own situation. Hence it is not surprising to come across parents and children alike who believe that their inability to be 'literate' or 'get an education' is more a function of their caste/community identity (for instance, *adivasis*/STs are considered to be backward and uncivilized, etc.) than the larger political economy. Casual and sometimes sarcastic comments on the worthwhileness of education in the lives of children who will 'end up' doing what their parents do – agricultural wage labour, sweeping, cobbling, etc. – also dampen children's self-esteem and aspirations.

What This Qualitative Study Tells Us

As stated earlier, our sample is skewed in favour of the very poor. Therefore, the observations need to be read contextually and may not be completely applicable to other socio-economic groups. However, this sample enables us to understand the situation of 10–15% of the poorest households in a village or slum, and provides insights into how children from these households continue to be deprived or marginalized. The rest of this section elaborates on the emerging themes that characterize the interplay of proximal and distal factors for determining the educational outcomes of children living in diverse poverty situations, especially with specific reference to gender, caste, community, and region.

Cumulative impact on child development

This research, by focusing on children in different age groups along the integrated child development continuum, illustrates the continuous and cumulative nature of social and economic exclusion that children face from the moment of their conception. Health, nutrition, and education are no doubt the three main sectors that impact a child's development. However, as we have seen, the impact of each these sectors is not discrete or merely additive; rather, it is far more complicated, varying in its intensity and manifestation according to the sub-stage under consideration. For example, the health and nutrition status of both mother and child is clearly significant in the early years. The continuous and cumulative nature of this impact also means that, although the age groups are discrete in nature, the impact of the non-attainment of appropriate developmental milestones, health and nutritional outcomes, or learning capacities, implies that these will necessarily accompany the child to the next stage – leading to more failures than successes and, in some cases, the intergenerational transfer of handicaps, resulting in a downward spiral of poverty, ill health, malnutrition, and poor learning outcomes. Our research also indicates that, despite their potential, current social policies are unable to effectively capture and tap the positive synergy of interventions into each of these different sectors. They are also unable to proactively harness the family and community to meet the objectives of reaching each child, especially the poorest ones, and creating a supportive environment for their development.

Intermeshing poverty and social status

Poverty impacts the overall health status of children. Lack of easy access to water affects the overall hygiene of children. Even when poor families have access to safe water, storage and usage practices affect the quality of the water they drink. The level of poverty significantly impacts food and nutritional security, and is closely related to the availability of work.[8] Even during 'normal' times, most poor (i.e., below-poverty line) families eat less nutritious food than households above the poverty line. Discussions on intra-household

[8] 'The issue of acute food shortage faced by poor people in drought affected areas in the midst of gigantic "surplus" food stocks rotting in godowns has been an area of debate and discussion [...] What is important to point out in this respect is that the poor do not face food insecurity only during drought or drought-like situations. In many places, periodic food crises are an annual phenomenon, the intensity of the crisis and length of the period increases during years of bad or no crop [...] Nearly one third of these families usually face a food crisis for more than four months in a year, the rest go through this for about two-three months' (Jha and Jhingran 2002).

food distribution also revealed that adult men are given precedence, followed by boys, girls, and finally women.

However, economic explanations *per se* rarely explain the persistent bottlenecks in attaining universal primary education. The reality is that economic factors are inextricably entwined with social variables that also determine children's ability to both access and continue in schools – often making it difficult to distinguish between the two. The in-depth study of the elementary education of the poorest and other deprived groups by Jyotsna Jha and Dhir Jhingran (2002) provides perhaps one of the more exhaustive databases in recent years on this issue, as well as shedding light on the competing factors that frame educational decision-making in poor households. They argue that enrolment and attendance is not only determined by a family's economic situation, but also by the social status of certain groups.

The same is reflected in our study, in which the caste and community of children, along with their economic status, influence their educational participation and outcomes. While an improvement in a family's economic situation certainly makes a difference, financial ability alone cannot explain why socially and economic backward or disadvantaged communities sometimes forgo access or only irregularly attend school. Other factors such as discrimination probably explain why SC, ST, and other minority groups in our sample emerge at the bottom of the educational ladder – in addition to financial difficulties. The attitudes and prejudices of teachers and children regarding the social and community identities of marginal groups also play an important role in defining possible educational outcomes (this is discussed at length in the next chapter). As a result, schools that cater to a specific disadvantaged group often record better attendance/performance for children from these groups, provided the quality of schooling is satisfactory. In many areas of the country, the teacher-pupil ratio and overall infrastructure/facilities in schools that exclusively cater to marginal groups are not comparable to the regular government schools that cater to mixed social groups. This phenomenon, known as hierarchies of access, has come to be accepted as a significant factor in explaining children's access to schooling as well as the quality of the schooling they receive – thereby influencing the ability of children to successfully complete primary schooling (Ramachandran 2004).

The interplay of gender with social and economic status adds another important dimension to this issue. Gender relations in the family, community, and society have a significant influence over the ability of girl children to access services, nutrition, immunization, healthcare (especially during bouts of illness), and schooling. The good news, however, is that if girls do reach school and are able to cross that initial barrier, they are highly motivated and struggle against all odds to remain in school. Similarly, if regularity can be assured, girls perform as well as, if not better than, boys.

Another significant factor that emerged during the course of this study – as well as in previous studies (Ramachandran 2004; Jha and Jhingran 2002) – is the changing social norms regarding schooling. Discussions in rural Karnataka and Andhra Pradesh revealed that sending children to school has become a norm in the communities. This transformation of social norms and practices has occurred over the last eight to ten years. Government-initiated campaigns (*Chinnara Angala*, *Ma Marali Shale*, *Chaduvula Pandaga*), intensive NGO-led mobilization against child labour (notably the contribution of the M.V. Foundation in Andhra Pradesh), and sustained efforts to make school a joyful experience (*Nalikali* and *Kalinali* of Karnataka) have made a major difference. The governments of Karnataka and Andhra Pradesh have ensured that teachers reach the schools and have tried to address cadre management concerns to rationalize teacher deployment. Conversely, the lack of sustained mobilization and corresponding governmental efforts to ensure functioning schools in Uttar Pradesh is perhaps responsible for the persistent apathy with respect to education, especially for the poor who rely on government schools.

Health, nutrition, and education

The relationship between health and education is often perceived as a one-way street, with most discussions focusing on the role education can play in facilitating health awareness and improving the health status of individuals and communities. Usually left out of this debate is the critical and reciprocal nature of the link between health and education, specifically in relation to children, whereby poor health and nutrition can actually prove to be a barrier to school attendance and educational attainment/achievement. The most obvious medical evidence indicates that childhood disability, chronic illness, and extreme malnutrition are direct impediments to children's participation in school. Further, the frequency, duration, and severity of illness also influence the regularity and attendance of an enrolled child. However, apart from the anti-polio drive, there is little energy in the health sector. What remains fuzzy is the less-than-tangible link between endemic poverty, malnutrition, and educational attainment and achievements.

Further, inadequate nutrition, manifested as short-term hunger, impacts a child's ability to learn and retain new concepts.[9] Hence, the common complaints by teachers that children lack the ability to concentrate, are listless, etc., take on a new urgency. Several studies, especially those on the Tamil Nadu midday meal programme, have stressed the benefits of a hot and sumptuous meal in school, particularly for children in poverty situations. As discussed earlier, dry rations seem to have little impact. No wonder the Supreme Court's judgment on serving cooked meals in school instead of distributing dry rations was welcomed by parents and children. The experience of Karnataka was quite impressive, not just for ensuring regularity in school, but also for making a significant difference for the children from the surveyed households.

[9] A review of the literature on the subject notes that the capacity for processing, structuring, and classifying information, asking and answering appropriate questions, and students' short-term memory, levels of alertness, attention, and concentration are crucial for success in school. These are all known to be adversely affected by nutritional and health deficits, which therefore limit the ability of the child to benefit from classroom instruction or later learning opportunities (Levinger 1994 and Del Rosso and Marek 1996, both cited in New Concept Information Systems 2003). Malnutrition, even in a milder form, can dull motivation and curiosity, and reduce the child's exploratory play and interactions with caregivers and the environment, thus restricting the child's psychosocial development (Martorell 1997, cited in New Concept Information Systems 2003).

Is There a Way Out?

The school system

An important insight of this study (as well as the documentation of good practices in Jandhyala and Ramachandran 2007) is that a well-functioning and attractive school, with basic facilities and motivated teachers, makes a major difference in the lives of poor children. *Access and quality thus need to be seen as part of the same continuum: one without the other is meaningless.* This is of particular importance in both government schools, which cater to the majority of poor children in rural and urban areas. The shift that we are seeking is the mobilization of the school system to ensure better learning outcomes for children. If this is taken as a non-negotiable principle, then the schools need to function as an integrated whole where a range of inputs/activities converge to create a non-intimidating and a creative learning environment.

Equally, a well-run midday meal programme positively impacts enrolment, retention, and learning. While Karnataka and Andhra Pradesh have a good programme, Uttar Pradesh continues to be inefficient – even in 2017.

The teacher is the fulcrum around which an effective schooling system revolves. While this appears to be cliché, the real task before us today is to break this down into doable steps, for instance by making teachers accountable for measurable learning outcomes. Actual teaching time and teacher-pupil ratios (TPR) remain important areas of concern. In the schools visited in Uttar Pradesh, the TPR in rural schools was as high as 1:173 in one case, and the actual teaching time was less than one hour per class per day! In the tribal school visited in Andhra Pradesh, only one of the three teachers were be present on any given day.

Another context-specific strategy worth exploring, especially in tribal areas where it has not been easy to ensure the posting and regular attendance of teachers, is to locate schools in a cluster (or *Mandal* in the case of Andhra Pradesh) that is easily accessible, and transport children from Class IV onwards to that school rather than having a school in every habitation. This recommendation is made in the light of evidence that tribal and SC children who are studying in residential schools (Ashram Shala, Residential School for SCs) come across as better equipped academically. This might enable the government to pool its resources, both human and financial, and ensure that children from poor families and disadvantaged communities get a good quality education. It would also facilitate good teacher-pupil ratios and ensure regular teaching. In the same vein, residential condensed programmes for girls in the 11+ age group could be particularly valuable in educationally backward regions.

The ICDS programme

Another major finding relates to the ineffectiveness of the preschool education component of the ICDS programme in all but one centre observed for this study. This has also been highlighted in other studies (NCAER, ICDS Field Study 1999). Our discussions in the communities raised the question of whether the ICDS programme should indeed be the nodal point for preschool education. One proposal worth considering is de-linking the preschool education component from the ICDS centre, and instead making it an integral part of primary schools. An important spin-off of this restructuring would be that children in the 3–5+ age group could benefit from the universal midday meal programme, thereby improving their nutritional status. This is of particular importance for children from poor families. For example, the Government of Assam has implemented a policy of opening a preschool section (Ka-Maan) for under-6 children. Similarly, the Shishu Shiksha Kendra in Madhya Pradesh caters to the preschool education needs of children in the 3–5 age group. There are several such initiatives across the country, underscoring the importance of preschool education in preparing children for primary schooling. The Balwadis of PRATHAM (in Mumbai, Delhi, etc.) and the Uttarakhand Sewa Nidhi of Uttar Pradesh are examples of successful innovations in the non-governmental sector.

This shift would enable the ICDS worker to concentrate on the nutrition and health of children in the 0–3+ age group, pregnant and lactating mothers, adolescent girls, and more importantly, the nutritional education of the community. This will also ensure that nutrition supplements and other health interventions reach those most in need. Since not all poor children have access to an ICDS centre (especially in Uttar Pradesh, where each village has only one centre), the government could consider making it a universal programme for all children in the appropriate age group as well as all pregnant and lactating mothers in the catchment area of the ICDS centre. It is important to emphasize that the provision of dry rations to pregnant and lactating mothers defeats the very purpose of supplementary nutrition. Given the precarious economic situation of the households studied, these supplements go into the food for the whole family.

Health and nutrition

Public education has received a setback over the last 30 years. It may be worthwhile to revisit earlier nutrition education and preventive health programmes. One disturbing feature is that every new programme that is introduced tends to diminish the validity of earlier efforts. For example, when food distribution through ICDS was made universal, older initiatives for community education fell by the wayside.[10] In all three states, it was rather disconcerting that simple messages (kitchen gardens, eating leafy vegetables, universal planting of common fruits like guavas and berries, the nutritional value of coarse grains, etc.) that enable people to harvest whatever local resources they have to improve their nutritional status are no longer seen as important.

More effective public education is also needed to highlight the link between *complete* immunization and disease prevention. Discussions in all three states revealed big gaps in people's appreciation of immunization. Unfortunately, as pointed out in the Mid-Term Review of the Ninth Five Year Plan of the Government (Planning Commission, GOI 2001), routine immunization has suffered, with serious implications for the health of children.

[10] The phenomenon of new projects/programmes overriding the priorities of earlier ones is not specific to the health or child development sector. In the education sector several valuable components of DPEP (like mainstreaming gender in textbooks, continuous training, and support for district gender coordinators) were dropped when SSA was introduced. When Lok Jumbish was merged into SSA, many block- and cluster-level activities were discontinued. This issue merits detailed research.

> It is a matter of concern that there has been a fall in routine immunization. It is obvious that the target of 100% coverage of all six Vaccine Preventable Diseases has not been achieved. Several states have reported substantial decline in routine immunization [...] Reported factors responsible for poor coverage range from vacancy at ANM level (40% in Bihar), poor mobility, poor access, problem in distribution and storage of vaccines, lack of supervision and monitoring, poor cold chain maintenance and ongoing campaign mode programmes disrupting routine activities. (Planning Commission, GOI 2001: 47)

Role of community groups and listening to the voices of children

There is today a multiplicity of village-level groups and committees with competing goals and overlapping roles. To seriously address the crucial factors that facilitate or impede the overall development and growth of children, it is vital to re-examine the roles and mandate of these committees and look for ways to empower specific groups of people to play a more positive, interventionist role in ensuring that benefits accrue to those most in need. The enthusiasm with which these committees are constituted is rarely matched by the resources or activities provided to educate and empower them. It is thus necessary to pay special attention to how these committees are constituted and to strengthen the training/capacity-building activities that are necessary to give them *teeth*. This applies to all of the committees established for primary education/child development at the village level.

Afterword

It has been fifteen years since this study was conducted. If we are to go by recent newspaper articles based on the National Sample Survey 66th Round (NSSO 2012), child malnutrition has gone up – especially among the rural poor, tribal communities, and the urban poor. The HUNGaMA Report (Naandi n.d.), which was based a sample survey, confirmed that more that 42% of children under 5 years old were severely or moderately underweight and that 59% of them suffered from moderate to severe stunting. Equally disturbing is that children from diverse poverty situations across the country constitute the majority of out-of-school/dropout children, and that they are the ones who do not make it to secondary schools. Innumerable research studies, sample surveys, and press reports continue to highlight the situation of children in poverty, and yet it looks like no one is really listening. My hunch is that if we were to repeat this study in 2017 the results may not be very different.

When children are malnourished and caught in a vicious cycle of poor health and poor nutrition, it is not surprising that the poorest families are busy trying to keep their children alive. While the midday meal may be a necessary intervention, the continued neglect of teaching and learning does not give our children a fighting chance to better their life. This, in essence, is the tragedy of India.

8 What it Means to be a Dalit or Tribal Child in Our Schools

As discussed in the preceding chapters, children from marginalized social groups drop out of school in greater numbers than other children. It is also fairly common knowledge that the schooling experiences of thousands of children from diverse poverty backgrounds are not positive. There have been a large number of press reports about caste discrimination, community-specific and gender-related exclusion, sexual harassment, and violence against children in schools.[1] Researchers, too, have repeatedly pointed to the problem of exclusion and segregation in schools. And yet, the education departments of both the Central and state governments do not officially admit the existence of this phenomenon.

For this reason, it was a positive development for Sarva Shiksha Abhiyan (SSA) to commission a research study looking at inclusion and exclusion in schools. This qualitative study was done in 2011–12 and involved a team of researchers from six states: Andhra Pradesh, Assam, Bihar, Orissa, Madhya Pradesh, and Rajasthan. The findings from these states may not be radically new, but the study

This chapter is based on a study commissioned by Sarva Shiksha Abhiyan, Ministry of Human Resource Development, Government of India in 2012 (Ramachandran and Naorem 2012). A shorter version of this chapter was published in *Economic Political Weekly* (Naorem and Ramachandran 2013).

[1] Notably articles published over the years by P. Sainath in *The Hindu*, reports and articles by Navsarjan Trust (Gujarat) and the Dalit Solidarity Network, working papers of the Indian Institute of Dalit Studies, and writings by Geetha Nambissan.

does make it possible for policymakers to officially acknowledge the prevalence of exclusionary practices in schools and the urgent need to address them.

The ability of children to access a school, participate in school activities, learn, grow, and – most importantly – develop a sense of self-worth, is influenced by a wide range of issues related to the school, family, and society. With this in mind, the 1986 National Policy on Education considered universal education to be a powerful tool for neutralizing 'distortions of the past' and giving children a level playing field. The Right to Education (RTE) Act of 2009 attempted to take this forward by guaranteeing all children the right to go to school, the right to be treated with love and care, and most importantly, the right to be treated equally and with dignity. However, despite these good intentions and policies, we are far from creating a level playing field for all of our children.

Education is not only about reading and writing; it is about preparing for life. It is in this context that the issue of inclusion and exclusion or, to put it more starkly, discrimination and unequal treatment, acquires a new meaning. In this study, the researchers explored the blatant, subtle, and hidden practices that influence the ability of children to learn, grow, and become confident young people.

Overarching Insights

It is now fairly well known that in most areas elementary schools run by the government primarily cater to the very poor and that the not-so-poor opt for private schools – even in rural and remote areas. There are, however, significant variations among the states. In Rajasthan and Madhya Pradesh, government elementary schools are not as exclusive and serve children from almost all social groups (castes/communities), especially girls from forward castes. In Andhra Pradesh, on the other hand, most upper or forward caste children do not attend government schools, which primarily cater to Backward Caste (BC), Dalit (SC), and *Adivasi* (ST) children. There has been a significant growth in residential schools, and many Dalit and *Adivasi* children who do well academically shift to residential programmes after primary school (Class V).[2] In Assam and Orissa, each social group's habitations/villages are distinct, and so are the schools that the children attend. In Assam, the major distinction is between the tea garden communities, the non-Assamese migrants, and the specific tribal areas for Bodos, Karbis, Mishings, and Chakmas. As each of these communities live in distinct habitations/villages, their children often go to schools that are fairly homogeneous.

Given the social norms in Indian society, overt discrimination against Dalits was generally found to be more prevalent than that against *Adivasis*, with Rajasthan a notable exception (especially with respect to children from Bhil and Sahariya communities). This may be attributed to the fact that most ST communities live in exclusive habitations/villages where they are often the majority. On the other hand, Dalit communities are spread across the state and are often a minority in villages. Where the Dalit community was the majority in a village, the dynamics were noticeably different.

One of the overarching insights from this study is that we need to view inclusion and exclusion from different vantage points: from the outside (who goes to what kind of school), from the inside (what happens inside the school), and from society (who is visible and who is not visible: for example, seasonal or new migrants are often invisible in data on out-of-school children). Equally significant is the influence of the larger society and social norms on what happens inside a school, the attitudes and practices of teachers, and the involvement or lack of involvement of parents and community leaders. Political and social assertion of the rights of Dalits and *Adivasis* also influences practices and attitudes. The differences between Andhra Pradesh, Bihar, and Rajasthan are a case in point. While the societies in all three states are highly stratified in terms of social and community identities, explicit (blatant) caste-based discrimination is not as evident in schools in Andhra Pradesh or Bihar as it is in schools in Rajasthan. A possible explanation

[2] The Andhra Pradesh Social Welfare Residential Institutions Society was created in 1987 and has been operating 289 residential schools for SCs. The Kasturba Gandhi Balika Vidhyalayas were extended to Class X, and there are also 48 residential schools for ST children. In total, all of these schools cover 15,252 students.

could be the impact of the political environment of the state. Both Bihar and Andhra Pradesh have experienced social and political movements led by Dalits and other backward classes. In fact, in the last two-to-three decades the social mobilization of Dalit and backward communities has led to them playing a significant role in the politics of both states. Rajasthan, on the other hand, has not witnessed similar social mobilization of Dalits and backward communities. On the other hand, the composition of the student community is more homogeneous in Andhra Pradesh, while in Rajasthan children from forward castes also attend government schools, leading to more diversity.

Language

Language is an important site of exclusion. This is particularly evident when the teachers' or official state language is different from the mother tongue of the children. In Orissa, teachers from coastal districts do not know either the language spoken in the districts in the western part of the state or the tribal languages, and are therefore unable to communicate with the children in these areas. Equally, the Odiya language spoken in the coastal areas is considered the standard official language and is used in textbooks and for classroom transactions. As a result, students from rural western Orissa and tribal students (both of whom speak different dialects) usually face difficulties. Language is therefore an important caste marker in the state.

In other states, language is a big factor in the exclusion of migrant children. For example, Assamese is the main medium of instruction in Assam, but is very different from the mother tongue of tea-garden workers, who migrated from Hindi-speaking areas of the country to work in the tea plantations. While the children go to Assamese-medium schools, the language they speak at home is different. Similarly, in the border areas of Andhra Pradesh the medium of instruction is very different from the mother tongue of the children. In Rajasthan the local dialects vary by district, and children may not be familiar with the standardized Hindi used in schools.

The issue of language as a site of exclusion has been highlighted in India for over 60 years. Hence, this study does not purport to explore this issue in detail, but instead tries to keep the language issue on the radar while exploring other factors of inclusion and exclusion.

Midday meals

Midday meals were one area in which social and community issues invariably surfaced. In some areas, the dominant practices of the village dictated certain norms, including: the appointment of cooks belonging to a caste/community that would enable all children to eat the food (mostly non-SCs); particular seating arrangements while eating the meal; access to water to wash plates and hands; and access to drinking water. Importantly, however, wherever there was a dynamic and strong headmaster or head teacher who was committed to equality, the school was able to swim against this current.

Infrastructure

Infrastructure facilities were poor in almost all of the states. There were not enough adequate classrooms. Usable toilets were few and far between, highlighting the fact that simply providing toilets means little unless equal attention is paid to ensuring that they are usable, have running water, and are cleaned regularly. A miniscule proportion of the schools had usable and functional toilets for girls – and where there were such toilets, girls from the poorest or most socially disadvantaged groups in the village were asked to clean them.

Teachers

The availability of adequate teachers was also a big issue. This was not only due to teacher shortages; it was also caused by the fact that not all of the teachers were present in the school every working day: some were either routinely absent or refused to take up their appointment while lobbying for a transfer. The ability of the educational administration at the district/block level to make sure that all posted teachers attended school every day was found to be uneven.

Another important similarity in the six states was that the present structure of the School Management Committees (SMCs) or erstwhile Village Education Committees (VECs) was not effective for ensuring the involvement of parents, especially those from the most backward communities. The parents were apathetic and admitted that their voices were rarely heard, and they did not regularly participate in school-level committee meetings. They were not aware of exclusionary practices in schools, often accepting the situation as a given and something that could not be changed. They were also unaware of the Right to Education Act of 2009.

Stereotypes and perceptions

Discussions with teachers from the six sample states highlighted a common perception: children from very deprived social groups did not perform well in school. Interestingly, information from the same schools revealed that this was a misconception; many children from deprived social groups *were* actually performing well academically. There was a disjuncture between teachers' perceptions and reality: it is noteworthy that the teachers themselves pointed out children who were 'bright' and keen on studies (many of them Dalits/*Adivasis*), while at the same time continuing to hold onto these prejudices and stereotypes. When asked about this seeming contradiction, they said that the bright children were exceptions.

Across all of the states, regular attendance was a serious issue among children from extremely poor and marginalized communities. This could be due to various reasons, including poor health and recurring illnesses, short-term migration by parents for work, taking on household responsibilities (especially by girls), parental absence because of daily wage labour, and neglect. Frequent absences essentially meant that the children were not able to keep pace with learning. When children fell behind, the teachers ignored them. In all six states, teachers preferred to focus on the 'bright' children who sat in the front rows. This process results in a self-perpetuating mechanism that leads to further exclusion.

Enrolment and attendance

In all six states, the proportion of children from socially disadvantaged groups, mainly BCs, SCs, and STs, was much higher in the sample schools than in the general population. *This confirms the available evidence that government elementary schools essentially cater to the poor and marginalized and that children from forward castes do not access government schools to the same extent.* However, there were state-wise variations in the proportion of children from different social groups who were enrolled in these schools (Table 8.1).

Equally significant is the finding that attendance rates were fairly uneven. There was a gap between the attendance reported in school records and the number of children present on the three days of field visits (i.e., the highest attendance observed over the three days). While this phenomenon is fairly well established, the main issue here is that children from extremely poor families (landless daily wage labourers, seasonal migrants) tended to miss school more often. Girls from very poor families were also irregular. Many children from such families reached the school late in the morning. They were, therefore, not involved in activities like leading the morning assembly or participating in school functions.

For the purposes of this study, it was therefore important to understand what happened to those children who were frequently absent when they *did* attend school. The study found that, in all of the states, such children were excluded from active participation in classes because they were not able to keep pace with the lessons

Table 8.1: Attendance in sample schools

State	Class	Total		SC		ST		Muslim		BC		General		CWSN[3]	
		B	G	B	G	B	G	B	G	B	G	B	G	B	G
Andhra Pradesh	IV	316	290	81	72	60	54	5	11	160	153	-	-	-	-
	VII	433	340	97	56	50	29	14	13	272	242	-	-	-	-
Assam	IV	91	103	27	28	15	14	13	15	33	42	3	4	0	0
	VII	215	135	49	27	44	12	22	55	34	40	66	1	0	0
Bihar	IV	456	516	75	98	11	13	38	41	267	294	41	49	2	2
	VII	666	785	122	101	44	60	73	97	351	409	51	57	1	5
Madhya Pradesh	IV	172	247	61	69	30	48	2	3	66	116	13	27	3	2
	VII	310	357	109	123	64	73	3	1	121	136	26	25	4	6
Orissa	IV	140	155	34	38	62	60	2	4	39	51	3	2	1	0
	VII	226	210	62	48	76	64	4	1	83	96	1	1	3	2
Rajasthan	IV	105	160	33	42	21	32	2	2	31	59	18	25	-	-
	VII	184	217	48	48	54	69	12	9	44	62	26	26	-	-

Source: Survey data.

[3] Children with Special Needs.

that were taught. They also tended to sit in the back rows and were often ignored. The combination of being poor, a first-generation school goer, SC or ST, and frequently absent was perhaps the most compelling reason for exclusion from school activities.

Infrastructure, Facilities, and Children's Access to Them

Despite compelling evidence spanning over five decades that the mere availability of a building, toilet, or hand pump/bore well does not necessarily ensure access, the school-level data that is compiled every year continues to record the availability of these facilities without any other information on their status (see Table 8.2). This study found a huge gap between the availability of such infrastructure and its maintenance, cleanliness, and even usability.

Another important point that was confirmed by this study is that water, toilets, and the mechanisms for cleaning them are the site of caste- and community identity-based exclusion. It is important to clarify at this stage that schools are not the only sites of such exclusion.

Table 8.2. Infrastructure facilities* in schools (%)

States	Available Toilets		Drinking	Playground	Kitchen	Library
	Common	Girls	Water			Books
Andhra Pradesh	63**		78	67	67	78
Assam	95	75	55	95	80	20
Bihar	75	80	100	40	95	60
Madhya Pradesh	23	5	80	53	65	13
Orissa	80	55	90	10	100	90
Rajasthan	50	50	90	60	85	60

Source: Survey data.
Note: * Facilities available but not necessarily accessible, especially in the case of toilets and libraries.
** Where the toilets were usable, teachers mainly used them; in a few schools, girls were also allowed to use them.

Water Woes

The study found a range of both exclusionary and inclusionary practices in the sampled schools. In Rajasthan, most of the forward caste children would drink water or wash their midday meal plates before the SC and ST children. In one school, the pitcher of drinking water was not accessible to SC and ST children; other children would pour water from it into separate glasses for the SC and ST children to drink. Interestingly, in three schools the children from one ST community (Meena) did not use the hand pump because other ST children who were perceived to be of a different status had touched it. Rajasthan perhaps revealed one end of the spectrum of exclusion, with rigid, caste-based societal norms playing out in the schools.

In Andhra Pradesh, almost all of the children brought water bottles from home, mainly because the water from the school pump was considered unhygienic. In three of the 20 schools visited, SC and ST children were seen standing at a distance from the hand pump/tap, even when they needed to wash their hands or plates. They had to wait for the other children to pour water for them.

In Orissa, 90% of the schools had some drinking water facility in the form of a tube well, open well, or piped water. The study did not find any water-related exclusion or discrimination in Orissa. Water potability is a big issue in this state, and the study found that in two schools the water was contaminated with fluorine.

In Assam, six of the 20 schools visited did not have a drinking water source. However, like in Andhra Pradesh, almost all of the children carried water bottles from home and no discrimination/exclusion was observed while using the hand pump or tap to wash their hands, plates, etc.

In Madhya Pradesh, most of the schools had a proper water source, mostly a hand pump. In five of the 20 schools studied, children had to fetch water from a short distance. In almost all of the schools, all of the children could drink water from the hand pump, but in one or two schools upper caste children brought water from home. However, only children from the upper castes were asked to fetch water for teachers and guests. In most of the schools, water was not stored; the children drank water directly from the hand pump. While all children drank the same water, there was subtle discrimination. For example, SC children often had to wait their turn by standing at a distance. In two schools where water was stored in steel tanks, a bucket was used to carry the water. SC children were not allowed to touch these vessels. OBC and general category girls were found to be practicing untouchability with their schoolmates more than the boys did in two schools. During focus group discussions, the girls informed us that they had been asked by their parents to conform to prevalent social norms. The state report notes that the teachers, who should ideally be leaders of change, got their own water or asked upper caste children to get it from their homes or from the home of a forward caste family near the school.

In Bihar, drinking water from hand pumps was available in all the sampled schools and UDISE 2014–15 figures for the state show that 92.74% of the schools had access to drinking water (NUEPA 2015a). However, it was observed that children from economically better-off families brought their own water bottles from home. In regard to drinking water directly from the source, it was observed that girls often drank water after the boys and children from deprived communities often got the last chance. Further, in the Muslim-majority district of Katihar, it was observed that in a majority of the schools the girls generally used the hand pump to drink water after the boys. A similar trend was found in 50% of the schools in Bhojpur and Muzaffarpur. Thus, there was discrimination not just on the visible basis of access (that is, who can touch the pump and who can drink water from it), but also more subtly in the order of access.

Toilet Tribulations

This study found overwhelming evidence of gender- as well as caste-based discrimination in the cleaning of toilets. To begin with, very few schools actually had usable toilets with proper water facilities. Therefore, wherever toilets were being used, the children cleaned them by fetching water from the hand pump or tap and pouring it in. In Andhra Pradesh and Madhya Pradesh, although people were hired to clean the toilets, girls (mostly from the SC community) also had to bring water from the hand pump and pour it in. In several schools, the usable toilets were locked and only used by teachers – but even in these schools, the girls were asked to clean these toilets.

In Orissa, none of the sample schools had separate usable toilets for boys and girls. In Assam, all of the toilets were unusable; where there were toilets for teachers, the girl students were expected to clean them. This was also the case in Bihar, where 75% of the schools had usable toilets, but most were locked and reserved for teachers. In Rajasthan, the situation was similar and the teachers said that they kept the toilets locked because the children would 'spoil and dirty them'. In Andhra Pradesh, parents said that their children were used to going in the fields and therefore toilets were not that essential. This arrangement was in fact preferred by many of the children, if the alternative meant they had to clean the toilets.

Sports Facilities and Libraries

One of the most notable findings of this study is the glaring absence of sports facilities for children in the sampled schools. Games and sports seemed to be notable in their absence. Where sports materials were available, they were made available only to 'select' boys who were favoured because of their perceived academic abilities. In most cases, teachers considered these boys 'bright' because they attended school regularly and sat in the front rows. The study did not find any caste-based discrimination in the access to and use of sports facilities, probably because sports was not a priority in schools and children were seen playing with their friends.

Like sports facilities, libraries were also noticeable in their absence. Even if the schools had books, they were not given to the children. Therefore, it was not possible to see if there was any exclusion. Only one school in Andhra Pradesh was an exception to this rule: although library books were issued to the students at this school, the teachers did not allow SC students to take the books home or use them because they were considered dirty and likely to soil the books.

Participation in School Activities and Tasks Assigned by Teachers

Observing what the children did, the activities they participated in, and the tasks that the teachers assigned to them were important elements of this study. School and classroom observation schedules tried to capture the various roles and responsibilities of the students and the conditions in which they performed various academic and classroom-oriented tasks.

Democratic practices were not the norm in the sample schools. Teachers decided the roster of tasks to be performed both in the classroom and around the school. These included routine duties like sweeping the playground, classrooms, veranda, toilets, and the place where the midday meal was served. Then there were high-profile tasks such as leading the morning assembly and participating in school functions by dancing, singing, giving a speech, or reading from the newspaper; and personal tasks such as fetching water or the register for the teacher, making and serving tea, cleaning the blackboard, collecting notebooks/papers, and carrying desks/chairs. These tasks were allotted to students depending on the teachers' perceptions of their acceptability. This made it possible for discrimination to arise from the teachers' own biases and prejudices.

An overwhelming finding of this study is that teachers across all the six states categorized students as being either 'bright' or 'passive'. These labels served to identify students who would be called upon to do various tasks. The 'bright' students had high visibility in the school community and performed high-profile and personal tasks. It was observed that the hierarchy of the chores that the children performed was also often an unspoken marker of caste, gender, and economic status.

School Functions and Morning Assemblies

Every school day began with a morning assembly. The teachers selected students to lead the morning assembly on the basis of two criteria: gender and perceived academic abilities. Girls were given precedence over boys, as the teachers felt that they sang in tune and were more disciplined.

While the criterion of gender was straightforward, the second criterion of perceived academic abilities demonstrated class and caste biases. Teachers perceived those students who attended school regularly as 'bright'. Because these children performed better in studies and spoken English compared to those who were not able to attend school regularly, they were given opportunities to lead the morning assembly. Therefore, irregular students generally lost out on these opportunities. Given these conditions, it was not surprising that upper caste students were identified as 'bright' and eligible for more prestigious responsibilities.

An exception to this rule was one school (out of 20) in Orissa, where an SC boy was seen leading the prayers. Again, in schools where a majority of the students belonged to the ST category, bright girls were asked to lead the prayers. Parents in almost all the states did not seem too concerned about the assignment of tasks and were largely silent on this form of discrimination.

This pattern of leading the morning assembly was repeated for all the major school functions, like those for Independence Day, Republic Day, and Teachers Day. Similar patterns were also found during the celebration of festivals. For example, whenever a formal *puja* was performed, it was led by teachers with the help of the 'bright' students.

Tasks Assigned by Teachers

The study observed clear gender- and caste-based behaviour in the classroom and school in the chores that the teachers assigned to the children (see Table 8.3). For tasks involving home-like chores such as filling water for teachers, making tea, or washing the teachers' lunch boxes, there was an overwhelming preference for girls – especially those from forward castes or OBCs. In tribal areas where almost all of the children belonged to the ST category, teachers asked any girl to do these tasks. There also seemed to be a clear hierarchy of tasks from menial to educational; teachers invariably called those students who (according to them) were the best and the brightest to perform the better tasks, while the poor and marginalized children were asked to do the menial ones.

Subtle and not-so-subtle gender and community/caste-based distinctions made by the teachers were quite obvious. Teachers did not make any effort to 'hide' their biases, and in fact they took them to be an extension of societal values and practices. The fact that teachers, administrators, and others involved with the schools did not perceive gender and community/caste-based distinctions as an issue was worrying.

In Andhra Pradesh, it was observed that the appearance of children played a significant role in determining which tasks would be allocated to them. Teachers differentiated between the neat and clean children and the ones who were untidy or 'dirty'. The colour of a child's skin too seemed to play an important role when special duties were assigned, such as speaking in an assembly or leading morning prayers. With respect to appointing class monitors, boys were given preference.

Cleaning-related Tasks

Across all three states, cleaning tasks were assigned according to caste, gender, or both. For example, for cleaning classrooms the preference was given to girls; boys were preferred for cleaning playgrounds. However, for cleaning toilets both gender and caste were important.

Discipline- and School Activity-related Discrimination

Interestingly, discipline was closely linked to being a 'good student' across all states. Only children perceived as 'good' or 'bright' were selected to be class monitors, speak in the morning assembly, participate in national day functions, and so forth. Although this form of discrimination was not blatant, it was a consequence of certain conditions. Teachers had clear views of what boys and girls could do. When it came to leading prayers, teachers preferred girls. Unfortunately there were no special school functions during our fieldwork, so we could not observe whether all of the children participated in these functions.

Corporal Punishment and Sexual Harassment

Scolding children was common across all of the states. Physical punishment was not that common – or it could be that after RTE teachers were more careful about using overt forms of punishment

Table 8.3. Tasks assigned by teachers

Task	Gender-based	Gender and caste-based	Caste-based	Ability-based ('bright', 'intelligent')	All or randomly assigned
Cleaning the classroom	Assam (G), Orissa (G)	Rajasthan (G, SC/ST), Madhya Pradesh (G, SC)			Andhra Pradesh
Cleaning the playground	Assam (B), Madhya Pradesh (B), Orissa (B)	Rajasthan (G, SC/ST)			Andhra Pradesh, Madhya Pradesh
Writing on the blackboard, coming forward to read, recite, or lead the class in an activity				Assam, Andhra Pradesh, Rajasthan (FC, OBC), Orissa, Bihar	
Class monitor	Orissa (B)	Rajasthan (B, FC)		Assam, Andhra Pradesh, Rajasthan	
Cleaning the toilets	Orissa (G)	Madhya Pradesh (G, SC), Rajasthan (B, SC/ST)	Andhra Pradesh		Andhra Pradesh, Assam
Fetching water for the teacher	Andhra Pradesh (B), Bihar (B)	Rajasthan (G, FC)	Orissa (non SC), Madhya Pradesh (FC, BC)		Madhya Pradesh, Assam

(contd.)

(Table 8.3 continued.)

Task	Gender-based	Gender and caste-based	Caste-based	Ability-based ('bright', 'intelligent')	All or randomly assigned
Preparing tea for the teacher/guests	Bihar (G)	Rajasthan (G, FC), Orissa (G, FC), Madhya Pradesh (G, FC)			
Arranging books	Assam (B), Bihar (G)	Rajasthan (B, FC)		Andhra Pradesh, Bihar (G)	Assam
Ringing the bell	Andhra Pradesh (B), Orissa (B)	Rajasthan (B, FC)			
Washing the teacher's lunch box		Andhra Pradesh (G, SC), Orissa, Rajasthan (G, SC/ST)			
Cleaning the floor after the midday meal	Orissa (G)	Rajasthan (G, SC/ST)			
Leading the assembly		Rajasthan (G, FC), Madhya Pradesh (G, FC, BC, Muslim)		Assam (G), Bihar (G), Andhra Pradesh (G)	Andhra Pradesh, Madhya Pradesh
Cultural activities	Assam (G)	Bihar (G, FC, BC)		Bihar (G), Andhra Pradesh	

(contd.)

(Table 8.3 continued.)

Task	Gender-based	Gender and caste-based	Caste-based	Ability-based ('bright', 'intelligent')	All or randomly assigned
Sports activities	Assam (B), Bihar (B), Rajasthan (B)			Andhra Pradesh (G & B)	
Seating arrangements	Separate rows for boys and girls, or girls occupying the first few rows in Andhra Pradesh, Assam, Bihar, Madhya Pradesh, Orissa, Rajasthan			Front Benchers: Assam (G & B), Bihar, Rajasthan (FC), Orissa (FC), Madhya Pradesh	

Source: Survey data.
Note: G: Girl; B: Boy; FC: Forward Caste; BC: Backward Caste; SC: Scheduled Caste; and ST: Scheduled Tribe.

when the research team was in the schools. In Orissa we found a small stick in almost every class and with every teacher, even though the government had painted 'No Punishment Zone' in all of the schools. In 20% of the schools, students reported the use of some kind of punishment. The children in a few schools in two districts reported verbal abuse, with caste or community identity often being used as a peg for the abuse.

During discussions with teachers and parents in Assam, it emerged that both groups blamed the recent ban on all forms of corporal punishment for the lack of discipline in schools. In particular, the parents complained that the teachers were not strict. In Rajasthan, teachers carried sticks and some were observed hitting/ slapping the children. In Andhra Pradesh, children reported that physical punishment was common; for example, teachers would pull girls' hair and twist the ears of the boys.

In Madhya Pradesh, discussions with teachers and parents revealed a complex understanding of discipline and learning. According to the parents, the difference between a 'good' teacher and a 'bad' teacher had to do with how much s/he punished children. Teacher quality was also measured in terms of her/his teaching 'well', which meant explaining lessons to the children, assigning work in class and as homework, conducting classes on time, coming to school daily, and being punctual and maintaining discipline. The Right to Education Act notwithstanding, both teachers and parents were quite vocal about the need for punishment.

We did not find any direct connection or link between caste discrimination and physical punishment in any of the states, although verbal abuse using students' caste/community identity was observed in several schools. However, there seemed to be a visible link between the attitude towards women and girls and the kind of punishment that was meted out to them: as it was considered taboo to touch a girl after puberty, teachers usually only scolded the girls.

One disquieting matter was that there were at least three schools in Andhra Pradesh where sexual harassment was mentioned by the parents. Girl students complained that some male teachers had touched them and abused them both verbally and physically. Their parents corroborated these statements. In one school, a complaint was registered with the District Collector and the teacher was reportedly suspended.

Midday Meal

It is no surprise that the midday meal (MDM) emerged as the single most important site for inclusion and exclusion. In this study, we came across a range of practices, as described below

Who eats and who does not?

In all of the states, children from better-off families (regardless of caste/community) did not eat the MDM and instead went home for lunch.

- In Andhra Pradesh, where a majority of the children in government schools were from very poor families, we found that extremely poor children from the Gondi and Lambada tribal communities did not eat MDMs because the food was cooked by a BC or forward caste person.
- In Assam, many forward castes, especially Brahmins and OBCs, went home during MDM.
- In Rajasthan, children from the Rajput, Brahmin, Jat, Bishnoi, and Gujjar communities did not eat the MDM. Children from the Meena (ST) community ate only if the cook was from their community. Children from relatively well-off families also did not eat MDMs in the sampled schools.
- In Orissa, adolescent girls did not eat the MDM because of socio-cultural taboos that restrict adolescent girls from eating food made outside their homes.

Who cooks?

By and large, the MDMs were made by cooks belonging to the OBC category while SCs worked as helpers.

- In Orissa, 20% of the sample schools had SC cooks; in the rest, they were either from OBC (50%) or ST communities (40%). This was a common practice to help avoid any social tension in schools. SC women worked as helpers.

- In Assam, caste identity was observed to be an important criterion in the selection of cooks.
- In Rajasthan, most of the cooks belonged to FC, OBC, or ST (Meena) communities. Older children from the relatively forward communities in the village served the food.
- In Andhra Pradesh, the cooks were mostly from BC (in five of the 20 schools) or FC (in 15 out of the 20 schools) communities. Although SC and ST women were appointed as helpers, they were not permitted to touch the food.
- The situation in Madhya Pradesh was quite ironic. Although the state government had taken a decision to allot MDM work to women self-help groups (SHGs) from the SC community, given the caste dynamics in the villages almost all of the cooks belonged to the OBC community.
- In Bihar, cooks belonged to FC or BC communities in a majority of cases.

Who sits where?

- In Orissa, children were observed sitting in their own community groups. While teachers did not ask them to sit separately, they also did not encourage them to sit together.
- In Assam, there was no pattern as to where the children sat during MDM. They were found sitting in small, single-sex groups of twos or threes – that is, girls with girls and boys with boys. In one school, the teachers said that Hindu and Muslim children sat separately.
- In Andhra Pradesh, the children sat in rows segregated by gender; in some schools they were seen sitting community-wise, sharing food brought from home along with the MDM. Children from the upper caste went home for lunch.
- In Bihar, the MDM was irregular. Children were observed sitting with their friends in caste-specific groups.
- In Rajasthan, children sat in caste groups, with a clear demarcation between boys and girls.
- While children were found sitting in caste-specific groups, what was noteworthy in Madhya Pradesh was that even teachers from different castes sat separately. For example, SC teachers

ate separately. In one school, children from the SC community were not allowed to eat from the plates provided by the school. When a child forgot to bring his/her own plate, food was served on a roti.

Seating Arrangements in Schools and Teaching-learning Practices

The study found three different classroom seating patterns:

- The brightest/more active and engaged students sat in the front rows, with boys and girls in separate rows. In Assam, roll numbers were assigned according to academic performance and children sat according to their roll numbers. Similarly, in Andhra Pradesh bright students or those who were active and regular sat in front. Other children sat at the back, along with students who were irregular. A similar situation was found in Orissa as well.
- Children from forward castes, from better-off families, or those who were better dressed sat in the front rows. Children who were poor or poorly dressed sat in the back. In Rajasthan, children from *Adivasi* and Dalit families sat in the back rows and remained passive and silent throughout the class. In one school in Andhra Pradesh, one girl was made to sit separately on the last bench because she did not wear proper clothes (due to her poor economic status; she also had little time to bathe regularly as she had to perform household chores). She was ignored by both the students and the teachers. Regular absentees and CWSN were normally ignored by the teachers. Similarly, in Nuapada and Keonjhar districts in Orissa, poor and shabbily dressed children were backbenchers who were ignored by the teachers during classroom transactions.
- There was no pattern and children sat wherever they found space. This was most common in overcrowded and multi-grade classrooms.

In all six states, boys and girls sat in separate rows or in clearly demarcated spaces. In observations of Classes IV and VII, it was seen that boys and girls did not interact much with each other and sat separately. The teachers did not encourage interaction between boys and girls. In all states, women teachers also did not interact much with male teachers.

Peer Interactions among Students

Seating arrangements were also reflected in peer interactions. In all states, children seemed to sit and play with their own 'kind,' as defined by their environment. While caste identity was not so significant for peer interactions in Andhra Pradesh, the children did not necessarily eat together. Children seemed to draw a line between eating together and playing together.

Teachers' Perceptions and Practices

Interviews and group discussions with teachers revealed some startling biases:

- Andhra Pradesh: Teachers did not voice any caste-based prejudices, but they said that 'some communities' were uneducated, unaware, unclean, only interested in employment, and 'not bothered about the performance of their children in school'.
- Assam: Most of the teachers did not articulate any caste- or community-specific biases. However, almost all of them said that they focused on academically bright children. They also said that girls were not interested in sports and that they should be trained in household duties. Some non-ST teachers attributed lack of cleanliness to community practices.
- Bihar was not very different from Assam. Teachers did not voice any caste-specific biases and were very careful in what they said. However, they were dismissive of children from extremely deprived/ poor families and believed they would not be able to learn.
- Madhya Pradesh: Teacher-pupil interaction mostly formal, related to assignments and tasks. Teachers were observed favouring children who were 'active' and expressed the opinion that children who were not 'bright' were left behind in class. According to them,

it was difficult to motivate those children because education was not considered a need in their homes.

- Orissa: General perceptions, including among teachers from the SC community, was that children from SC, ST, and some OBC communities were 'dull and lagged behind in studies'. They attributed this to a lack of interest among the parents. A majority of the teachers exhibited a strong bias against some SC children (especially very poor ones) and called them 'unhygienic and dull'.
- Rajasthan: The reticence visible in Assam and Bihar was absent in Rajasthan, where the teachers freely talked about caste/community patterns and stereotypes. All of them said that they paid greater attention to children who were bright and attentive in class. Teachers were found to put in extra effort to reach out to all children academically in only two schools. A majority of the teachers said that the parents of SC/ST children were often not educated and that they eventually pulled their children out of school or did not ensure regular attendance. They were aware of the caste hierarchies that played out during the MDM and in accessing water resources.

Children's Perceptions

This study captured children's perceptions through structured activities with those studying in Classes IV and VII and through discussions with adolescent boys and girls (in separate groups) who had studied in the sample schools.

- Andhra Pradesh: In all the schools, children said that the girls did the cleaning and the boys carried and dumped the collected dirt. The children did not refer to toilet cleaning duties; however, they said that sometimes the teachers asked girl students to throw water in the toilets. In all of the schools, the children said that they brought water from home in bottles. In one school in Adilabad district, SC children were not allowed to drink water from the well. Girls led both the prayers and the daily pledge that was taken by the children.
- Assam: In almost all of the schools, children said that the teachers

paid attention to bright children. In a few schools, ST children said that the teachers used derogatory language when referring to tribal children or to those from tea garden areas. The children also said that the boys cleaned the playground and the girls cleaned the classrooms, that girls led the morning assembly, and that all of the children had equal access to the hand pump. Overt forms of exclusion were not an issue for the children, and they did not have anything to say about those who cooked the MDM or where they sat. The children also said that everyone participated in school functions on national days, and that no child was excluded.

- Bihar: A majority of the children said that the cleaning tasks in the schools were mostly assigned to children from deprived communities; children from the upper castes generally came up with some excuse to get away from such tasks. Though all children had equal access to drinking water, during rush hours, such as break time and during the MDM, children from upper castes and boys got first preference in using the hand pump. It was mostly children from the upper castes, particularly girls, who got more opportunities to lead the assemblies.
- Madhya Pradesh: The children were quite articulate about the tasks assigned to them and were also aware of the caste that they belonged to. They said that by and large none of them cleaned the toilets or classrooms. If and when the children had to do the cleaning, the teachers did not ask forward caste girls to do it; instead, they asked students who were disadvantaged and poor to do the cleaning. While in most schools all of the children had access to hand pumps/taps, SC children were not allowed to touch water that was stored in a bucket or tank. However, adolescent girls in one village said that SC children could not touch the hand pump and that the forward caste children pumped water for them. Like in several other states, girls who could sing well and were doing well academically led the morning assembly.
- Orissa: The children were quite clear that only girls did the cleaning, with older girls cleaning the veranda while the open space used for the MDM was cleaned by the cook and helper. Toilet cleaning was not a regular activity. However, when they had to clean the toilets, the boys poured the water and the girls

swept them with a broom. In one school, SC boys and girls said that they were not allowed to touch the hand pump. During Saraswati and Ganesh Pujas, boys took the lead.

- Rajasthan: The children were quite articulate about the tasks assigned to them. In almost all schools they said that sweeping was invariably done by girls from the SC or ST communities. While most of the children used the hand pump, there was an order in which the children could access it – boys from forward castes used it first, while girls from the backward caste/community and CWSN used it last. Most of the children washed the mouth of the hand pump before they drank water. However, when water was stored in clay pots, SC and ST children were not allowed to touch it. Older girls (from senior classes) led the morning assembly, and the teachers picked girls who could sing well.

What the Parents Had to Say

Discussions with parents from the most deprived/poor communities in the villages were illuminating. Like other studies have shown (Singh 2011), parents were by and large not aware of what happened inside the school, particularly the tasks assigned to the students by the teachers. In Rajasthan, they were not aware that cleaning duties were assigned to the children; in Assam, they were aware that the girls swept the classrooms and the boys cleared the playground. In Madhya Pradesh, parents had almost nothing to say about what their children did in school; they believed that, barring a few forward caste children, the others ate their midday meal and used the hand pump and water source without discrimination. Awareness levels were also very low in Orissa, where discussions with parents on what the children did in school did not yield much except that the children did not use the toilets in the schools, and that those from better-off families were given priority in school activities like the morning assembly and special functions. Parents in Rajasthan had more information about cleaning duties being assigned to girls, but did not have much to say about it. They believed that forward caste children would naturally get more attention from the teachers. In Andhra Pradesh, the parents participated in school-level committee meetings (School

Development and Management Committee, School Management Committee, etc.) at random, particularly male members and those from the BC community. They were largely ignorant of what was happening in the schools. The parents were not aware of issues like the sexual abuse of girl students, and seemed to be unaware of such incidents reported in the media.

Inclusion and Positive Practices

This study also tried to understand and identify inclusive and positive practices in schools and classrooms. A few inclusive practices were observed by field investigators across all of the states. These included the head teacher being interested in and committed to certain issues, or teachers taking the initiative in promoting the inclusion of students regardless of caste, gender, or physical disabilities.

The observed positive practices include:

- Midday meal: Out of the 20 schools visited in Andhra Pradesh, only one school was setting an example of inclusive practices in terms of access to food and water for all of the children. During the MDM, all of the children were observed eating together and no one was seen skipping a meal or sitting in segregated caste groups. The teachers also monitored the MDM to ensure that good food was served and its quality maintained. Similarly, during classroom transactions it was observed that the teachers paid attention to all of the students. During school activities, equal participation of children was ensured through rotation; students shared responsibilities during the morning assembly, special days/events, and school functions. Similarly, in Orissa 30% of the schools had a MDM that was managed by women self-help groups to ensure the quality of food and the maintenance of proper hygiene, thereby leading to greater participation of children in the midday meal.
- SMCs and the community: Cooperation and communication between the local community, teachers, and the school management committee was another factor that led to promoting inclusion in schools. For example, in Assam it led to better

infrastructure and classroom transactions and 100% participation during the midday meal, irrespective of caste and religious backgrounds. In this context, the role played by SMCs was quite crucial. In Orissa, it was found that the SMCs played an active role in appointing voluntary teachers to address teacher shortages, leading to better classroom transactions.

- Good infrastructure: Even though improving infrastructure facilities in schools has been a major focus of SSA, a majority of the sample schools still did not have adequate infrastructure. However, some schools did have adequate and good facilities. In Madhya Pradesh, good infrastructure facilities were seen in some schools; these also opened on time and had active classroom transactions and teachers who were supportive of the needs of students.
- Promising state-level initiatives: The Bihar government's initiatives like the Tola Sewak (a volunteer in a village or habitation who takes the responsibility of getting children to school and interacting with them in school) and Hunar programmes have gone a long way toward promoting inclusiveness and cohesiveness in schools. Tola Sewaks have been appointed at the school level and their prime responsibility is to provide handholding support to the students, and more importantly bring them to school. Their second major responsibility is ensuring that no discrimination takes place in the participation of children in school activities.

Children with Special Needs

It is important to note that in every one of the schools that was visited there were very few or no children with special needs. This raised a serious question: Were these children actually being included in schools or not? The SSA framework clearly mandates that the programme ensure that every child with special needs, irrespective of the kind, category, or degree of disability, be provided with education in an appropriate environment. SSA is also supposed to adopt a zero rejection policy so that no child is left out of the education system. In all six states, however, factors such as the availability of CWSN-friendly infrastructure and the lack of a positive and supportive attitude from the teachers and students towards

CWSN did not seem to reflect this mandate. There have also been instances of such children being ignored, denied admission, and verbally abused by teachers and peers, such as in Andhra Pradesh. Moreover, even though ramps have been provided in some schools and there was a CWSN-friendly toilet in one school in Andhra Pradesh, their usability still remains a big concern. However, there is also a silver lining: in one school in Andhra Pradesh, the teachers and headmaster were observed taking an active interest in school activities and being supportive of the needs of CWSN and children from deprived social groups.

Can We End All Forms of Exclusion/Discrimination in Schools?

As the research teams travelled across the sample states, one question kept cropping up: 'When and how will the situation change?' There are no easy answers and, given our track record over the last 65 years, some prejudices have only become more palpable. Issues such as caste, religion, economic status, place of residence, and knowledge of English have become significant markers in our social and political life. Many elder teachers admitted that the feeling of national unity is fading. Many of them attributed this to the way our electoral system has evolved to mobilize political support on the basis of social and cultural identities.

However, despite the fact that the larger socio-political environment is becoming more stratified and divisive, there are islands of hope across this vast and diverse country.

Taking the Constitution of India as a guiding spirit, teachers, administrators, and community leaders need to be told that any violation of the right to equality and the constitutional right to freedom from discrimination will invite strict penal action. A non-negotiable code of behaviour needs to be communicated to all of those who are involved in school education. This needs to be done in writing and prominently displayed in all schools and educational institutions. Simultaneously, children need to be involved in activities that enable them to understand and appreciate diversity, respect differences, and formulate school-level norms of behaviour.

Involving children in creating an egalitarian atmosphere could bring moral pressure on teachers, administrators, and local leaders to not differentiate or discriminate.

As it stands today, teacher orientation and training is often limited to administrative requirements and subject knowledge; there is also growing evidence of training fatigue among teachers. Equally, short-duration training programmes are not able to do justice to basic subject knowledge and pedagogy issues. In this situation, creating alternative forums/platforms where teachers and community leaders can come together to communicate the need of keeping schools free of discrimination may well be a worthwhile exercise.

What is required is that we start looking seriously at inclusion and exclusion in our schools and work at all levels to bring about lasting change on the ground. There are no shortcuts; the government and civil society organizations need to take this issue seriously and address it in every context.

9 The Teacher Conundrum

In this chapter I draw upon four studies that I conducted between 2005 and 2015 to investigate why Indian school teachers are demotivated and disheartened, why they are targeted by the media and the government for the poor learning outcomes of their students, and most importantly, what we as a nation have done to not only improve teachers' social status but also to make them accountable.

Teacher Motivation

In the year 2005, I asked a group of teachers in Rajasthan what the characteristics of a motivated teacher were, in their opinion (Ramachandran et al. 2005). After thinking for a while, one of them said: 'A "motivated" teacher comes to school every day, does what he is told and provides information the higher ups want!'

I was puzzled by this answer and probed further. Almost all of the teachers believed that daily attendance and complying with orders and requests for information were reasonable indicators of teacher motivation. A headmaster with a sense of humour said:

> We have five teachers in our school. One of them is a *dakia* [postman], who responds to enquiries that come from above and dispatches data/information to the district or block office. The other is a *halwai* [sweetshop owner], who manages the midday meal. The third one is perpetually in training and the fourth is a clerk who has to maintain

accounts and pay salaries. Who is left to manage five classes and teach around 200 children? One teacher!

I then asked the administrators the same question. The first group was empathetic toward the teachers and talked about the systemic issues that impact regular teaching and learning in schools. They did not see teachers as villains, but rather as cogs in a giant wheel that are trapped in a hierarchical system rife with corruption. One officer admitted, 'Teachers are victims of the transfer and posting phenomenon. They have no security of tenure and are constantly haunted by the fear of being transferred. They have to acquire godfathers for protection. Once they develop these contacts and linkages, then there is no need for them to do their job seriously!'

The second group saw teachers as shirkers and placed the entire blame for poor quality education on them. They described a motivated teacher as one who was regular, did what she or he was told, and was, by and large, compliant. One of the more sensitive members of this group lamented:

> Job satisfaction and morale are highly interrelated. To my mind, job satisfaction and morale require a certain basic seriousness about what one is doing. Thereafter, the level of satisfaction or morale may decrease due to prevailing work conditions or other factors. But, government schoolteachers lack this basic seriousness. There is a lot of dissatisfaction and frustration among them but it does not generate any action. Surprisingly, they do not seem interested enough to do anything about it.

It is worth noting here that the notion of 'quality' is linked to efficient management for all of the interviewed groups. As a result, obedience and predictability become pervasive, sought-after values in the system. Actual teaching transaction time, classroom processes, or the learning outcomes of children do not appear in their first response to this question. However, the percentage of children clearing the terminal examination was considered an important indicator of quality.

Parents had a different view. Teachers who were regular, stayed in the school for the stipulated time, did not use excessive force (beatings, abusive language, shouting, and other punishments), and taught with interest were considered motivated. I then talked to several educationists of Rajasthan (Sharada Jain, Nagendra Nagpal, and the faculty members of Sandhan, an educational resource and training centre located in Jaipur). They argued that a motivated teacher was one who could communicate with the children, who drew energy from interactions with children, and who was concerned about what and how much the students were learning.

Parents alone said that their children's ability to read, write, and pass examinations was important. Ironically, administrators brushed learning aside: it was incidental to the mountain of data they gathered and fed into the system. Enrolment, attendance, midday meal distribution, and teachers' participation in training programmes have become the indices of education.

Discussions about motivation led to comparisons between government and private schools. Teachers, administrators, and parents quickly pointed out that private schools attached great importance to discipline, regularity, and successful results in both the yearly and public examinations (administered in Classes V, VIII, X, and XII). Almost all of the teachers I interacted with sent their own children to private schools. They admitted that the irregular attendance of teachers was uncommon in private (both aided and unaided) schools, and that teachers there taught for the stipulated hours/periods. When asked why government schools were different, most could not give us a convincing answer. They ended up blaming the system, in which the dice is loaded against teachers in primary schools.

Most people squarely blame the teachers, citing absenteeism, the politicization of teachers' unions and, most importantly, a lack of professional ethics. Teachers, on the other hand, argue that the system has placed them in a position where if they do not take the initiative and just plug along, nothing can go wrong; but if they do and get noticed, they can get into trouble. For this reason, they lie low and do what the higher-ups tell them to do. The more politically savvy and active teachers spend time and energy cultivating politicians and local networks.

Quite unlike the popular perception, teachers do have a nuanced understanding of motivation. Almost all of them admitted that 'motivation' is a dynamic feeling that changes from time to time. They linked it to the larger environment in which they worked and how it had affected their sense of self-worth. One *panchayat* leader narrated an interesting experience:

> If the family or community considers teaching a noble profession, the teacher will naturally be more motivated. When someone does their job well, their status in society increases. Kalyanpura village is in Chaksu block. Girls here were hardly sent to school. Their numbers were minimal. We started Pehchan Shala. The appointed teacher identified all the girls herself. She built a rapport with the community. Everyone was full of praise for her work. The villagers themselves organized her salary. She had to leave when she got married. A huge party was thrown for her and that day was called Balika Utsav (festival for girls). About 700 girls from all the nearby Pehchan Shalas participated. The villagers organized everything themselves. The function cost almost Rs. 20,000!

Teachers complained about feeling demeaned when they were sent out to collect data or conduct door-to-door campaigns. They argued that their job was not to do research surveys and campaigns for the government, and that doing so affected their social status. They valued justice and fair play, and were ready to work with administrators and political leaders who they felt were just, and who appreciated and rewarded hard work.

The government's decision to hire contract and para-teachers in the mid-1990s was a further blow to regular teachers' self-worth. They felt they were no longer discharging a unique, special duty: apparently, even untrained hands could do what they had been doing. It was clear that their motivation hinged on their emotional energy. Teachers' skills and unique strengths, especially their status as professionals, are not appreciated, and there is no positive affirmation or encouragement.

Teacher motivation is a complex issue in India. There is virtually no incentive for teachers to go beyond the call of duty and empower their students to learn or move on in life. On the other hand, teachers who network with political leaders and local bureaucrats receive plum postings and, if they are lucky, teachers' awards too!

The answer to the question of poor motivation lies buried, perhaps, in the labyrinth of a complex educational system. The issue of teacher motivation is framed in an intricate matrix of cause and effect in which one cannot really discern a clear, one-to-one linear correlation between motivation and any one issue, such as salary, preferred posting, or the location of the school.

First, while the education system has expanded rapidly and enrolment rates have shot up, the growth rate in the number of teachers has not kept pace. Classrooms have become very complex. Children from extremely poor families and first generation school-goers account for an overwhelming majority of the students in government schools. Most rural schools are multi-grade, with one, or at most two, teachers managing five classes. Teacher-pupil ratios are also high in such schools.

Second, the social distance between teachers and students is wide in government schools (which cater to the very poor). Social attitudes and community prejudices play an important role in determining the ability and willingness of teachers to empathize with the children. Press reports reveal cases of sexual exploitation of girls in both rural and urban (municipal) schools. We hear of teachers and even headmasters being arrested for raping girls and sexually abusing young boys. Senior police officials who had arrested teachers for sexual misconduct said that teachers used abusive language when they talked to/about children from very poor or socially disadvantaged communities. It was as though they were doing a big favour by teaching children from erstwhile 'untouchable' or very poor migrant communities from other parts of India.

Third, teachers lack the skills to manage so much diversity in the classroom. Training programmes for teachers are designed for the situation in large urban schools, where each teacher manages one class. The problems faced by teachers in multi-grade situations and where teacher-pupil ratios are high are rarely covered in training programmes. Labels like child-centred learning do not mean anything to teachers who have to deal with social diversity, different levels of students, and most importantly, children who are undernourished, hungry, and frequently ill. Focus group discussions with teachers in several states over the last few years revealed that teachers wanted subject-specific training for multi-grade situations. But most training programmes focus on generic skills. The mismatch between the problems faced by teachers inside the classroom and the training programmes designed by administrators and teacher educators (who have very little idea of how to manage a multi-grade class) is stark.

Fourth, systemic issues of corruption (paying for/to prevent transfers, deputations, appointments, promotions, and special assignments) have vitiated the larger teaching environment in the country. Teachers say this has politicized the environment: actual teaching is rarely monitored, while building networks with patrons and supporters is more important. Teachers who are in leadership positions in trade unions or affiliated with the political parties in power rarely attend school. Teachers' continuation in the job and/or in preferred posts depends on their ability to strike the right chord with the people in power. As a result, highly motivated and honest teachers end up transferred to difficult areas, saddled with a number of non-teaching duties, and made scapegoats when the need arises. Even though there are no incentives for performing better in the classroom, it certainly pays to build networks and cultivate political godfathers.

Fifth, teachers' unions and block- and district-level administrators claim that teachers are asked to do a range of non-teaching tasks that take them away from the classroom. For example, in 2005 the Rajasthan Government asked teachers to motivate couples to use terminal family planning methods. This led to a series of protests by teachers in February of that year. In 2001–03, the Rajasthan state government directed teachers to maintain the books of women's self-help groups and monitor whether loan repayments were made on time. District Magistrates rely on teachers to distribute drought or flood relief supplies and to identify beneficiaries for government welfare schemes. Discussions with teachers revealed that, while the task of meeting family planning targets (for example) may be given to all of the teachers, the more difficult and time-consuming non-teaching duties go to the teachers who are considered dedicated. Teachers with political links or who are active in trade unions are not given additional duties.

Both the central and state governments contest this finding. Senior administrators in the Government of India point out that less than 5% of teaching days are taken up by non-teaching duties

(Ramachandran, Bhattacharjea, and Sheshagiri 2008). The DISE collected information on non-teaching duties and the days spent on them. While state-wise data has not been published by NUEPA, an interview with Arun Mehta (who was the person heading the DISE and UDISE unit of NUEPA) revealed that non-teaching duties accounted for about 1.6% of working days. Teachers' unions and local administrators disagree. They argue that the government may expect teachers to do such work after school hours, but invariably teachers spend teaching time performing these non-teaching assignments. The problem becomes particularly severe during elections, the census, and other government surveys. This is not captured in official statistics on non-teaching duties of teachers during school hours.

Sixth, teacher training has increased since 1994, with almost all teachers expected to attend a range of training programmes every year. Many of these workshops are held during the academic session, and teachers are eligible for compensatory leave if they attend these workshops during vacations. This reduces teaching days. While the training programmes are intended to improve teachers' knowledge level and skills – especially in child-centred teaching processes – teachers claim that they add little value when the overall teaching environment, examination system, and other aspects of schools remain unchanged. Nearly all of the teachers I interacted with in several states in 2005, 2007, and 2009 said the training was a burden that was neither well planned nor catered to their needs. Another disturbing issue came to the fore. Teachers in several states revealed that training is a ritual; they often reach a training venue by 11.00 am and leave by 3.00 pm after attending two or three lectures. In some remote districts where the state government is not able to monitor whether the training programme actually happens, teachers reach the venue, collect their travel allowance, and are asked to disappear for three to four days – while ticking off the activity as done.

Seventh, teachers and administrators are continuously embroiled in court cases concerning promotions and placements, arrears due to them, and disciplinary action-related issues. Administrators explain that a lot of their time is spent attending to court cases filed by teachers. Teachers argue that they have no option but to go to court if they are to receive justice. Teacher cadre management is highly politicized: both administrators and ordinary teachers are caught in a web of allegations and counter-allegations. This has affected the recruitment of new teachers in several states.

The silver lining is that during the course of my research work I came across teachers who loved children and were highly motivated, regardless of where they were posted. These were exceptional people. It was, indeed, humbling to meet teachers who worked hard despite all odds. I came across situations where good teachers received tremendous community support, leading to an improvement in teaching and overall results. The reverse was also true. There were also villages that had had a wonderful teacher in the past, but that could do little to motivate/support a new teacher who just refused to teach.

Interestingly, everyone we spoke to were of the opinion that 25% to 30% teachers are highly motivated and work very hard, regardless of their personal circumstances. Another 30% comply with all of the formal requirements – regularity, attendance, data on enrolment and retention, midday meal distribution, and so on. These teachers have potential, but the system has worn them out. The remaining 40% to 45% can be categorized as 'indifferent': they are just not motivated and really do not care.

India is a large country. It is possible that the observed gap between the educationists' perception of motivation and that of teachers, administrators, and the larger community may be lower in educationally advanced states like Kerala, Tamil Nadu, and Himachal Pradesh. Similarly, the nature of the problem in Rajasthan may be a far cry from the issues bothering West Bengal.

One thing is certain. Teacher motivation is a huge issue in education, and unless we, as a nation, decide to take it seriously, the future of our children is shaky. This is particularly grave when it comes to children living in diverse poverty situations and those who depend on government schools.

Becoming a Schoolteacher

In 2008, I travelled across Maharashtra, Kerala, West Bengal, Tamil Nadu, and Karnataka to talk with schoolteachers, teacher-education institutions, and administrators in an attempt to unravel the daily life of an ordinary schoolteacher (Ramachandran, Bhattacharjea, and

Sheshagiri 2008). Teachers do not enter the profession knowing what to do; they learn what is expected of them – and it is the education system itself that teaches them their role within it, not only through training programmes but in many other ways both obvious and subtle. In other words, what teachers do in the classroom reflects a set of conditions *on the ground* that establish the real parameters within which they work. I emphasize 'on the ground' because policies on paper and their implementation in the field can often look very different.

Contradictions that frame the world of teachers

- If beginning teachers are required to have completed Class XII, but are never screened for teaching aptitude (either initially or at any time during their careers), what message does the 'system' transmit about the importance of teaching ability?
- If teachers are told in training programmes that they must use creative teaching techniques to promote 'joyful learning', but are evaluated by supervisors based on whether they have covered the textbooks, which set of instructions are they likely to follow?
- If students' learning levels make absolutely no difference to teachers' access to a better job or a higher salary, why should they bother to try to teach better?
- If a local politician can have a teacher transferred simply because he wants to, can we really expect her to put the needs of her students before her own need for job security?

I therefore look at how the conditions on the ground affect how teachers perceive and implement their work. These conditions are sometimes the result of stated, formal policy measures (such as those that specify criteria for promotions), and sometimes the outcome of unstated but nevertheless powerful informal processes and mechanisms (such as local politicians' ability to transfer teachers). Taken together, these conditions reflect the reality, as opposed to the theory, of 'teacher development' in India.

Meet Malati, who was just appointed as a regular schoolteacher in a rural two-teacher school

Malati is 20 years old. She comes from a middle class family and lives in a small town. She completed Class XII two years ago with

58% marks, and then enrolled in a primary teacher-training course that lasted one year. She decided to become a teacher because the salary is good and she likes being with children. Also, she wants to get married soon and feels that the hours are convenient: she will be able to look after her household duties also.

Malati is among the lucky few to get a job as a regular teacher. She had hoped for a school within her town, but was appointed to a rural district school about an hour away by bus. On her first day of school, Malati leaves home early to take a bus to the village where her school is located, then walks from the bus stand to the school. As she nears the school, her heart sinks. In front of her is a dilapidated two-room structure with broken windows. A few children playing in a playground strewn with bricks, stones, and garbage greet her; when she asks where the teacher is, they tell her that he comes by the 10 am bus. She goes into the building, where there are no benches for children, the rooms are dusty and unkempt, and two chairs lie in the far corner of the veranda. There is no water in the pot, no toilet anywhere in sight, and no sign of charts or books.

Eventually, her co-teacher arrives. A middle-aged man who has worked in this school for three years, he has nothing much to say to her about how she should go about her work, how she should manage a group of children from different classes in the same room or, most importantly, what it is that is expected of her as a teacher. She is just asked to handle Classes I to III: more than 60 children crammed into a dark, airless, colourless room. Nothing she had learnt in her teacher-training programme helps her understand how to deal with this reality.

She decides that she should get to know her students and tries to ask them some questions. But only those sitting in the first few rows are paying attention. The children at the back are talking among themselves and she worries that they are making too much noise. Confused, she decides that she ought to try to teach them something. So she leads them through a recitation of the alphabet, interrupted many times by her telling first one child then another to keep quiet, pay attention, and sit down. Somehow she gets through the day.

Within a week of joining, Malati is informed that the textbooks have to be collected from the local cluster office. She makes the journey, hoping to meet other teachers like her. When she reaches the cluster school she is handed a number of forms to fill out before she can collect the books. Since she did not bring the data with her, she trudges back to her school and returns four hours later with the registers. As she fills out the forms she notices that the official enrolment is far higher that what she has seen so far. Brushing it aside, she picks up the books and notices that the one on Environmental Studies (EVS) is missing. The cluster resource centre (CRC) resource person tells her that he will inform her when the EVS books arrive.

Malati tells the resource person that she is having trouble dealing with such a large group, especially since they are of different ages, and asks what is expected of her as a teacher. He looks surprised and says: follow the textbook, teach, and make sure you furnish all the information that the government asks for.

Malati attends a monthly meeting in the nodal school at the cluster level and collects her salary slip (her salary is deposited in a bank). Life goes on and she quickly realizes that no one visits her school – there is no inspection. Sometimes the block resource centre (BRC) or cluster resource centre (CRC) teachers come for a little while, fill out forms, look at the registers, scrutinize the midday meal register, and then leave. There is little discussion about teaching and learning. No one asks her what she has been doing in the classroom or how she is managing a group of children who are technically in different classes. She notes that even the children in Class III cannot read or write.

Soon Malati is informed that she must help conduct the household survey in the village. As the survey date approaches, she seeks out her co-teacher to take his advice on how to conduct the survey. He asks her to meet the local *panchayat pradhan* (village head) and fill out the forms with his help. He hands her a copy of the previous year's survey and tells her to make sure that the number of out-of-school children comes down. The next few days are spent filling out the forms and compiling the data; Malati has little time to teach.

As the months go by, Malati settles into a routine. She takes the 10 am bus and reaches the school around 11 am, teaches a group of students for about 40 to 50 minutes, supervises the midday meal, does some paperwork, and returns by the 3 pm bus. She is home by 4 or 5 pm. Every month she goes with her co-teacher to the cluster or block centre for the monthly meeting and to collect her salary slip.

At the end of the academic year, she purchases question papers from the market and organizes an annual examination for the children. She writes the questions on the blackboard one by one (as the blackboard is not very good and the children cannot read if anything is written on the periphery of the board). Sometimes, she also writes the answers. Some children are able to copy from the board, a few are able to answer the questions but most of them just hang around not knowing what to do. The children who go to tuitions do better, but many others do not even understand the language used for teaching. In one sense it hardly matters: the no detention policy ensures that they will all pass. But what are they learning? Malati dwells on this thought for a bit, and then the moment passes. The days go by swiftly and before she realizes it the summer vacation arrives.

Malati is asked to attend a six-day training at the district headquarters. She seeks out other women teachers to find out where they stay during the training. She does not get any specific answer; many of them do not attend the mandatory training, as there is no place to stay. She arranges with a distant relative to stay in his home. On the first day of training she reaches the venue at 9 am and waits for a long time before a few more teachers arrive. A resource teacher teaches a class on 'hard spots' in mathematics teaching for Class III. Later, another resource person talks about joyful learning and the Play Way method – a lecture that she finds interesting, but which provides no solutions for her situation. How do you use the Play Way method with one teacher and 60 children in one room? At no point do the resource persons or teachers ask her or any other teacher about the problems they face in school. She wants to ask many questions, but is afraid to speak in front of her seniors. Malati attends all six days and collects some photocopied material. The attendance in the training varies from 60 on the first and last day (when registers are signed and travel forms are filled) to 20 on the other days.

During the training, Malati meets a few older relatives who have been teachers for a long time. She is asked to become a member of a teacher's union and is informed that once she is a member (or for that matter once anyone becomes a regular school teacher) no one can touch her. The job she has is secure and for life. Her relatives tell her not to worry about the rural posting; it can be sorted out in due course, but she will have to bear with it for the first few years. She experiences a sense of pride at becoming a permanent government employee. Her work has turned out to be not so difficult, since she only needs to do what her seniors tell her to do. Her salary is fairly good and she is ready to wait.

It did not take Malati very long to understand what was expected of her. She must come to school for at least a few hours for a few days every week. She must read the textbook aloud to the students. As far as possible, she must keep them quiet and obedient. By the end of the year she must cover the syllabus, because if an inspector comes this is what he will check. She must supervise the midday meal, fill out innumerable forms, and do whatever other duties she is assigned.

Malati received no induction training. There was no one to guide or help her understand how to manage a multi-grade classroom or how to teach a chapter in a textbook that she herself didn't understand. In the monthly meetings and the annual training, nobody talked or asked about the real problems that teachers face. Nobody knew or seemed to care whether the children were learning or not.

Malati came to teaching with an open mind, eager to learn and grow. Her experience is not unique. Discussions with teachers across the country reveal a similar picture, of course with notable exceptions. Young and not-so-young people become 'regular government school teachers' and start off like most young people do – with energy and hope. But as the years roll by they change.

What does Malati's journey tell us? Conceptually, we can delineate three domains of questions:

- How do the environment and administration of the school shape what teachers do?

- How do formal institutional systems for supervision and support affect teaching practices?
- How do these ground realities affect teachers' perceptions of their professional roles and objectives?

I now try to trace these domains through the experiences of the teachers, teacher educators, and administrators with whom we interacted, as well as the many studies and reports that I have reviewed.

How the school environment shapes teachers and vice-versa

As discussed in the preceding chapters of this book, school infrastructure has received a great deal of attention from policymakers in the last 25 years. On the ground, the uneven implementation of this policy is clearly perceptible. At one end of the spectrum are Kerala and Tamil Nadu, where the school buildings look fairly neat and clean and there are adequate classrooms. The schools are secure behind boundary walls and have proper drinking water arrangements. Maharashtra and Rajasthan come next: the availability of buildings and basic amenities is fairly good, except in remote or interior areas and in tribal villages. West Bengal, on the other end, is a picture of decay and neglect: the buildings are old and often crumbling, they look as if they have not been painted in decades, and the overall environment is depressing. Similarly, if we focus on the number of single- and two-teacher schools or on the academic infrastructure (the availability of library and reference books, charts, maps, basic science equipment, and so on), there is again a huge difference between the states. Overall, at the ground level our primary/elementary schools are nothing like what our policymakers conceptualize as an ideal school environment.

Although the environment of a school affects the teachers, whether we look at infrastructure, the availability of TLM, or the number of teachers appointed, it is equally true that teachers influence the overall learning environment: **it is a two-way process**. What then determines whether this two-way process produces a positive, creative learning environment or a negative spiral of indifference and decay?

Case studies of exceptional government schools suggest that a highly committed head teacher and other teachers can make a big difference. In 2004, SANDHAN did in-depth case studies of ten schools in Rajasthan as part of a larger quantitative study covering over 660 schools (SANDHAN 2004). The overall understanding that emerged through school observations, community dialogues, and teacher interviews is revealing:

- If teachers are hardworking and sensitive towards the students, it ensures positive community support for the school.
- Mutual understanding and coordination among the teachers within a school has a positive impact on the overall educational environment of the school.
- If teachers are creative and committed, they always find a way out, even in difficult circumstances.
- If teachers are sincere, the budget allocated by the District Primary Education Programme (DPEP) is optimally utilized.
- The teacher-student ratio affects the educational environment and students' learning achievements. This one thing is beyond the control of the teachers.

We are confronted with this reality: regardless of the 'system', there are always teachers and head teachers who are able to elicit parental and community support to improve the overall learning environment of a school. Equally, notwithstanding high budget allocations and the best programmes, a group of indifferent administrators, head teachers, and teachers will become a stumbling block for improving the learning environment. The agency of the head teacher and teachers at the school level, and that of the administrators who directly supervise the school, emerge as the critical factors.

Why are most teachers unable or unwilling to provide this leadership? The crux of the problem lies with what the 'system' expects from the teacher. Let us revisit Malati's story. When she joined the school she was not given an induction workshop, and no one told her what she should do in class or what was expected of her. The few times she interacted with her superiors, all they wanted her to do was fill out forms and take the textbooks; later, she was asked to do a household survey. When she wanted to know what she should do or how she should tackle a particular educational situation or problem, she had nobody to turn to. In this situation, Malati had no way of judging whether she was teaching well or whether she could help students learn more by doing things differently. Nor did she have any reason to even bother to think about it.

Discussions with teachers in the five states we visited confirm Malati's experience. Academic support institutions in Kerala, Maharashtra, and Tamil Nadu were engaged in some degree of academic monitoring and support – although these were, for the most part, designed to ensure teachers' compliance with predefined tasks and procedures, rather than to encourage them to analyse their teaching practices in light of their students' learning outcomes. In West Bengal and Rajasthan, on the other hand, the signals from the top were often contradictory and limited to administrative matters and data collection. Let us take a quick look at these last two states.

In **West Bengal**, primary schools are 'managed' by a number of different bodies. Teachers were appointed at the district level, but come under the administrative control of the West Bengal Board of Primary Education (WBBPE); their training and supervision comes under the Directorate of School Education (DSE) and the District Inspector (DI) of Schools; and the funds are controlled by the DI of Schools. Under the Sarva Shiksha Abhiyan (SSA), the SSA directorate and the district office handled the flow of funds, monitoring (data gathering), and training. Further, the District Primary School Council (DPSC) and the *panchayats* do not speak with one voice: the DPSC is partly nominated, and the elected *panchayat* has no administrative role in the schools.

These various bodies were created at different points of time for specific reasons. According to a DPSC Chairperson we interviewed and the SSA coordinators at the state and district levels, the existence of multiple chains of command implies that the teacher is at the receiving end of instructions from all, but effectively accountable to none in particular. Each 'school' also comes under different bodies, and therefore there is no coherent/coordinated body that takes care of all aspects of schooling. No one can be held responsible for this sorry state of affairs and everyone passes the buck. Organizations working in the state, such as Pratichi Trust, say that there is no effective support or supervision and the schools are in a state of decay.

> Pratichi Trust's strategy is clear. We believe that teachers are a huge resource and they can contribute a lot and only they can break the cycle of poor quality and high drop out rates. We need to demand involvement of teachers in policy-making and in formulation of curriculum and textbooks. Simultaneously we need to list what the teachers can do and what they cannot do. This could be followed by a series of workshops on teaching methods, English and science teaching. We can see visible change in Bribhum and some work has also started in Malda, Jalpaiguri, and Gopiballabganj. (Interview with Kumar Rana of the Pratichi Trust, July 2008)

In other words, there is a fundamental need to improve teachers' capacity to be effective in the classroom as well as their ability to shape policies that affect schools.

Discussions across different directorates/boards and councils reveal that this lack of coordination and coherence means that important issues are lost sight of. The most pressing problem, highlighted in almost all discussions, is the high pupil-teacher ratios. At a meeting in the Birbhum district DPSC, for example, council members and teachers said that the situation was extremely worrying.

Class sizes (especially for Classes V to VIII) were huge; there were almost 80 children per teacher. This is the main reason for the high dropout rates. While some children resort to private tuitions, poor families cannot afford to do so and their children barely learn anything. Teachers get frustrated; they cannot do anything about it, and just cannot handle 80 children. Worse, the children are cramped in small, dark, and dingy rooms. Given the no-detention policy until Class IV, children enter Class V without any knowledge because they are promoted from one class to the next without ever ascertaining how much they have learned. Teachers know that remedial classes alone are not enough, and that these students have to start from the basic alphabets and numbers. Given the fact that teachers are expected to complete the prescribed curriculum, however, they do not have the time to go back to the basics.

In **Rajasthan**, teachers flagged the politicization of education and its impact on the school and larger educational environment as a key issue. Teachers are vulnerable to regular mass transfers and can be assigned any duty by the district administration. Teachers with clout or political linkages can even be posted in the district collector's office for years on end! While civil work and the construction of buildings have been quite effective in the state, rural and remote areas continue to have an overwhelming proportion of single-teacher and two-teacher schools.

There are two principal problems with the issue of teacher transfers. The first problem is that of institutional corruption: teachers can manipulate the system when it permits 'informal' (or corrupt) practices to exist. The second and equally important problem is that there are no *educational* criteria by which teachers can apply for a desired transfer. In other words, if effective teaching practices (however measured) were linked to teachers' ability to access desired postings, it could have the effect of both catalysing better teaching and diminishing the importance of 'informal channels' within the system. This, of course, requires that educational outcomes be placed at the centre of the decision-making criteria that affect teachers.

The situation is not as dismal everywhere. In **Tamil Nadu**, teachers' non-teaching duties are restricted to election and census duties. The District Magistrate (DM) does not have overarching power over the teachers, and in 2006 non-teaching duties accounted for only 2.66% of working days spent on educational work. The government started transfer counselling in 1988, and for the past 20 years transfers have occurred only when there was a vacancy. The chain of command is clear and teachers are not expected to report to multiple authorities.

This is echoed in **Kerala**, where numerous school-level support structures such as Parent Teacher Associations, School Resource Groups, School Support Groups, and Local Resource Groups work actively with teachers, the District Institutes of Education and Training (DIETs), and BRCs to promote public debate on educational processes and outcomes at the community level. The long history of public involvement in education in the state has been catalysed by the Kerala Sastra Sahitya Parishad (KSSP) through its work on science popularization since the 1960s. In the last 10–12 years, the 'People's Planning Process' has brought *panchayats* to the forefront. Significant fiscal decentralization has taken place, and the three-tier Panchayati Raj Institutions (PRIs) have played a key role in education as part of the local-level planning and development processes. In Thrissur, for instance, the Gram and District *Panchayats* have come together to organize training for teachers in different areas, and the Panchayati Raj bodies pay resource persons and bear the travel costs of the teachers.

Administrators and teachers in Kerala felt that there was an increasing recognition of the need to focus on the environment in which the teacher works, especially given that the nature of this environment clearly transmits what the education system 'expects' of him or her. A similar change has been reported in **Karnataka** and **Andhra Pradesh**, where growing social pressure for good quality education has initiated serious thinking about the factors that enable teachers to work to their full potential. However, it is too early to tell whether these changes will be allowed to flourish and deepen to the point where their effects are visible in improved learning outcomes among students.

How Formal Monitoring and Supervision Systems Affect Teachers' Practice

It is said that *we learn what we are tested on*, and *we do what is monitored*. This is the case with teachers too. If we look back over the last 30 years – since the National Policy on Education in 1986 – the Government of India has produced an impressive array of data. DISE, for example, was developed during DPEP to enable the government to track educational progress. What does DISE capture, and how is this information collected?

According to the National University of Educational Planning and Administration (NUEPA), the institution that houses DISE (now known as UDISE), it was felt that a sound information system was essential for the successful monitoring and implementation of DPEP. It was stressed that DPEP's focus on decentralized planning required up-to-date and reliable school-level information. In 1994, the Ministry of Human Resource Development (MHRD) decided to design and develop a school-based computerized information system as a part of the DPEP national endeavour, the main responsibility for which was entrusted to the National Institute of Educational Planning and Administration (NIEPA), New Delhi. Information on the following key variables over a period of time is now available at the school, cluster, block, district, state, and national levels:

- Key data on elementary education in terms of the number of schools, enrolment, and teachers, classified by school category and school management (also in respect to a few variables concerning rural/urban areas).
- Grade-wise and level-wise enrolment in each state.
- Examination results for the previous academic session for the terminal classes at the primary (Class IV or V) and upper primary (Class VI or VII) levels.
- Classrooms categorized into good condition, or requiring minor or major repairs, by school category.
- Number of schools by category and type of building.
- Sex-wise enrolment of children with disabilities at the primary and upper primary levels.
- Gender and caste distribution of regular and para-teachers, and the proportion of teachers undergoing in-service teacher training during the previous year.
- Distribution of regular and contract teachers by educational and professional qualifications and by school category.
- Enrolment by medium of instruction and by school category.
- Sex-wise number of students benefited by various incentive schemes at the primary and upper primary levels.
- Performance indicators in terms of school category: the ratio of primary to upper primary schools/sections; enrolment distribution by total, Scheduled Castes and Scheduled Tribes, and females; gender-parity index; number of classrooms; single-teacher schools; schools with attached pre-primary classes; percentage of under-age and over-age children in primary and upper primary classes; apparent survival rate within the school system (up to Class V), drop out rate, retention rate, and transition rate from primary to upper primary level.

- Quality indicators according to school category: teacher-pupil ratio; schools without female teachers; availability of blackboards and proper buildings; percentage of schools that received and utilized the school development and TLM grant; student-classroom ratio; availability of drinking water, common toilets, and girl's toilets in the school; etc.

By any reckoning, this list is quite impressive. During August and September every year, schools are expected to send this information in a specified format to their cluster. This information is collated and sent to the block, and then to the district, the state, and finally to NUEPA.

Almost everyone engaged in education has found this information valuable. Yet, during DPEP there was a lot of debate about the reliability of the data; comparisons between the results of 'system statistics' and those of sample surveys like the National Sample Survey Organisation (NSSO) and the National Family Health Survey (NFHS) led to growing cynicism about the validity of this information. As early as 1996, Madhya Pradesh conducted a state-wide house-to-house survey (known as the Lok Sampark Abhiyan) covering 53,460 villages, 6.1 million households, and 10 million children, which revealed that the situation of primary education in the state was nowhere near as positive as commonly believed. The problem of non-enrolment was found to be quite severe, especially in tribal hamlets with highly dispersed populations and for children from socially disadvantaged communities in non-tribal areas. The state government decided to act on this information. More importantly, the survey data was alleged to have 'called the bluff' of the statistics, and teachers were blamed for fudging the data.

Very soon other states followed Madhya Pradesh. For example, Andhra Pradesh carried out a similar survey to ascertain the number of out-of-school children and of child workers. Gradually, by 2000, the household survey gained official recognition as a tool for educational planning; when SSA was launched in 2001–02, household surveys were built into the programme as precursors to the 'bottom-up' planning process. Ironically, teachers were made responsible for conducting these surveys. Remember Malati? She was expected to go from house to house to gather data, but was advised by her superior to meet the village *pradhan* (village head) and fill out the forms, making sure the number of out-of-school children in the official records of the government went down every year (in order to meet the government target of reducing the number of out-of-school children).

Until the late 1960s, schools were visited by a School Inspector (SI) who was supposed to interact with teachers and students, observe classroom sessions in progress, look at the notebooks to ascertain how much of the syllabus had been covered, maybe teach one lesson, and so on. The older teachers we interviewed said that the SI's visit was an important event that impelled teachers to make an effort to cover the syllabus. But things changed in the 1970s, and today there is no inspection: cluster and block officials visit the schools just to collect data and look at the registers. For example, a study done by the Indian Institute of Management (IIM), Calcutta found that the SIs in **West Bengal** were overloaded and that it was impossible for them to visit all of the schools even once a year. The report noted that 'management and control at the ground level was almost non-existent and nothing was done even after the Ashok Mitra Committee recommended immediate measures to improve management of schools and supervision' (Chattopadhyay and Reddy 2001). Teachers in **Rajasthan** said that schools located in rural areas, especially those that were not on the main trunk routes, were rarely visited. With the coming of the mobile phone, block level officials now call to get the information, or give out the forms during monthly meetings. There are many schools that have not been visited for over a year; even when the BRC or CRC personnel visit, all they do is look at the registers – especially those related to midday meals.

This observation corroborates the findings of an earlier study that teachers in sample schools spent a lot of time filling out forms, and that teachers in single-teacher schools devoted very little time to teaching. When the cluster resource persons who are meant to provide academic support to teachers visited the schools, they ended up merely collecting information:

> The supervision of schools indicated the same dynamic. On the one hand, supervisory officers told interviewers that they were engaged

> in so many tasks that they could not find time to supervise. On the other hand, discussions with teachers showed that when they did supervise, they rarely went beyond checking records. A teacher's comment on the last visit of the supervisor was typical: '*He chatted with the teachers and checked the records regarding midday meals. He did not ask for any information about teaching or children's learning levels.*' [...] Concerns about TLM centred on the misuse of money, and not on obtaining and using relevant materials in the classroom. In Andhra Pradesh, a 'handbook' that listed various important questions and answers had been prepared at the district level. This was obviously a signal to the paper setters as well as teachers to focus on certain questions to ensure that too many children did not fail. (Sharma and Ramachandran 2009: 180, original emphasis)

When formal supervision and monitoring systems do not engage with teaching and learning processes, the message is that these do not matter.

There are, of course, islands of hope. There are instances where stricter monitoring along with effective on-site academic support has made a difference. A review (Jandhyala and Ramachandran 2007) of promising interventions from different parts of the country revealed that most of the successful interventions for improving the quality of education revolved around some generic principles, including:

- *Coming to terms with the reality on the ground through a realistic assessment of the situation with respect to learning* by both teachers and their immediate superiors. This implies that the teachers and administrators acknowledge the problem and do not brush it aside. Acknowledging the problem in all of its complexity and with all of its imperfections represents a cognitive break with the past.
- *Creating open communication channels between education administrators/implementers and local officials, teachers, parents, and children.* This is more difficult in a hierarchical system, yet there have been fledgling initiatives that have demonstrated some promise. Assessment has always been a contentious issue

feared by both teachers and administrators, given its potential to 'expose' the real picture. At the same time, the assessment of learning outcomes is integral to monitoring quality.

- *Energizing the school system by supporting teachers to improve classroom practices and assessment processes.* The ability of a teacher to teach creatively and to make the classroom child-friendly does not only come from teaching-learning materials and textbooks. It is inextricably linked to how the teacher is treated in the system, the autonomy she or he enjoys, and his or her sense of accountability to the children and parents. An important criticism of initiatives like DPEP is that they were implemented in a self-limiting, modular fashion. For example, teacher training has not always been linked to textbook development. The various dimensions of quality are rarely woven together. Treating each 'input' as a self-contained unit and not seeing the entire education process as an organic whole is a big handicap. Working with teachers as cornerstones in the process of change demands going beyond formal training and material production – it starts with efforts to kindle teachers' self-motivation and enthusiasm by appealing to their creativity and desire for satisfaction in their work.
- *Building a nurturing system of assessments and developing monitoring systems that create a continuous link between the school and academic and administrative support structures.* The classic, well-documented example of a nurturing monitoring system was the Shiksha Karmi Project (SKP) in Rajasthan.[1] This project was designed in 1987.

[1] The Rajasthan Shiksha Karmi Project (SKP) was launched in 1987 with financial assistance from the Swedish International Development Agency (SIDA). This was the second externally aided basic education project in India. The genesis of SKP can be traced to the night school experiment by the Social Work and Research Centre (SWRC) in Tilonia (Rajasthan), under which local youth with some basic education (Class X for young men and lower for young women) were identified, trained, and provided with continuous educational support to teach children in Shiksha Karmi Day Schools, Prehar Pathshalas (schools with convenient timings), and Angan Pathshalas (courtyard schools). SKP's concept rests on the assumption that barefoot teachers belonging to the local community, once intensively trained and enjoying support from the local community, can overcome a lack of formal educational qualifications. The project evolved a structure of training, support, and

> The key to SKP's effectiveness lay in providing academic resource support to teachers and teacher educators in a manner that fostered creativity and respected the teachers' experience, skills, and knowledge. The resource persons were drawn from NGOs working in the education sector, especially as teacher trainers. Central to the success of the Shiksha Karmi Project and process was the training–review-monitoring system, which was adopted in Lok Jumbish as the Review-Planning-Monitoring (RPM) cycle. Training programmes were contextual and need-based. The 40-day induction training was residential and participatory in nature. The conventional didactic training method, still used in the formal school system, was rejected in favour of discussions, self-learning, and co-learning through exercises and a range of activities including songs, role-play, and practice teaching. This induction training was followed by two 30-day training programmes in the first two years, a 20-day training after two years, and a 10-day training programme every year. In addition, the Shiksha Karmis (SKs) also participated in a two-day monthly review, planning, and difficulty removal meeting.

SKP was probably the first significant attempt to involve research centres as co-participants (rather than external evaluators) in an ongoing and concurrent process to provide assistance with course correction and problem solving. What is noteworthy is that their monitoring function was not seen as an isolated activity, but instead closely linked to the selection of Shiksha Karmis, training, documentation, and field support. This enabled the monitoring and support agency to tailor the training programme to the needs of the Shiksha Karmis and also to follow evaluation findings with concrete inputs into training and other field supports.

---

monitoring by involving Shiksha Karmi Sahyogi (SKS) drawn from both the formal school system and NGOs. To address the gender imbalance in the appointment of Shiksha Karmis, the project also created Mahila Prashikshan Kendra (residential training schools). Marking a departure from the formal school system, this project initiated school mapping to involve the community in identifying children who were not in school. Partner NGOs provided ongoing training support and also participated in community mobilization.

Interestingly, the SKP model of ongoing and on-site teacher support was not adopted by the government in large-scale education programmes like DPEP and SSA. Why such an effective model was not scaled up remains one of the puzzles of educational planning in India. I have tried to explain possible reasons in a recent paper on the legacy of the Shiksha Karmi and Lok Jumbish Projects (Ramachandran 2016).

How Ground Realities Affect Teachers' Perceptions of Their Roles and Objectives

Discussions with teachers around the country reveal that many experience nostalgia for the 'good old days' when they enjoyed high social status and were respected and looked up to by the community. Today, teachers are acutely aware of the negative perceptions about them, as government functionaries who draw good salaries and, more importantly, as shirkers who take no responsibility for the learning of children (Synovate and Azim Premji Foundation 2005). Their reflections on this situation generate a series of disturbing conclusions.

Teachers generally hold an idealized vision of the importance of the teacher's role in shaping future generations. At the same time, teachers today rarely consider helping children learn to be their main objective. Rather, their objective is to follow orders.

> [The teacher] is like a sculptor who moulds the future of students, backbone of society; they transmit cultures and traditions to students, inculcate moral values, provide guidance, etc. and have a significant role in preparing students as complete human beings in society. (Synovate and APF 2005: 11)

Let us contrast this statement with the one discussed in the preceding section of this chapter: 'A motivated teacher comes to school every day, does what he is told and provides information the higher-ups want' (Ramachandran et al. 2005). Teachers have no control over what they should teach and how they should teach. These decisions are taken at higher levels. For this reason, one teacher said, 'Our role is to follow orders and do as we are told.'

> Critically looking at themselves, as normal government servants, those who suffer from political interference and non-teaching duties, multipurpose government workers, who blindly follow rules and responsibilities laid down and who do not receive any training and capacity building for improving their teaching skills. (Synovate and APF 2005: 11–12)

'Orders from above', as we have seen earlier, typically have little to do with what children are learning – or how teachers are teaching. Thus teachers feel no responsibility for learning outcomes among their students.

Reflecting on who is a good teacher, H. K. Dewan noted:

> During discussions teachers said that a teacher was seen as being 'good' as long as she was sincere, regular, punctual and seen to be teaching the class. How she taught and engaged the children or what material and activities she used to make learning a fun experience was immaterial! This 'good teacher' could get away with almost anything if she abstained from excessive violence (beating). (Dewan 2009: 198)

If teachers are just doing what they have been told to do, it becomes easy to blame poor educational outcomes on other factors – like the family background of their students. Despite their idealized view of teachers and teaching, government schoolteachers often express resentment at having the traditionally high status of teachers jeopardized by teaching poor and underprivileged children:

> Another reason why quality is poor is because we have no control over who we can admit – every riffraff is allowed to be enrolled. (Interview with a *Madhyamik* schoolteacher in Birbhum, West Bengal, July 2008)

> The biggest difference between teachers in government and private schools is that the private school gets the cream while government teachers have to deal with '*kachara*' – we have to teach children from *kachi basti* and slums and from the poorest classes. (Interview with a primary schoolteacher from Jaipur district, Rajasthan, February 2008)

In a separate study that came to similar conclusions, Caroline Dyer sums up the situation as follows:

> A majority of the teachers interviewed showed no critical awareness of the relationship between their pedagogical practices, conditions in schools, and teaching problems. Some teachers understood teaching as imparting the content of the textbook, regardless of whether children learn. In some schools, children who had attended classes for four years could still neither read nor write. The teachers did not accept responsibility for this situation and blamed the local environment: they did not feel that their own pedagogical practices required attention. (Dyer 1996: 16)

Thus, despite years of government policy statements promoting education for all, school systems transmit exactly the opposite message. The words of the teachers quoted above offer damning evidence of the failure of school systems to put into practice what educational policy so ardently preaches. Rhetoric apart, teachers do what their day-to-day professional environment demands. The focus on rules rather than outcomes, the treatment of teachers as the bottom rung of government servants, and the absence of mechanisms for teachers to even discuss the reality of their working conditions – let alone develop solutions to local problems – all contribute to the attitude among many teachers that learning outcomes are not their responsibility, and that if poor or disadvantaged students do not learn, it is because of their own or their families' shortcomings.

Current trends suggest that many states have started holding teachers responsible for learning outcomes. Yet, as we have seen, they are given neither the tools nor the support needed to do so successfully. This puts enormous pressure on teachers; at the same time, the way the system treats them ends up undermining their position in the school, their self-esteem, and their confidence.

Summing Up

As we saw in Malati's example, new teachers join the profession with hope. Their daily experiences either energize them to reach out to every child, or they gradually become disillusioned and indifferent. We have placed the blame at the doorstep of teachers for too long without trying to understand what makes them into what they are today. Yes, there are thousands of teachers who absent themselves from school, do not teach, and discriminate against children from certain social and economic groups. There are also hundreds of instances where a supportive environment has nurtured individuals who have struggled against all odds to be true to their vocation as teachers. As Professor Punam Batra (2005) says:

> Access to schooling, an adequate teaching-learning environment, an appropriate school curriculum and an empowered and inclusive teaching community are four crucial prerequisites of a school system that seeks to enable social transformation. While educational reform since the 1980s was strongly focused on the first two elements, the late 1990s brought the role of the curriculum into national focus. The critical link that binds these four critical elements together – the agency of the teacher continues to be cast aside, by political ideologies of most hues, contemporary curriculum reform efforts and the professional practices of much of the teacher education community. (2005: 4348)

The teacher conundrum has been with us for a long time. Unless the government takes a holistic perspective and makes simultaneous efforts to fix all dimensions of the school system, teachers will continue to be caught in the unenviable situation discussed in this chapter.

References

ASER Centre and PRATHAM. 2005. *Annual Status of Education Report (Rural) 2005*. Mumbai: Pratham Resource Centre. http://www.asercentre.org/p/85.html.

———. 2007. *Annual Status of Education Report (Rural) 2006*. Mumbai: Pratham Resource Centre. http://www.asercentre.org/p/84.html.

———. 2007b. *Annual Status of Education Report (Rural) 2007*. http://www.asercentre.org/p/83.html.

———. 2011. *Annual Status of Education Report (Rural) 2010*. Mumbai: Pratham Resource Centre. http://www.asercentre.org/Annual/Status/of/Education/Report/The/Annual/p/80.html.

———. 2012. *Annual Status of Education Report (Rural) 2011*. http://www.asercentre.org/p/79.html.

———. 2015. *Annual Status of Education Report (Rural) 2014*. New Delhi: ASER Centre. http://www.asercentre.org/Keywords/p/234.html.

———. 2016. *Annual Status of Education Report (Rural) 2015*. New Delhi: ASER Centre.

———. 2017. *Annual Status of Education Report (Rural) 2016*. New Delhi: ASER Centre. http://www.asercentre.org//p/289.html.

Azim Premji Foundation. 2004. 'The Social Context of Elementary Education in Rural India.' http://righttoeducation.in/sites/default/files/TheSocialContextofElementaryEductaioninRuralIndia.pdf.

Azim Premji Foundation Research Group. 2017. 'Teacher Absenteeism Study.' Azim Premji University Field Studies in Education series, March. http://azimpremjiuniversity.edu.in/SitePages/pdf/Field-Studies-in-Education-Teacher-Absenteeism-Study.pdf.

Batra, Poonam. 2005. 'Voice and Agency of Teachers.' *Economic and Political Weekly* 40(40): 4347–56. http://www.epw.in/journal/2005/40/special-articles/voice-and-agency-teachers.html.

Bhog, Dipta, Saswata Ghosh and Disha Mullick. 2011. 'Secondary education in the context of Rashtriya Madhyamik Shiksha Abhiyan: A desk review.' Report prepared for Nirantar Centre for Gender and Education, New Delhi. http://www.nirantar.net/uploads/files/RMSA%20June%202011.pdf.

Bourdieu, Pierre and Jean-Claude Passeron. 1977. *Reproduction in Education, Society and Culture*. London: Sage Publications.

Chandramouli, V. 2013. 'Release of Primary Census Abstract – Scheduled Castes and Scheduled Tribes.' Handout from presentation, 28 October.

Chattopadhyay, R., and V.N. Reddy. 2001. 'Role of Panchayat in Primary Education in West Bengal.' Unpublished draft, IIM-Calcutta, Kolkata.

Chaudhuri, Hom and Nikhil Mathur. 2014. 'Report of the Working Conditions of Teachers in Mizoram.' Unpublished report prepared for IASE and NUEPA.

Chitnis, Suma. 2013 [1972]. 'Education for Equality: The Case of Scheduled Castes in Higher Education.' In *Higher Education in India: In search of Equality, Quality and Quantity*. Jandhyala B. G. Tilak, (ed). New Delhi: Orient Blackswan.

Cohen, Joel E. 2010. 'Goals of Universal Basic and Secondary Education.' In *International Perspectives on the Goals of Universal Basic and Secondary Education*. Joel E. Cohen and Martin B. Malin, (eds.). New York: Routledge.

Colclough, Christopher, Pauline Rose and Mercy Tembon. 2000. 'Gender inequalities in primary schooling: The roles of poverty and adverse cultural practice.' *International Journal of Educational Development*, 20(1): 5–27. doi:10.1016/S0738-0593(99)00046-2.

Das, Vidya. 2004. 'More unequal than others: Evidence from Surguja District, Chhattisgarh.' In *Hierarchies of Access*, Ramachandran (ed.).

De, Anuradha, Reetika Khera, Meera Samson and A. K. Shiva Kumar. 2011. *PROBE Revisited: A Report on Elementary Education in India*. New Delhi: Oxford University Press.

De Haan, Arjan. 2004. 'Disparities within India's Poorest regions: Why do the same institutions work differently in different places?' Equity & Development: World Development Report 2006 Background Papers, World Bank. http://siteresources.worldbank.org/INTWDRS/Resources/477365-1327693659766/8397901-1327773323392/Disparities_within_Indias_Poorest_Regions_Why_Do-the_Same_Institutions_Work_Differently_in_Different_Places.pdf.

Delors, Jacques, et al. 1998. *Learning: The Treasure Within: Report to UNESCO of the International Commission on Education for the Twenty-first Century.* Paris: UNESCO Publishing.

Deshpande, Satish. 2013. Introduction to *Beyond Inclusion: The Practice of Equal Access in Indian Higher Education*. Satish Deshpande and Usha Zacharias (eds.). New Delhi: Routledge.

Dewan, Hriday Kant. 2009. 'Teaching and Learning – The Practices.' In *The Elementary Education System in India: Exploring Institutional Structures, Processes and Dynamics*. Rashmi Sharma and Vimala Ramachandran (eds.). New Delhi: Routledge.

Dreze, John and Amartya Sen. 2002. *India: Development and Participation*. Oxford: Oxford University Press.

Dyer, Caroline. 1996. 'The Improvement of Primary School Quality in India: Successes and Failures of Operation Blackboard.' Edinburgh Papers in South Asian Studies No. 4. University of Edinburgh, Edinburgh. http://www.san.ed.ac.uk/__data/assets/pdf_file/0009/38169/WP04_DYER.pdf.

EdCil (India) Ltd. and SRI (Social and Rural Research Institute). 2010. 'All India Survey of Out-of-School Children of Age 6–13 Years and Age 5.' http://www.educationforallinindia.com/aii_india_survey_of_out_of-school_cghildren_2010.pdf

Education Reforms Commission, Mizoram. 2010. 'Towards an Enlightened and Inclusive Mizo Society: Report of the Education Reforms Commission, Mizoram, 2009–10.' Report prepared for the Government of Mizoram. https://schooleducation.mizoram.gov.in/uploads/attachments/c059bc7e27e0d2bc7a99bce2084a79dc/erc-edited123.pdf.

Educational Initiatives, Pvt. Ltd. 2006. 'Quality Education Study.' Unpublished report commissioned by Wipro.

———. [2011]. 'Quality Education Study.' Commissioned by Wipro Applying Thought in Schools. http://www.ei-india.com/wp-content/uploads/2012/07/Executive_Summary-Low_Resolution-QES1.pdf.

G. B. Pant Social Science Institute. 2011. 'Children's Education in Conflict and Crisis Situation of Salwa Jhudum in Chhattisgarh.' Unpublished report prepared for NEGFIRE, New Delhi.

Government of Arunachal Pradesh. 2006. 'Arunachal Pradesh Human Development Report 2005.' Itanagar: Department of Planning, Government of Arunachal Pradesh. http://planningcommission.gov.in/plans/stateplan/sdr_pdf/shdr_ap05.pdf.

Government of Meghalaya. 2008. 'Meghalaya Human Development Report.' Shillong: Government of Meghalaya. http://megplanning.gov.in/MHDR/Human_De.pdf.

Goyal, Sangeeta and Priyanka Pandey. 2009. 'How do Government and Private Schools Differ? Findings from Two Large Indian States.' Report No. 30, South Asia Human Development Sector, The World Bank. Available at https://openknowledge.worldbank.org/handle/10986/17962.

IIPS (International Institute for Population Sciences). 1993. 'First National Family Health Survey (NFHS-1)' [Dataset]. Commissioned by the Ministry of Health and Family Welfare, Government of India. Available from http://rchiips.org/nfhs/index.shtml.

———. 1999. 'Second National Family Health Survey (NFHS-2)' [Dataset]. Commissioned by the Ministry of Health and Family Welfare, Government of India. Available from http://rchiips.org/nfhs/index.shtml.

———. 2006. 'Third National Family Health Survey (NFHS-3)' [Dataset]. Commissioned by the Ministry of Health and Family Welfare, Government of India. Available from http://rchiips.org/nfhs/index.shtml.

Jandhyala, Kameshwari. 2004. 'So close yet so far: Primary schooling in Warangal District, Andhra Pradesh.' In *Hierarchies of Access*, Ramachandran (ed.).

———, and Vimala Ramachandran. 2007. 'Energising Government Schools for Meaningful Access.' *Economic and Political Weekly,* 42(48). http://www.epw.in/journal/2007/48/special-articles/energising-government-schools-meaningful-access.html.

———. 2015. 'Why Women Teachers Matter in Secondary Education.' *Economic and Political Weekly*, 50(32): 48–54. http://www.epw.in/journal/2015/32/special-articles/why-women-teachers-matter-secondary-education.html.

Jejeebhoy, Shireen J. and Zeba A. Sathar. 2001. 'Women's Autonomy in India and Pakistan: The influence of Religion and Region.' *Population and Development Review,* 27(4): 687–712. doi:10.1111/j.1728-4457.2001.00687.x.

Jenkins, Robert and Eimar Barr. 2006. 'Social Exclusion of Scheduled Caste Children from Primary Education in India.' New Delhi: UNICEF India. https://www.unicef.org/files/Social_Exclusion_of_Scheduled_Caste_Children_from_Primary_Education_in_India.pdf.

Jha, Jyotsna and Dhir Jhingran. 2002. *Elementary Education for the Poorest and Other Deprived Groups.* New Delhi: Centre for Policy Research.

Kabeer, Naila and Ramya Subrahmanian. 1996. 'Institutions, Relations and Outcomes: Framework and Tools for Gender-Aware Planning.' Discussion paper no. 357. Institute of Development Studies, Sussex, UK. Available at http://www.ids.ac.uk/publication/institutions-relations-and-outcomes-framework-and-tools-for-gender-aware-planning.

Kremer, Michael, Nazmul Chaudhury, F. Halsey Rogers, Karthik Muralidharan and Jeffrey Hammer. 2005. 'Teacher Absence in India: A Snapshot.' *Journal of the European Economic Association* 3(2–3): 658–667. doi:10.1162/jeea.2005.3.2–3.658.

Lall, M. C., and S. Srinivasa Rao. 2011. 'Revisiting the equity debate in India and the UK: Caste, Race and Class Intersections in Education.'

In *Education and Social Justice in the Era of Globalisation: Perspectives from India and the UK*, M. C. Lall and Geetha B. Nambissan (eds.) New Delhi: Routledge.

Lewis, Maureen and Marlaine Lockheed. 2007. *Inexcusable Absence: Why 60 Million Girls Still Aren't in School and What to Do About It.* Washington D.C.: Centre for Global Development. Available at https://www.cgdev.org/publication/9781933286143-inexcusable-absence-why-60-million-girls-still-arent-school-and-what-do-about-it.

Madhulika, Sahoo. 2013. 'Impact of Displacement on Tribal Children: Cases of Child Labour and Abuse in Odisha.' *Afro Asian Journal of Anthropology and Social Policy*, 4(2): 148–153. doi:10.5958/ j.2229-4414.4.2.024

Mahajan, Vandana. 2004. 'The hidden picture: A case study from Hisar District, Haryana.' In *Hierarchies of Access*, Ramachandran (ed.).

Majumdar, Manabi and Jos Mooij. 2011. *Education and Inequality in India: a classroom view*. London: Routledge.

Meetei, Arambam Noni. 2011. 'Armed Conflict and its impact on children and education: A study with reference to Manipur.' Unpublished report prepared for NEGFIRE.

New Concept Information Systems Pvt. Ltd. 2003. 'Integrated Child Development – A Conceptual Framework.' New Delhi.

Ministry of Tribal Affairs, GOI (Government of India). 2014. 'Report of the High Level Committee on Socio-Economic, Health and Educational Status of Tribal Communities in India.' New Delhi: Ministry of Tribal Affairs. https://ruralindiaonline.org/resources/report-of-the-high-level-committee-on-socio-economic-health-and-educational-status-of-the-tribals-of-india/.

MHRD (Ministry of Human Resource Development), GOI (Government of India). 1994. 'Guidelines for the District Primary Education Programme (DPEP).' New Delhi.

———. N.d. [2001]. 'Selected Educational Statistics 2000–2001.' http://www.educationforallinindia.com/selected-educational-statistics-2000-2001.pdf.

———. 2005. 'Abstract: Selected Educational Statistics, 2005–2006.' http://www.educationforallinindia.com/SES2005-06.pdf.

———. N.d. [2008]. 'Tables of Statistics of School Education, 2007–2008.' http://www.educationforallinindia.com/selected-education-statistics-2007-08-MHRD.pdf.

———. 2008b. 'National Evaluation-II: Kasturba Gandhi Balika Vidhyalaya (KGBV).' Available at http://ssa.nic.in/research-studies/KGBV%20-%20NPEGEL%20Evaluation%20Report%20Feb%2008.zip/view.

———. 2010. 'Report of the Committee on the Implementation of the Right of Children to Free and Compulsory Education Act, 2009 and the resultant revamp of Sarva Shiksha Abhiyan.' New Delhi, Department of School Education and Literacy.

———. 2012. 'Statistics of School Education, 2010–2011 (As on 30th September 2010).' New Delhi: MHRD, GOI. http://www.educationforallinindia.com/Statistics_of_School_Education_2010-11,%20MHRD.pdf.

———. 2013. 'National Report: Second National Evaluation of KGBV Programme of GOI, November-December 2013.' Report prepared for the MHRD, GOI. http://ssa.nic.in/girls-education/kasturba-gandhi-balika-vidyalaya/Final%20revised%20National%20Report%20of%20KGBV%20Evaluation%2012%20Dec%202013%20.pdf/view.

———. 2014. 'Educational Statistics at a Glance.' New Delhi: MHRD, GOI. http://www.educationforallinindia.com/educational_statistics_at_glance-MHRD_2014.pdf.

———. 2014b. 'Statistics of School Education, 2009–10 (As on 30th September 2009).' New Delhi: MHRD, GOI. http://www.educationforallinindia.com/Statistics_of_School_Education_2009-10,%20MHRD.pdf.

———. 2014c. 'Statistics of School Education, 2011–12 (As on 30th September 2011).' New Delhi: MHRD, GOI. http://www.educationforallinindia.com/Statistics_of_School_Education_2011-12,%20MHRD.pdf.

———. 2016. 'National Policy on Education 2016: Report of the Committee for Evolution of the New Education Policy.' http://www.nuepa.org/new/download/NEP2016/ReportNEP.pdf

———. 2016b. 'All India Survey on Higher Education (2014–15).' New Delhi: MHRD, GOI. http://aishe.nic.in/aishe/viewDocument.action?documentId=206.

Mooij, Jos. 2014. 'Redressing Poverty and Enhancing Social Development: Trends in India's Welfare Regimes.' In *Persistence of Poverty in India*, Nandini Gooptu and Jonathan Parry (eds.). New Delhi: Social Science Press.

Muralidharan, Karthik. 2013. 'Priorities of Primary Education Policy in India's 12th Five-Year Plan.' NCAER-Brookings India Policy Forum. https://pdel.ucsd.edu/_files/paper_2013_karthik.pdf.

———, and Michael Kremer. 2006. 'Public and Private Schools in Rural India.' Revised version of a paper prepared for the 'Mobilizing the Private Sector for Public Education' Conference, Kennedy School of Government, Harvard University, 5–6 October 2005. http://dise.in/Downloads/Use%20of%20Dise%20Data/Karthi%20Muralidharan%20and%20Michael%20Kremer.pdf.

Naandi Foundation. N.d. 'HUNGaMA: Fighting Hunger and Malnutrition, The HUNGaMA Survey Report.' Hyderabad: Naandi. http://www.hungamaforchange.org/HungamaBKDec11LR.pdf.

Nambissan, Geetha B. 1994. 'Language and Schooling of Tribal Children: Issues Related to Medium of Instruction.' *Economic and Political Weekly*, 29(42). http://www.epw.in/journal/1994/42/special-articles/language-and-schooling-tribal-children-issues-related-medium.

———. 1996. 'Equity in Education – Schooling of Dalit Children in India.' *Economic and Political Weekly,* 31(16–17). http://www.epw.in/journal/1996/16-17/special-articles/equity-education-schooling-dalit-children-india.html.

———. 2000. 'Dealing with deprivation.' Paper presented at 'Redesigning Curricula' Seminar, September. http://www.india-seminar.com/2000/493/493%20geetha%20b.%20nambissan.htm.

———. 2001. 'Social Diversity and Regional Disparity in Schooling: A Study of Rural Rajasthan.' In *Elementary Education in Rural India: A Grassroots View*, A. Vaidyanathan and P. R. Gopinathan Nair, (eds.). New Delhi: Sage Publications.

———. 2009. 'Exclusion and Discrimination in Schools: Experiences of Dalit Children.' 'Children, Social Exclusion and Development' working paper series 1(1), Indian Institute of Dalit Studies and UNICEF. http://dalitstudies.org.in/wp/wps0101.pdf.

Naorem, Taramani and Vimala Ramachandran. 2013. 'What it means to be a Dalit or Tribal Child in Our Schools: A Synthesis of a Six-State Qualitative Study.' *Economic and Political Weekly,* 48(44): 43–52. http://www.epw.in/journal/2013/44/special-articles/what-it-means-be-dalit-or-tribal-child-our-schools.html

Nawani, Disha. 2013. 'Corporal Punishment in Schools.' *Economic and Political Weekly*, 48(24). http://www.epw.in/journal/2013/24/commentary/corporal-punishment-schools.html.

NCAER (National Council for Applied Economic Research). 1999. 'ICDS Field Survey' Unpublished Report. New Delhi.

NCERT (National Council of Educational Research and Training). 2005. 'National Curriculum Framework 2005.' http://www.ncert.nic.in/rightside/links/pdf/framework/english/nf2005.pdf.

———. 2006. 'National Focus Group on Gender Issues in Education.' Position Paper 3.2. http://www.ncert.nic.in/new_ncert/ncert/rightside/links/pdf/focus_group/gender_issues_in_education.pdf.

———. 2015. 'What Students of Class V Know and Can Do: A Summary of India's National Achievement Survey, Class V (Cycle 4), 2015.'

New Delhi: NCERT. http://www.ncert.nic.in/departments/nie/esd/pdf/NAS_Class_V_(Cycle%204)_Summary_Report_National.pdf.

———. 2016. 'Eighth All India School Education Survey (As on 30 September, 2009): A Concise Report.' New Delhi: NCERT. http://aises.nic.in/8th%20AISES%20Concise%20Report.pdf.

NSSO (National Sample Survey Office). 2010. 'Education in India: 2007–08: Participation and Expenditure: NSS 64th Round.' Report no. 532, prepared by the National Sample Survey Office, National Statistical Organisation, Ministry of Statistics and Programme Implementation, Government of India. http://www.mospi.gov.in/sites/default/files/publication_reports/532_final.pdf.

———. 2011. 'Employment and Unemployment Situation in India, 2009–10: NSS 66th Round.' Report no. 537, prepared by the National Sample Survey Office, National Statistical Organisation, Ministry of Statistics and Programme Implementation, Government of India. http://www.mospi.gov.in/sites/default/files/publication_reports/NSS_Report_No_537.pdf.

———. 2012. 'Nutritional Intake in India: NSS 66th Round (July 2009–June 2010).' Report no. 540 prepared by the National Sample Survey Office, Ministry of Statistics and Programme Implementation, Government of India. http://www.mospi.gov.in/sites/default/files/publication_reports/nss_rep_540.pdf.

———. 2013. 'Status of Education and Vocational Training in India: NSS 66th Round (July 2009–June 2010).' Report no. 551, prepared by the National Sample Survey Office, Ministry of Statistics and Programme Implementation, Government of India. http://www.mospi.gov.in/sites/default/files/publication_reports/nss_report_551.pdf.

———. 2016. 'Education in India: NSS 71st Round (January-June 2014).' Report no. 575, prepared by the National Sample Survey Office, Ministry of Statistics and Programme Implementation, Government of India. http://www.mospi.gov.in/sites/default/files/publication_reports/nss_rep_575.pdf.

NUEPA (National University of Educational Planning and Administration). 2007. *Elementary Education in India: Progress towards UEE: Flash Statistics, DISE 2005–06*. New Delhi: NUEPA and Department of School Education and Literacy, Ministry of Human Resource Development, Government of India. http://udise.in/Downloads/Publications/Publications%202005-06/Flash%20Statistics.pdf.

———. 2008. *Elementary Education in India: Progress towards UEE: Flash Statistics, DISE 2006–07*. New Delhi: NUEPA and Department of School Education and Literacy, Ministry of Human Resource Development,

Government of India. http://udise.in/Downloads/Publications/Publication%202006-07/Flash%20Statistics%202006-07.pdf.

———. 2009. *Elementary Education in India: Progress towards UEE: Flash Statistics, DISE 2007–08*. New Delhi: NUEPA and Department of School Education and Literacy, Ministry of Human Resource Development, Government of India. http://udise.in/Downloads/Publications/Publications%202007-08/Flash%20statistics2007-08.pdf.

———. 2010. *Elementary Education in India: Progress towards UEE: Flash Statistics, DISE 2008–09*. New Delhi: NUEPA and Department of School Education and Literacy, Ministry of Human Resource Development, Government of India. http://udise.in/Downloads/Publications/Publications%202008-09/Flash%20Statistics%202008-09.pdf.

———. 2012a. *Elementary Education in India: Progress towards UEE: Flash Statistics, DISE 2010–11*. New Delhi: NUEPA and Department of School Education and Literacy, Ministry of Human Resource Development, Government of India. http://udise.in/Downloads/Publications/Publications%202010-11/Flash%20Statistics-2010-11.pdf.

———. 2012b. *Secondary Education in India: Progress towards Universalisation: Flash Statistics, SEMIS 2010–11*. New Delhi: NUEPA and Department of School Education and Literacy, Ministry of Human Resource Development, Government of India. http://udise.in/Downloads/Publications/SEMIS/2010-11/Flash%20Statistics%20%202010-11.pdf.

———. 2013. *Elementary Education in India: Progress towards UEE: Flash Statistics, DISE 2012–13*. New Delhi: NUEPA and Department of School Education and Literacy, Ministry of Human Resource Development, Government of India. http://udise.in/Downloads/Publications/Documents/Flash%20Statistics%202012-13.pdf.

———. 2013b. *Elementary Education in India: Progress towards UEE: Flash Statistics, DISE 2011–12*. New Delhi: NUEPA and Department of School Education and Literacy, Ministry of Human Resource Development, Government of India. http://udise.in/Downloads/Publications/Publications%202011-12/Flash%202011-12.pdf.

———. 2014a. *Elementary Education in India: Progress towards UEE: Flash Statistics, DISE 2013–14*. New Delhi: NUEPA and Department of School Education and Literacy, Ministry of Human Resource Development, Government of India. http://udise.in/Downloads/Publications/Documents/Flash%20Statistics2013-14.pdf.

———. 2014b. *Secondary Education in India: Progress towards Universalisation: Flash Statistics, U-DISE 2013–14*. New Delhi: NUEPA and Department of School Education and Literacy, Ministry of Human Resource

Development, Government of India. http://udise.in/Downloads/Publications/Documents/SecondaryFlash%20Statistics-2013-14.pdf.

———. 2015a. *Elementary Education in India: Progress towards UEE: Flash Statistics, U-DISE 2014–15.* New Delhi: NUEPA and Department of School Education and Literacy, Ministry of Human Resource Development, Government of India. http://udise.in/Downloads/Publications/Documents/ElementaryFlash2014-15.pdf.

———. 2015b. *Secondary Education in India: Progress towards Universalization: Flash Statistics, U-DISE 2014–15.* New Delhi: NUEPA and Department of School Education and Literacy, Ministry of Human Resource Development, Government of India. http://udise.in/Downloads/Publications/Documents/SecondaryFlash%20Statistics-2014-15.pdf.

———. 2015c. *School Education in India: Flash Statistics, U-DISE 2014–15.* New Delhi: NUEPA and Department of School Education and Literacy, Ministry of Human Resource Development, Government of India. http://udise.in/Downloads/Publications/Documents/U-DISE-SchoolEducationInIndia-2014-15.pdf.

———. 2016. *Secondary Education in India: Progress towards Universalisation: Flash Statistics, U-DISE 2015–16.* New Delhi: NUEPA and Department of School Education and Literacy, Ministry of Human Resource Development, Government of India. http://udise.in/Downloads/Publications/Documents/Secondary_Flash_Statistics-2015-16.pdf.

———. 2017. *Elementary Education in India: Progress towards UEE: Flash Statistics, U-DISE 2015–16.* New Delhi: NUEPA and Department of School Education and Literacy, Ministry of Human Resource Development, Government of India. http://udise.in/Downloads/Publications/Documents/Flash_Statistics-2015-16_(Elementary).pdf.

Office of the Registrar General & Census Commissioner, GOI (Government of India). 1961. 'Census 1961' [Dataset].

———. 1971. 'Census 1971' [Dataset].

———. 1981. 'Census 1981' [Dataset].

———. 1991. 'Census 1991' [Dataset].

———. 2001. 'Census 2001' [Dataset].

———. 2011. 'Census 2011' [Dataset]. Available from http://www.censusindia.gov.in/2011-common/census_2011.html.

Ota, A B. 2006. 'Livelihood restoration of the displaced.' In *The Economics and Politics of Resettlement in India*, Shobhita Jain and Madhu Bala (eds.). New Delhi: Dorling Kindersley (India) Pvt. Ltd.

Periodi, Vani. 2004. 'Second generation issues in equity and education: Learning from Kolar District, Karnataka.' In *Hierarchies of Access* (ed.).

Planning Commission, GOI (Government of India). 2001. 'Mid-term Appraisal of the Ninth Five Year Plan (1997–2002).' New Delhi: GOI. Available at http://planningcommission.nic.in/plans/mta/index.php?state=ap9702cont.htm.

———. 2013. *Twelfth Five Year Plan (2012–2017): Social Sectors.* New Delhi: Sage Publications. http://planningcommission.gov.in/plans/planrel/fiveyr/12th/pdf/12fyp_vol3.pdf.

PROBE Team, The. 1999. *Public Report on Basic Education in India.* Delhi: Oxford University Press.

Ramachandran, Vimala. 1996. 'Critical Consciousness, Credit and Productive Assets: Key to Sustainable Livelihood [Women's Mobilisation in Nellore and Anantapur Districts of Andhra Pradesh]'. Report prepared for UNICEF, New Delhi and Commissioner Rural Development, Government of Andhra Pradesh, Hyderabad. https://www.researchgate.net/publication/237777791_Critical_Consciousness_Credit_and_Productive_Assets_Key_to_Sustainable_Livelihood.

Ramachandran, Vimala, (ed.). 2003. *Getting Children Back to School: Case Studies in Primary Education.* New Delhi: Sage Publications.

———. 2004. *Hierarchies of Access: Gender and Equity in Primary Education.* New Delhi: Sage Publications.

Ramachandran, Vimala. 2005. 'Democratic inequalities: The dilemma of elementary education in India'. Paper prepared for CSDS, New Delhi.

———. 2009. 'Towards Gender Equality in Education.' Paper prepared for the 'Education for All – Mid-decade Assessment'. New Delhi: National University of Educational Planning and Administration (NUEPA). http://www.teindia.nic.in/efa/Vimla_doc/EFA_Paper_Vimala_2009.pdf.

———. 2016. 'Exploring the legacy of three innovative programmes: Lessons learnt from Lok Jumbish, Shiksha Karmi and Mahila Samakhya.' In *Education and Empowerment in India – Policies and Practices,* Avinash Kumar Singh (ed.). New Delhi: Routledge India.

——— and Tara Naorem. 2012. 'Inclusion and exclusion of students in schools and in the classroom in primary and upper primary schools: Nation Synthesis Report.' Report commissioned by Sarva Shiksha Abhiyan, MHRD, Government of India. Technical Support Group, EdCIL India Limited. http://ssa.nic.in/research-studies-document_old/survey-report-on-out-of-school-children/list-of-studies/National%20Synthesis%20Inclusion-Exclusion%20study.pdf.

———, and Prerna Goel Chatterjee. 2014. 'Evaluation of Gender and Equity Issues Under Sarva Shiksha Abhiyan.' *Indian Journal of Gender Studies,* 21(2): 157–178. doi:10.1177/0971521514525081.

———, Madhumita Pal, and Vandana Mahajan. 2003. 'Balika Shikshan Shivir, Lok Jumbish.' Report prepared for DFID, India. http://www.eruindia.org/files/Balika%20Shikshan%20Shivir%20Rajasthan%20-%20a%20review.pdf.

———, Madhumita Pal, Sharada Jain, Sunil Shekar and Jitendra Sharma. 2005. 'Teacher Motivation in India.' Report prepared for Knowledge and Skills for Development, Brighton, UK. https://www.researchgate.net/publication/23778093_Teacher_Motivation_in_India.

———, Niti Saxena, and ERU Consultants Pvt. Ltd. 2017. 'Formative study to enhance the understanding about reasons for smooth transition among boys and girls to secondary schools: National Synthesis Report, Vol. 1.' Report prepared for UNICEF India. file:///Users/chelseamcgill/Downloads/Formative%20Study%20Final%2027%20Feb%202017%20Vol%201.pdf.

———, T. Linden, Tara Beteille, Sangeeta Dey, Sangeeta Goyal, and Prerna Goel Chatterji. 2015. 'Teachers in the Indian Education System: Synthesis of a nine-state study.' NUEPA Research Report. http://www.nuepa.org/New/download/Research/Teachers_in_the_Indian_Education_System.pdf.

———, Suman Bhattacharjea, and K. M. Sheshagiri. 2008. *Primary School Teachers: The twists and turns of everyday practice*. New Delhi: Educational Resource Unit. https://www.researchgate.net/profile/Vimala_Ramachandran/publication/254412564_Primary_School_Teachers_The_Twists_and_Turns_of_Everyday_Practice/links/00b49538c8a61172db000000.pdf.

——— et al. 2004. 'Snakes and Ladders: Factors Influencing Successful Primary School Completion for Children in Poverty Contexts.' South Asian Human Development Sector Report No. 6. South Asian Human Development Sector, South Asia Region, The World Bank. Available at http://hdl.handle.net/10986/17813.

Ramalingaswami, Vulimiri, Urban Jonsson and John Rohde. 1996. 'Commentary: The Asian Enigma.' Paper presented at 'The Progress of Nations' conference, UNICEF, Paris, France. https://www.unicef.org/pon96/nuenigma.htm.

Rana, Kumar, Abdur Rafique and Amrita Sengupta. 2002. 'The Pratichi Education Report: with an introduction by Amartya Sen.' New Delhi: Pratichi (India) Trust.

Ratnam, Aruna. 2004. 'The weft and the warp of public education: A tale of two primary schools in Cuddalore District, Tamil Nadu.' In *Hierarchies of Access* (ed.).

Reardon, Sean F. 2011. 'The widening academic achievement gap between the rich and the poor: New evidence and possible explanations.' In *Whither Opportunity? Rising Inequality, Schools, and Children's Life Chances*. Greg J. Duncan and Richard J. Murnane (eds.). New York: Russell Sage Foundation.

Research, Evaluation and Studies Unit, Technical Support Group. 2009. 'Attendance of Students and Teachers in Primary and Upper Primary Schools: Synthesis Report of the Study Conducted in 20 States.' Report prepared for Sarva Shiksha Abhiyan, EdCIL (India) Limited. http://www.educationforallinindia.com/Report%20of%20school%20attendance.pdf.

———. 2009b. 'Teachers' Absence in Primary and Upper Primary Schools (Synthesis Report of Study Conducted in Andhra Pradesh, Madhya Pradesh and Uttar Pradesh).' Report prepared by EdCiL (India) Limited. https://www.scribd.com/document/76992602/EdCIL-Report-of-Teachers-Absence-Study-Doc-by-Vijay-Kumar-Heer.

———. 2010. 'Teachers and Students' Time-on-Task in Primary and Upper Primary Schools: Synthesis Report (Assam, Haryana, Maharashtra, Karnataka, Orissa).' Report prepared for Sarva Shiksha Abhiyan, EdCIL, India. Available at http://ssa.nic.in/page_portletlinks?foldername=research-studies.

Rustagi, Preet, Sunil Kumar Mishra and Balwant Singh Mehta. 2011. 'Scheduled Tribe Children in India: Multiple Deprivations and Locational Disadvantage.' IHD–UNICEF Working Paper Series, Children of India: Rights and Opportunities no. 8. New Delhi: UNICEF and Institute for Human Development.

Saihjee, Aarti. 2004. 'Long live the alphabets! Reflections from Betul District, Madhya Pradesh.' In *Hierarchies of Access*, Ramachandran (ed.).

———, and Vimala Ramachandran. 2002. 'The New Segregation: reflections on gender and equity in primary education.' *Economic and Political Weekly*, 37(17). http://www.epw.in/journal/2002/17/review-womens-studies-review-issues-specials/new-segregation.html.

———, Kameshwari Jandhyala and Vimala Ramachandran. 2003. 'Through the Life Cycle of Children.' *Economic and Political Weekly*, 38(47). http://www.epw.in/journal/2003/47/special-articles/through-life-cycle-children.html.

SANDHAN. 2004. 'In-depth case studies of ten government schools in Rajasthan.' Unpublished research report. SANDHAN, Jaipur.

Sharma, Neerja and Nandita Chaudhary. 2003. 'From home to school: circumstances in the lives of children from poor families.' In Vimala Ramachandran (ed).

Sharma, Rashmi and Vimala Ramachandran, (eds.). 2009. *The Elementary Education System in India: Exploring Institutional Structures, Processes and Dynamics*. New Delhi: Routledge.

Singh, A. K. 2011. 'Study of the Role of VECs/PTAs/SDMCs/Urban Local Bodies in School Management and Supervision in the context of SSA.' Unpublished report. New Delhi: National University of Educational Planning and Administration.

SRI (Social and Rural Research Institute) and EdCil. 2014. 'National Sample Survey of Estimation of Out-of-School Children in the Age 6–13 in India.' http://ssa.nic.in/pabminutes-documents/NS.pdf.

SSA (Sarva Shiksha Abhiyan). 2002. 'Sixteenth Joint Review Mission Aide Memoire.' Sarva Shiksha Abhiyan, Government of India. Available at http://ssa.nic.in/monitoring/joint-review-mission-ssa-1/the-aide-memoire-of-16th-jrm-on-ssa.

———. 2007. '6th Review Mission of Sarva Shiksha Abhiyan, Government of India (16–26 July, 2007): Aide Memoire.' http://ssa.nic.in/monitoring-documents_old/jrm/6th%20JRM/Final%20Aide%20Memoire%20 6th%20JRM.pdf.

———. 2009. '10th Joint Review Mission of Sarva Sikhsha Abhiyan, Government of India (20th to 31st July, 2009): Aide Memoire.' Available at http://ssa.nic.in/monitoring/joint-review-mission-ssa-1/The%20 Aide%20Memoire%20of%2010th%20JRM%20on%20SSA.

———. 2010. 'Eleventh Joint Review Mission & Mid Term Review, 15th to 29th January 2010: Aide-Memoire.' http://ssa.nic.in/monitoring-documents_old/jrm/11th%20JRM/11th%20JRM%20Final%20 Aide%20Memoire.pdf.

Subrahmanian, Ramya. 2002. 'Gender and Education: A Review of Issues for Social Policy.' Social Policy and Development Programme Paper no. 9. United Nations Research Institute for Social Development. Available at http://www.unrisd.org/unrisd/website/document.nsf/(httpPublications)/0A8ADED14E7E1595C1256C08004792C4.

Sujatha, K. 1999. 'Education of Indian Scheduled tribes: A study of community schools in the district of Vishakhapatnam, Andhra Pradesh.' Working document in the series 'Strategies of Education and Training for Disadvantaged Groups.' Paris: International Institute for Educational Planning. http://unesdoc.unesco.org/images/0012/001202/120281e.pdf.

———. 2002. 'Education among Scheduled Tribes.' In *India Education Report: A Profile of Basic Education*, R. Govinda (ed.). New Delhi: Oxford University Press.

Synovate and Azim Premji Foundation. 2005. 'Understanding the views, perceptions and feelings of primary school teachers: A Study Report.' Bangalore: Azim Premji Foundation.

Tilak, Jandhyala B. G., (Ed). 2013. *Higher Education in India: In search of Equality, Quality and Quantity*. New Delhi: Orient Blackswan.

UNICEF (United Nations Children's Fund). 2009. *Equity in School Water and Sanitation – Overcoming Exclusion and Discrimination in South Asia*. Kathmandu: UNICEF Regional Office for South Asia (ROSA).

UNPF (United Nations Population Fund). 2014. *Programme of Action of the International Conference on Population Development*, 20th Anniversary Edition. Available at http://www.unfpa.org/publications/international-conference-population-and-development-programme-action.

Veerbhadranaika, P, Revathi Sampath Kumaran, Shivali Tukdeo and A. R. Vasavi. 2012. 'The Education Question from the Perspective of Adivasis: Conditions, Policies and Structures.' Bangalore: National Institute of Advanced Studies. http://eprints.nias.res.in/423/1/2012-R7-PV.pdf.

Visaria, Leela. 2014. 'Schooling of Tribal Children: Current Scenario and Challenges.' In *Tribal Development in Western India*, Amita Shah and Jharna Pathak (eds.). New Delhi: Routledge.

Wazir, Rekha (ed.). 2000. *The Gender Gap in Basic Education: NGOs as Change Agents*. New Delhi: Sage Publications.

Weisskopf, Thomas E. 2013. 'Impact of Reservations on Admissions to Higher Education.' In *Higher Education in India: In search of Equality, Quality and Quantity*. Jandhyala B. G. Tilak, (ed). New Delhi: Orient Blackswan.

World Bank. 2004. *Reaching out to the child: An integrated approach to child development*. Human Development Sector, South Asia Region, World Bank. Report No. 29695. Available at http //documents.worldbank.org/curated/en/743781468750325946/Reaching-out-to-the-child-an-integrated-approach-to-child-development.

———. 2009. *Secondary Education in India: Universalising Opportunity*. Prepared by the Human Development Unit, South Asia Region, World Bank. Available at http://hdl.handle.net/10986/3041.

Young India, Odisha. 2010. 'Plight of Children in Post Communal Violence Situation in Kandhamal District of Orissa.' Report of study commissioned by NEGFIRE.

Index